T0330546

BUSINESS ESSENTIALS FOR UTILITY ENGINEERS

BUSINESS ESSENTIALS FOR UTILITY ENGINEERS

RICHARD E. BROWN

CRC Press
Taylor & Francis Group
Boca Raton London New York

CRC Press is an imprint of the
Taylor & Francis Group, an **informa** business

CRC Press
Taylor & Francis Group
6000 Broken Sound Parkway NW, Suite 300
Boca Raton, FL 33487-2742

© 2010 by Taylor and Francis Group, LLC
CRC Press is an imprint of Taylor & Francis Group, an Informa business

No claim to original U.S. Government works

Printed in the United States of America on acid-free paper
10 9 8 7 6 5 4 3 2 1

International Standard Book Number: 978-1-4398-1196-2 (Hardback)

Library of Congress Cataloging-in-Publication Data

Brown, Richard E., 1969-
    Business essentials for utility engineers / Richard E. Brown.
        p. cm.
    Includes index.
    ISBN 978-1-4398-1196-2 (hbk. : alk. paper)
    1. Public utilities--Management. 2. Public utilities--Finance. I. Title.

HD2763.B684 2010
363.6068'1--dc22   2009044119

Visit the Taylor & Francis Web site at
http://www.taylorandfrancis.com

and the CRC Press Web site at
http://www.crcpress.com

Contents

Preface   *ix*

Author   *xv*

1. UTILITIES **1**
   1.1. Types of Utilities 2
   1.2. Natural Monopolies 15
   1.3. Utility Ownership 16
   1.4. Utility Regulation 20
   1.5. Utility Rates 23
   1.6. Service Standards 25
   1.7. Deregulation 28
   1.8. Summary 30
   1.9. Study Questions 31

2. ACCOUNTING **33**
   2.1. The Basic Accounting Equation 33
   2.2. Journals, Ledgers, and Accounts 38
   2.3. Accounting Principles 41
   2.4. Financial Statements 55
   2.5. Other Types of Accounting 74
   2.6. Summary 82
   2.7. Study Questions 82

3. ECONOMICS **85**
   3.1. Supply and Demand 86
   3.2. Market Pricing 90
   3.3. Producer Surplus 95
   3.4. Monopolistic Pricing 97
   3.5. Business Cycles 101
   3.6. Summary 106
   3.7. Study Questions 107

4. FINANCE **109**
    4.1. Time Value of Money 110
    4.2. Company Valuation 121
    4.3. Market Efficiency 124
    4.4. Capital Structure 136
    4.5. Tax Shields 140
    4.6. Bankruptcy 141
    4.7. Summary 145
    4.8. Study Questions 145

5. RISK **147**
    5.1. Probability and Statistics 148
    5.2. Stock Price Movement 158
    5.3. Diversification 160
    5.4. Portfolio Theory 165
    5.5. Capital Asset Pricing Model 173
    5.6. Financial Options 180
    5.7. Real Options 191
    5.8. Summary 192
    5.9. Study Questions 194

6. FINANCIAL RATIOS **195**
    6.1. Profitability Ratios 196
    6.2. Activity Ratios 200
    6.3. Leverage Ratios 203
    6.4. Liquidity Ratios 204
    6.5. Market Ratios 205
    6.6. DuPont Analysis 209
    6.7. Residual Income and EVA 211
    6.8. Summary 213
    6.9. Study Questions 214

7. RATEMAKING **215**
    7.1. Regulatory Goals 216
    7.2. Revenue Requirements and the Rate Base 219
    7.3. Rate Design 222
    7.4. Rate Cases 237
    7.5. Rate Base Misconceptions 240
    7.6. Summary 244
    7.7. Study Questions 245

8. BUDGETING **247**
   8.1.   Top-Down Budgeting 248
   8.2.   Bottom-Up Planning 256
   8.3.   Types of Budgets 261
   8.4.   Project Prioritization 264
   8.5.   Business Case Justification 280
   8.6.   Summary 300
   8.7.   Study Questions 301

9. ASSET MANAGEMENT **303**
   9.1.   What Is Asset Management? 304
   9.2.   History of Utilities 306
   9.3.   History of Asset Management 309
   9.4.   Organizational Functions 311
   9.5.   Asset Management Process 313
   9.6.   Summary 316
   9.7.   Study Questions 317

Index *319*

Preface

This is a book for utility engineers. Typical readers will have studied engineering in college, received an engineering degree, and somehow ended up pursuing a career within a utility or taken a job associated with utilities. Academic credentials for most of these readers will include advanced mathematics, probability, statistics, chemistry, physics, and materials science. Most will have further specialized in a specific area such as electrical engineering, mechanical engineering, or civil engineering. These types of readers are well-educated and intelligent, an assumption made by the author when presenting material, sometimes difficult material, throughout the book.

Utilities have many challenging engineering problems to be solved. New customers must be served. Old equipment must be maintained. New technologies must be assessed and adopted. To solve these challenges, engineers find themselves responsible for planning, engineering, system analysis, system design, equipment specification, maintenance management, operations, and a host of other functions.

Whatever their role, utility engineers make many decisions. Some of these decisions result from extensive and careful analyses. Others are made quickly during everyday activities. In virtually all cases, decisions have cost and other implications for the utility. Some options are cheap. Others options are expensive. Some options spend as little as possible now. Others options spend money now in order to save money later. Some options result in high safety margins. Other options are more risky. With so many choices, it is valuable for utility engineers to understand the criteria for deciding which decisions are best from a business perspective.

Typical businesses prefer engineering decisions that result in higher profits. A cheaper engineering solution may produce higher profits if a resulting lower price causes an increase in sales. It is also possible that a more expensive engineering solution will produce higher profits if the resulting higher quality product can command a premium price. In both cases, the business objective is clear: while acting legally and ethically, maximize profits whenever possible.

Do utilities want their engineers to maximize profits whenever possible? Stockholders of publicly traded utilities would answer with a resounding "yes." However, the profits of utilities are essentially controlled by regulators, making profit maximization for utilities somewhat different than for other industries. There is a large body of literature on every imaginable business topic, but few address the application of these business topics to utilities in an effort to describe similarities, differences, and why these differences are important. *Business Essentials for Utility Engineers* hopefully fills this gap.

This book does not address a variety of topics that are typically covered in a business curriculum. Examples include strategy, marketing, organizational design, human resources, and business ethics. These topics are important in a general sense, but tend to be peripheral to most issues facing utility engineers. Their application to utilities is similar to other types of businesses, allowing the interested reader to effectively study these topics through general business publications that are not specific to utilities.

The author spent about six years writing this book. The material evolved from a large number of asset management and business consulting engagements between the author and utilities, including numerous seminars and short courses. Hopefully, this collaborative process has resulted in a book that is both interesting and practical for utility engineers. An exhaustive academic approach has been intentionally avoided so that the focus can be on essential material rather than on details and nuances. Even with this approach, the book is suitable for self-study, undergraduate study, graduate study, and desk reference. It will be most useful for utility engineers, but will also have value for utility managers, regulators, and others associated with utilities such as vendors, consultants, and for engineers and consultants working with non-utility companies with a utility-like infrastructure.

Although this book mostly focuses on practical issues, the reader should be prepared for a generous amount of business theory and the periodic use of advanced mathematics. The reader should also be prepared for an extensive vocabulary lesson. For the reader's benefit, the first instance of each important business term is presented in italics.

Like engineering, business has a large number of words and phrases with precise meanings. Many of these words and phrases represent simple concepts, and can be easily learned and understood. Although some terms represent difficult or confusing concepts, this book will remove much of the mystery and serve as a reference when necessary. For many utility engineers, much of the value of this book will derive from an increased ability to communicate with senior management using business jargon.

Chapters are organized, for the most part, to build sequentially upon each other. This is not always possible since topics overlap. For example, accounting sometimes uses financial concepts and finance sometimes uses accounting concepts, resulting in the periodic need for forward referencing. Many terms and concepts presented in earlier chapters are used in later chapters. Therefore, the first-time reader will benefit from reading each chapter in order, from Chapter 1 through Chapter 9.

This book starts in Chapter 1 by addressing the special aspects of public utilities that affect their characteristics as a business. Specific topics include the main types of utilities, the nature of monopolies, different utility ownership structures, regulation, rates, service standards, and utility deregulation. It does not address any business topics. To do so would be premature and not helpful to the reader. Instead, the information in Chapter 1 provides the reader with a general overview of utilities and introduces terminology and context that are required for a full comprehension of future chapters.

Accounting, covered in Chapter 2, is chosen as the first business topic because it is difficult to address other aspects of business without using accounting concepts and terminology. Accounting is covered from an engineering and technical management perspective; detailed treatment of accounting mechanics, rules, nuances, and processes is beyond the scope of this book and is avoided. The focus is to present the theory and principles underlying financial accounting, provide a sound basis for the interpretation of financial reports, and demonstrate an array of key accounting issues that are important for utility engineers to understand.

The subject of "engineering economics" typically deals with the time value of money and net present value calculations. In the business world, these topics are included in the subject of finance. Business economics, the subject of Chapter 3, may occasionally use these tools, but addresses topics very different from traditional engineering economics. Economics can broadly be classified into microeconomics and macroeconomics. Most of Chapter 3 focuses on microeconomics, as only a few aspects of macroeconomics are important for utility engineers. Specific topics include supply and demand, market pricing, producer surplus, monopolistic pricing, and business cycles.

Chapter 4 is the first of three chapters on the subject of corporate finance. It begins by discussing the time value of money, a subject most likely familiar to many readers. It builds on this concept by showing how investors determine the value of a business, including the ability of markets to efficiently determine the fair price of a publicly traded stock. Included in this discussion are the sources of financial capital, capital structure, tax shields, and bankruptcy.

Businesses are forward looking, and there is always uncertainty about the future. As such, Chapter 5 addresses the important topic of financial risk. Although most engineers have studied probability and statistics, this chapter begins by reviewing the core mathematics used in financial risk assessment. It then uses these tools to build a mathematical model of random stock price movement. With this foundation established, the following subjects are described and discussed within the context of utility finance: diversification, portfolio theory, capital asset pricing model, financial options, and real options.

Chapter 6 completes the subject of finance by addressing financial ratios. Although somewhat tedious from a pedagogical perspective, lists and definitions of important ratios are unavoidable. While discussing how utility values might be similar or different from other industries, this chapter works its way through profitability ratios, activity ratios, leverage ratios, liquidity ratios and market ratios. It ends with a discussion of the DuPont analysis, residual income, and economic value added.

Chapter 7 covers a subject of critical importance to public utilities: ratemaking. Keeping in mind that although each regulator and each utility will have many unique aspects, there are certain overarching principles that guide the ratemaking process. The focus of Chapter 7 is on this high level view rather than on nuances and exceptions. Specific topics include the regulatory goals of ratemaking, utility revenue requirements, the rate base, rate design, and rate cases.

Corporate budgeting, the subject of Chapter 8, addresses the "who, what, when, where, and why" of corporate spending. At the highest level, budgeting consists of identifying the total amount that a utility will spend over a fiscal year. At the lowest level, budgeting consists of identifying specific amounts of money for specific projects. Since budgets play a stronger role in utilities than in most other companies, the specifics of utility budgeting are emphasized, including the author's vision of an effective and credible budgeting process. As such, this chapter has a strong focus on spending justification, which goes well beyond the traditional utility budgeting function of spending allocation. Specific topics include top down budgeting, bottom up planning, project budgets, program budgets, blanket budgets, project prioritization, and business case justification.

Chapter 9 concludes this book with the topic of asset management. Although not all utilities are pursuing an asset management approach to managing their business, many are, and it is beneficial for business-savvy engineers to have a basic grasp of the subject. In addition, asset management provides a good mechanism to bring together many of the business topics covered in previous chapters, since it is essentially the harmonious interaction of business management and engineering management. In addition to the mechanics of asset management, this chapter addresses the traditional utility culture, and how this culture differs from the requirements of asset management. In many ways, the content of this book is designed to help overcome the typical utility engineering mindset from "we should do what makes good engineering sense" to "we should do what makes good business sense."

The breadth of subject matter covered in *Business Essentials for Utility Engineers* is necessarily extensive so that the reader can become exposed to the essentials of business. As such, this book should be viewed as foundational. It provides a robust framework for correct business thinking and a solid foundation for further learning. Graduate degrees are offered in accounting, finance, and economics. This book covers each subject in a single chapter. Completion of a master's degree in business administration requires thousands of hours of reading, homework, and lecture time. A thorough read-through of this book can be accomplished in several weeks. The reader should be under no illusion that this book is all encompassing. Rather, its goal is to distill and present the business essentials that will help utility engineers be more effective in their daily jobs.

Many people helped to make this book possible, both directly and indirectly. I would like to specifically mention and give thanks to the following people who took the time to review chapters and suggest improvements: Lee Willis, Dr. Le Xu, Dr. Julio Romero Agüero, and Majida Malki of Quanta Technology; Prof. Matt Clements of St. Edwards University; Prof. Gerry Sheblé of the University of Porto; Prof. Santiago Grijalva of the Georgia Institute of Technology; Kevin Dasso of Pacific Gas & Electric; Tony Hurley of FirstEnergy;. Mike Spoor of Florida Power and Light; and Larry Vogt of Southern Company. I also express many thanks to the people at Taylor & Francis (publishing under the CRC imprint) with special regards to Jessica Vakili and Karen Simon. Last, I would like to offer special thanks to my wife, Christelle, to our four children (Ashlyn, Colton, Landrey, and Selah), and to our two beloved Welsh Corgis (Guinevere and Lancelot), for their invaluable inspiration and support.

Richard E. Brown

Author

Richard E. Brown is the Senior Vice President of Operations and co-founder of Quanta Technology, a firm specializing in technical and management consulting for utilities and utility-related industries. Dr. Brown has been on the leadership team of three successful startup organizations, and has provided consulting services to most major utilities in the United States and many around the world. He is a frequent instructor, has taught courses in eleven countries, and is an adjunct professor at North Carolina State University.

Dr. Brown has published more than 90 technical papers related to asset management and performance management, and is also author of the book *Electric Power Distribution Reliability*. In 2007, he was elected to the grade of Fellow by the Institute of Electrical and Electronics Engineers (IEEE), which is conferred by the IEEE Board of Directors for an extraordinary record of industry accomplishments.

Dr. Brown earned his BSEE, MSEE, and PhD degrees from the University of Washington in Seattle, and his MBA degree from the University of North Carolina at Chapel Hill. He is a registered professional engineer.

1
Utilities

Public utilities provide essential services to society. Because of their importance, legal precedent has upheld the need for specialized government oversight of these businesses to ensure that safe and reliable utility services are widely available for rates that are reasonable and non-discriminatory.

The types of public utilities considered in this book require large investments in fixed infrastructure, typically extending to the premises of end-use customers. Examples of these *infrastructure utilities* include electric utilities, gas utilities, telephone utilities, water utilities, and wastewater utilities. Sometimes public transportation facilities are also considered public utilities (e.g., railroads, buses, subways), called *transportation utilities*. Much of this book is relevant to both infrastructure utilities and transportation utilities, but there are certain aspects of infrastructure utilities that require special consideration. Most people use the terms public utility and utility interchangeably. Therefore, unless otherwise stated, the remainder of this book uses the term *utility* to refer to a public utility that relies heavily on fixed infrastructure to provide an essential utility service.

Many of the business topics in this book apply to all industries. However, utilities have a large number of differences that make traditional business thinking potentially misleading or inappropriate, requiring this book to serve in dual roles. First, it educates the reader on the fundamentals of business theory, taking advantage of the analytical sophistication and deductive nature of typical engineers. Second, it discusses and applies each aspect of business theory within the context of utilities to describe what is similar and what dissimilar when compared to other businesses.

The remainder of this chapter describes the special aspects of utilities that affect their characteristics as businesses. It does not address any specific business topics, nor does it attempt to link these characteristics to business topics that have yet to be presented. To do so would be premature and unhelpful to the reader. Instead, the information in the remainder of this chapter provides the reader with a general overview of utilities and introduces terminology and context that are required for a full comprehension of future chapters.

1.1   TYPES OF UTILITIES

There are many types of utilities. The reader is probably familiar with the technical aspects of one or more, but may not be familiar with others. This section presents a general overview of the major types of infrastructure utilities as organized by the service provided to the end user. Emphasis is given to the physical system and service delivery process rather than industry or business structure.

A company that owns and controls the entire infrastructure required to deliver end-use utility service to retail customers is called a *vertically integrated utility*. Often customers do not receive services from a vertically integrated utility. In these cases, multiple utilities must coordinate to provide all vertical functions. The process is similar to other businesses where one company may produce raw materials, another may manufacture a product, and still another may sell the product to retail customers. The following sections present different types of infrastructure utilities in terms of vertically integrated infrastructure functionality. They also discuss the typical roles of major vertical business functions.

1.1.1   Electric Utilities

Electric utilities produce electric power and deliver this power to customers. The functions required to do this are generally categorized into generation, transmission, and distribution.

Most electric power is generated at large central stations that use fuel such as coal, oil, natural gas, or uranium (for nuclear reactors). Some generation facilities use non-thermal energy sources such as hydroelectric dams, wind turbine farms, and photovoltaic or solar-thermal facilities.

There are several fields of study that address the economics of electric power generation. This includes the optimal use of generation from a centralized planning perspective including real time generator output (called *economic dispatch*) and weekly generator scheduling (called *unit commitment*). It also includes the optimal market mechanics of generation and transmission from a competitive industry perspective. Both of these business topics are important to

Figure 1.1. Electricity meters. The picture on the right shows a typical residential meter that is fed from underground electric cables. The picture on the left shows a more advanced meter, typically used for large industrial customers.

electric utilities, but do not generalize well to other utilities. They are specialized fields rather than general business subjects and, therefore, are beyond the scope of this book.

Most electric power is generated at large facilities that are long distances from the customers they serve. Therefore, most electric power must be moved from generation facilities to high concentrations of customers via high voltage transmission lines. These lines allow electric power to be transported efficiently over long distances, but require large and expensive transmission towers.

Electric transmission systems deliver power to electric distribution systems, which operate at lower voltages and require smaller structures (typically wood utility poles). Electric distribution systems transport power to retail customers, where consumption is tracked by electricity meters (see Figure 1.1). Retail customers are typically billed based on electric energy consumption, measured in kilowatt-hours (kWh).

Electric utilities are required to provide electric power to their customers at an acceptable voltage. High voltages can harm electric appliances and cause potential safety hazards. Low voltages can cause electric appliances to function improperly and can also cause electric motors to overheat. Other aspects of voltage that utilities must consider are frequency, flicker (i.e. slowly oscillating voltage), transients (i.e., voltage spikes), harmonics, and others.

Electric utilities are increasingly expected to provide high levels of service reliability to customers. When switches are flipped, customers want their lights to turn on and their computers to boot up. Utilities typically measure customer reliability by the number of service interruptions and the length of these service interruptions. Utilities are also expected to respond well after major weather events that cause widespread infrastructure damage and extensive customer interruptions.

Electricity has an important quality that affects the economics of electric utilities; it is difficult and expensive to store large amounts of electric energy. Because of this, most electricity must be instantaneously produced when it is needed. Generation and transmission capacity must be sized to handle the highest period of electricity demand. Most electric energy cannot be economically stored up during times of low demand for later usage.

The difficulty of electric energy storage is becoming increasingly problematic as more wind power and solar power are connected to the electric system. Wind turbines only produce electricity when the wind blows. Solar panels only produce electricity when it is sunny. Since electric utilities cannot depend on these sources being available when needed, the installation of wind and solar does not typically eliminate the need for traditional sources of generation. For example, if peak electricity demand increases by ten gigawatts, it is not sufficient to add ten gigawatts of wind power; it is likely that six or seven gigawatts of traditional generation will also need to be added.

Historically, most utilities were vertically integrated and owned the generation, transmission, and distribution systems required to provide service to their customers. There are still many vertically integrated utilities, but the industry recently went through a wave of "vertical unbundling." There are now a number of dedicated generation companies (gencos), dedicated transmission companies (transcos), dedicated distribution companies (discos), and combined transmission and distribution companies (wires companies, or electric delivery companies).

The primary goal of vertical unbundling is to introduce competition to electricity generation. In a perfectly unbundled world, a disco will purchase required generation from a variety of gencos. This energy is then transported from the genco to the disco via the transco for a fee. An independent system operator (ISO) controls the transmission system to ensure reliability, allow fair access, and prevent market power abuses. Much of the electric energy in the US is purchased and delivered in this manner.

Retail electricity purchases amount to about 3% of the US economy. More importantly, society and economic activities essentially shut down when electricity becomes unavailable. Electricity has become one of the cornerstones of modern societies and electric utilities are the stewards of this essential service.

1.1.2 Gas Utilities

Gas utilities produce natural gas and deliver it to customers. Like electricity, the functions required to do this are generally categorized into generation, transmission, and distribution.

Most natural gas today comes from wells that drill into underground deposits in natural gas fields. Natural gas also occurs with oil deposits, coal deposits, and shale deposits (although it is often uneconomical to extract natural gas from shale deposits).

Figure 1.2. A typical residential gas meter. Natural gas is supplied from the vertical pipe on the left. The gas passes through a pressure regulator before entering the meter. The pipes to the right go on to supply gas appliances in the house such as stoves and hot water heaters.

Natural gas delivered to consumers primarily consists of methane. When first extracted from the ground, natural gas also contains significant quantities of ethane, propane, butane, pentane, and hexane, as well as non-fuel substances such as water vapor, carbon dioxide, hydrogen sulfide, nitrogen, and helium. After extraction, natural gas is transported via pipes to processing plants so that these impurities can be removed in a process often called *sweetening*.

Most natural gas is produced in remote areas that are long distances from the customers they serve. Therefore, most gas must be moved from production and processing facilities to concentrations of customers via high pressure transmission pipelines. These high pressure pipes allow large volumes of natural gas to be transported efficiently, but require compressor stations about every fifty miles to keep pressures sufficiently high.

It is not practical or cost-effective to build natural gas pipelines across oceans. For trans-oceanic transport, natural gas is cooled until it becomes *liquefied natural gas* (*LNG*). Liquefication occurs at about 260 °F below zero, and the resulting LNG occupies about six hundred times less volume than the gas. LNG can then be transported in special ships similar to oil tankers. LNG can also be transported in specialized tanker trucks and tanker railcars.

Transmission systems deliver natural gas to distribution systems. Up to this point, the natural gas has been colorless and odorless. Upon entering the distribution system, a tiny amount of odorant is added, typically with a rotting cabbage or rotten egg smell. The foul odor allows gas leaks to be easily detected. The gas distribution system operates at lower pressures than the transmission system and delivers gas to customer premises, where consumption is tracked by gas meters (see Figure 1.2). Customers are typically billed based on either gas volume or gas heat content. Gas volume is measured in thousands of cubic feet (Mcf). Heat content is measures in *therms*, with one therm defined as one hundred thousand British thermal units (BTUs). A therm is typically equal to about one hundred cubic feet of natural gas.

Gas utilities are required to provide natural gas to customers with sufficient pressure and heat content. Excessive pressure can cause leaks and ruptures. Insufficient pressure and insufficient heat content can cause gas appliances to function improperly. From a billing perspective, utilities that base bills on gas volume increase customer charges when heat content is reduced; more gas is required to perform the same amount of heating. Gas utilities must also place a high emphasis on safety, since gas leaks can lead to dangerous explosions. A safe gas system is ensured by regular inspections for leaks and timely repairs.

Unlike electricity, storage is an important part of the natural gas delivery system. Most gas storage occurs on the transmission system. Gas that is not needed by the local delivery companies is diverted to underground storage reservoirs, of which there are three major types. By far, the largest amount of storage occurs in depleted gas fields (86% of storage volume in the US). The remaining storage occurs in aquifer reservoirs (10% of storage volume in the US) and salt cavern reservoirs (4% of storage volume in the US). These storage facilities are drawn from when gas demand exceeds the rate of production or when large production facilities become unavailable.

Wholesale natural gas is sold and traded as a commodity in the US and in many other countries. Gas pricing, trading, and purchasing are important and interesting business subjects, but are narrow specialties and are therefore not covered in this book.

Methane is a natural by-product that occurs when organic matter undergoes anaerobic decay. Newly produced methane is commonly called *biogas*. It is becoming more common to capture biogas and use it to run electric generators, such as at a landfill or a hog farm. Since these facilities are not connected to gas transmission or distribution systems, they are not considered part of the gas utility infrastructure.

The major actors in the US natural gas industry are producers, pipeline companies, local distribution companies (LDCs), and natural gas marketers. A typical producer will sell gas to a combination of LDCs, power marketers, and end users. Gas is then transported through transmission pipelines for a federally regulated fee.

Many houses do not have natural gas service, and rely on electricity for all of their energy needs. Those with gas typically use it to fuel furnaces, hot water heaters, ovens, and clothes dryers. In the last decade, many private companies have built electric generation facilities powered by natural gas. These facilities have the advantages of fast start times and relatively low emissions. In terms of volume, gas sales are about 20% residential, 50% commercial/industrial, and 30% power generation. Revenue splits are somewhat different, since rates for these customers classes vary widely.

For-consumption gas purchases amount to about 1.5% of the US economy, which is about half that of electricity. Society is better able to handle gas interruptions than electricity interruptions, but many industrial processes are shut down when gas is not available. Natural gas is a critical element of our economy, and gas utilities are the stewards of this essential service.

1.1.3 Water Utilities

Water utilities produce drinkable (i.e., potable) water and deliver it to customers. The functions required to do this are generally categorized as treatment, transmission, and distribution.

Rainwater is naturally collected in geographic watersheds, where it drains into river systems, lakes, and underground aquifers. These serve as the sources for most of the water that is eventually consumed by people. River systems are typically tapped at reservoirs behind dams. Underground aquifers are tapped through wells.

Water taken directly from a watershed is not suitable for drinking. Utilities will therefore either gravity feed or pump this water through pipes to treatment plants where contaminants are removed. Treatment always addresses health-related issues such as the presence of biological and toxic contaminants. Treatment often also addresses aesthetic issues that affect taste, odor, and color.

After treatment, water is pumped through high pressure transmission pipes to local distribution systems, where pressure is reduced. These local systems contain elevated water tanks (see Figure 1.3), which serve dual purposes of water storage and ensuring that customers have sufficient water pressure.

Local distribution systems deliver drinking water to customers, where consumption is tracked by water meters (see Figure 1.4). These systems also supply water to fire hydrants; water utilities are responsible for ensuring that there is enough water for fire departments to effectively fight fires. The water is under pressure, but not nearly enough to fight fires. Fire trucks have dedicated pumps to draw water from hydrants and push it through hoses at high pressure.

Even in first-world countries, not all homes receive their drinking water from utilities. In the US, about fifteen percent of homes have private water wells. These wells are not connected to the utility distribution system and are not required to meet the same federal standards for purity.

Water utilities are required to provide drinking water to customers with sufficient pressure and purity. Insufficient pressure results in a dribble at the tap, miserable showers, and an inability to perform certain industrial processes. Excessive pressure can cause seals to leak and pipes to rupture. Insufficient purity can cause widespread public health concerns. Heath violations are rare but do occur. In 2008, health violations were reported on over five thousand US local water systems, representing about 11% of all systems and about 7% of all customers served by local water systems.

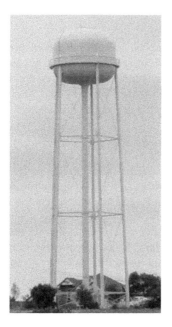 

Figure 1.3. Elevated water tanks, commonly called water towers. In addition to storing drinkable water, elevated water tanks ensure that customers have sufficient water pressure. The picture on the left is a typical water tower design, which communities often deem to be aesthetically undesirable. There are vast arrays of more pleasing designs. The picture on the right is a water tower designed to look like a peach, which is proudly displayed in Georgia, the Peach State.

Figure 1.4. A typical residential water meter. The picture on the left shows the closed water meter box. The picture on the right shows an opened box with the exposed meter. A water meter typically has a flow indicator and a total volume counter (gallons in the US).

Water utilities are required to have a high continuity of service. In first-world countries, service continuity is most threatened by an insufficiency of supply. This could be due to low water tables or a lack of water treatment capacity. Water may also be interrupted due to transmission or distribution pipe leaks, which typically require water flows to be stopped for repairs to occur.

Certain areas do not have sufficient water tables to supply drinking water to their populations. Some of these areas are near the ocean and use desalination technologies to convert abundant salt water to drinking water. Most desalinated water comes from distillation facilities, where salt water is boiled in reduced pressure chambers. Most new desalination facilities use reverse osmosis, where salt water is forced through a filter membrane using pressure. It costs much more to produce drinking water through desalination than through traditional methods that draw from water tables.

Compared to many other developed countries, the US uses a lot of drinking water. Much of this is due to the use of drinking water for outdoor uses such as irrigation and swimming pools. In the US, about 58% of water is used outdoors and 42% indoors. Since water for landscaping does not need to be treated, many communities are considering the use of dual water systems: one system to supply drinkable water for indoor use and a separate system to supply untreated water for outdoor uses such as landscape irrigation, fire hydrants, and street cleaning. The downsides of dual systems include the costs of building and maintaining two systems, and the possibility of cross-contamination.

Water and sewage bills are typically combined, and amount to about 0.7% of the US economy. This is about half the amount of natural gas purchases and about one quarter the amount of electric energy purchases. Despite being a smaller part of the economy, society is severely impacted by water supply interruptions. Without water, people cannot drink, bathe, or wash their dishes. People get annoyed when they cannot wash their cars or water their yards. Many industrial processes also depend upon a ready supply of water. Water is an essential service, is critical to all life, and is supplied to most of us by water utilities.

1.1.4 Telephone Utilities

Telephone service is different from electricity, gas, and water in that is does not generate a product and transport this product to customers. Rather, the job of a telephone system is to connect two customers so that they can communicate. Despite this difference, telephone utilities and telephone infrastructure are similar to other utilities in many respects.

Traditional telephone service is commonly called *plain old telephone service*, or *POTS*. POTS connects a *central switching office* to each customer through a pair of *twisted wires*, called a *subscriber loop*. The twisted pair terminates at a customer junction box, allowing additional twisted pairs to be routed from the junction box to the telephone jacks (see Figure 1.5).

Figure 1.5. The telephone junction box is where the local subscriber loop terminates and premise wiring begins. This junction box is sometimes called a network interface device. One twisted pair of wires is terminated for each line. These terminations are connected to additional twisted pairs that are routed to telephone jacks.

A central switching office can connect any two subscriber loops so that they can communicate. A *local exchange* consists of all subscriber loops within a local area; any calls within the local exchange are billed as local calls. The local exchange is roughly analogous to the distribution systems of electric, gas, and water utilities.

Each local exchange provides its own electric power, independent of the electric utility. Each subscriber loop is powered with forty-eight volts direct current. Since this system is independent, POTS typically remains in service when electric service becomes interrupted. However, many modern phones, such as those with cordless handsets, must be plugged into an electrical outlet, and will not function if electric service is unavailable.

POTS was originally designed as an analog system, and local exchanges are still primarily analog. These analog facilities support data transfers through the use of modems. Common examples are facsimile machines and computer modems. Even digital subscriber lines (DSLs) are just high speed modems that use the same wires as voice communication.

Local exchanges are connected together via long distance *trunk lines*. Historically, these trunk lines were based on analog technologies. Today, most trunk line capacity is provided through digital technology over fiber optic cables. The effective transmission distance of optical fiber is limited by signal attenuation and distortion, requiring the periodic placement of signal repeaters or optical signal amplifiers.

The use of mobile telephones has become pervasive in many parts of the world. These devices communicate to a network of cellular towers, which are connected to local exchanges. From the perspective of the local telephone company, cellular towers are customers within the local exchange; mobile telephone users access the local exchange through these towers.

The global system of interconnected local exchanges is called the *public switched telephone network* (*PSTN*). This system allows virtually all locations with telephones to connect and communicate to one another. Through cellular towers, mobile telephones can connect to both each other and to telephones on the POTS. The PSTN is roughly analogous to the transmission systems of electric, gas, and water utilities. There are a number of large private telephone networks that are not linked to the PSTN, but these are usually for military purposes.

The business aspects of telephone service are generally divided into *local exchange carriers* (*LECs*), who are responsible for the local exchange, and *inter-exchange carriers* (*IXCs*), who are responsible for calls between LECs. LECs are commonly called local telephone companies, and typically charge a fixed monthly fee for unlimited calls within a *local access and transport area*. IXCs are commonly called long distance companies, and have a variety of pricing approaches that typically assign a charge to each interexchange call.

There has been a steady progression from telephone service transporting voice conversations to telephone service transporting digital data. Initially, there was analog voice only. Then there was data encoded as audible signals. Next there was data transmitted simultaneously with voice. Most recently, voice has been encoded digitally and treated the same as any other data package. It is true that most subscriber loops still use analog voice signals, but this is due to legacy infrastructure and is not a limitation of available technology.

Since conversations can be treated as a form of generic data, telephone service can now occur through any digital data link, as long as this link can encode, transfer, and decode voice signals with sufficient speed. As such, telephone service can now be provided through the Internet and cable television systems. The reliability and voice quality of these systems are not as good as POTS. But they are getting better, and many businesses and households today no longer use POTS for voice communication.

In 2006, about 1.1% of US household spending was on local telephone service (including long distance charges) and another 1.1% was spent on cellular phone service. Total spending on local service has been slowly declining, while total spending on cellular service has been rapidly increasing. It is becoming more common for residential homes to rely exclusively on mobile phones and to not have a land line. It is also becoming more common for companies to use internet-based telephones, although this will often entail the use of high speed data lines supplied by the local phone company. There are rapid changes occurring in the business environment of telephone utilities, but the function they provide remains an essential service for both businesses and the general public.

1.1.5 Wastewater Utilities

A wastewater utility, often called a sewage utility, transports dirty water away from customers and treats the dirty water so that it is suitable to be discharged into the environment. A wastewater utility is different from other utilities in several respects. First, it transports something away from customers rather than delivering something to customers. Second, wastewater infrastructure is strictly local and there are no long-distance transportation functions.

Wastewater enters the sewer system through drainpipes. These drainpipes are combined into *lateral pipes* that feed into larger *collector pipes* under the street. Cleaning access to lateral pipes is through *sewer cleanups* located on the customer premises. Utility workers can access the collector pipes through manholes (see Figure 1.6).

Collector pipes eventually feed into larger *interceptor pipes*, which transport the collected wastewater to treatment plants. Together, collector pipes and interceptor pipes are called *sewer mains*. Sewer systems are designed to be gravity driven from the customer all the way to the treatment facility. As such, treatment facilities are typically located at low elevations. Sewer pipes often follow rivers and creeks, which naturally flow downhill. If a pure gravity system is not feasible, the sewer system must use pumping or lifting systems to transport sewage up and over hills.

A wastewater treatment plant will typically have several stages of treatment. The first stage (called *primary treatment*) screens out large particles as the wastewater enters a tank. The tank allows heavy solids to settle at the bottom (forming *sludge*), and light solids to rise to the top (forming *scum*). The second

Figure 1.6. The picture on the left shows a sewer manhole with the manhole cover removed. The manhole allows for human access to the collector pipes. The picture on the right shows the cover of a sewer cleanup, which allows cleaning equipment to access pipes than connect customers to the collector pipe. This sewer cleanup is located on a front lawn, but many are located in basements.

stage (called *secondary treatment*) uses bacteria to consume organic materials and other nutrients. At this point, the treated water is suitable for discharge into rivers or lakes, and may be suitable for certain agricultural uses.

A wastewater treatment plant may also have additional treatment stages, collectively called *tertiary treatment*. This includes a disinfectant process involving chorine or ultraviolet light. The disinfected water can then undergo additional filtration and treatment depending upon its intended purpose, typically agriculture reuse, non-drinkable reuse, or drinkable reuse. As water shortages become more prevalent, many communities are considering the conversion of wastewater into drinking water.

Some sewer systems are connected to storm drains. In these cases, rain and runoff water collected by storm drains are deposited into sewer mains. Some areas have chosen to have completely separate sewer systems and storm water systems. Two separate systems are costly. However, wastewater treatment plants can be smaller since they do not have to treat rain and runoff water.

Wastewater utilities are responsible for treating sewage to an acceptable level of purity. Some of the more important measures include acidity, organic materials, suspended solids, fecal bacteria, dissolved oxygen content, phosphorous, nitrogen, and chorine.

In the US, about 25% of homes are not connected to a utility wastewater system. These homes must use either a holding tank or a septic tank to handle their sewage. A holding tank, sometimes called a *cesspool* or a *cesspit*, is a sealed container that holds sewage and must be pumped out when it becomes full. A septic tank allows solids to settle, scum to rise, and the remaining liquid to flow into a drain field. Since liquids are not permanently stored in septic tanks, they require emptying far less frequently than holding tanks.

Many wastewater utilities would like to bill customers based on the volume of sewage introduced into the system. Though nice in principle, very few wastewater utilities actually measure this value. Most rely on water usage as a proxy for sewer usage. Of course, not all water goes into the sewer system. This is especially true of outdoor water usage such as landscape irrigation and swimming pools. It can also be true for industrial uses such as a water bottling company. Wastewater utilities understand these issues and try their best to compensate. For example, annual residential sewer usage may be based on winter water usage, which typically does not involve landscape irrigation or the filling of swimming pools. Some utilities allow customers to separately meter indoor water usage and outdoor water usage, and only assess sewer charges for indoor water usage.

Some customers are connected to utility sewer systems, but get their own drinking water from a private well. Even these customers are sometimes charged for sewer use based on volume, though there is not even a water meter. Usage in these cases is typically assumed based on the characteristics of similar customers who do have water meters.

Wastewater utilities have dramatically improved the quality of life in developed countries by removing sewage odor and greatly reducing diseases such as cholera, typhoid, paratyphoid, yellow fever, and dysentery. Sewer mains are highly reliable due to their gravity-fed design, allowing people to dispose of sewage down drains at virtually any time. Less visible to customers is the highly reliable treatment process, which is of equal or greater importance.

1.1.6   Combined Utilities

The previous sections have described public utilities according to the service they provide. Multiple companies may be involved in providing each utility service, but each were still classified as a specific utility type such as electric, gas, water, telephone, or wastewater. This approach is helpful for initial presentation but somewhat oversimplifies the real world.

It is common for a single company to provide multiple utility services. Often these services are managed by separate organizational divisions that in many ways resemble separate businesses. Other times utilities will have substantial overlap when providing multiple utility services. The following are brief descriptions of the most common types of combined utilities, where a single company provides more than one utility service.

Electric & Gas Utilities. Many utilities offer both natural gas and electric power to customers. This type of utility might serve some customers with electric power only, other customers with natural gas only, and still other customers with both electric and gas. Many of these utilities use "electric and gas" or "gas and electric" in their name and are easily identified (e.g., Pacific Gas & Electric, Public Service Electric & Gas). It is also common for combined electric and gas utilities to use the word "energy" in their name (e.g., Duke Energy, Puget Sound Energy). Still others use nondescript names without any of these terms (e.g., Exelon, Avista).

Electric & Heat Utilities. In many parts of the world it is common for buildings to be connected to a steam pipe system, called a *district heating system.* The steam is distributed and sold to customers for space heating, water heating, and potential other uses. Steam for district heating is often produced by using the waste heat from small local electric generators owned and operated by electric utilities. Often these electric utilities also own and operate the district heating system, making them a combined utility.

Electric & Water Utilities. Many parts of the world have watersheds that are not capable of providing enough drinking water for the surrounding population. In these cases, often on small islands, the waste heat of electric generation facilities is used to distill seawater into drinking water, making the combined electric and water utility a logical choice.

Water & Wastewater Utilities. It is common for water and wastewater utilities to be combined. Both deal with water pipe installation and maintenance. Both deal with water quality issues. Also, much of the drinkable water produced by the water utility system ends up in the wastewater utility system, requiring coordination for expansion planning. Last, many wastewater utilities rely on water metering for bill calculation; a combined water and wastewater utility does not have to share billing data or read the meter twice.

1.2   NATURAL MONOPOLIES

Most of the world economy is based on *free markets* where businesses are free to offer and price goods and services, and consumers are free to either purchase them or not. Free market theory, discussed further in Chapter 3, shows that consumer choice in a free market results in prices and consumption quantities that are optimal from a societal perspective.

But there is a catch. Free markets require competition to operate efficiently and to maximize societal welfare. A single supplier that dominates an industry without competition and is able to control prices is called a *monopoly*. Monopolies occur for several reasons.

A new industry may start with a single company that grows large before other companies can establish themselves. This incumbent company can then use anti-competitive behavior to deter the formation of new competition or to squash new competition if it does occur (think of Microsoft for computer software).

Another scenario starts with a large number of relatively small competitors. As one company grows large compared to others, it becomes more cost efficient, offers prices lower than the competition, increases market share, and repeats the cycle. In this way, industries with *increasing economies of scale* naturally tend towards a single dominant company (think of Wal-Mart for discount retailing).

Last, a product or service can require a huge upfront investment. Once an initial company makes this investment, a second company would have to repeat this investment to compete for customers. Since a large percentage of the cost is related to an initial fixed investment, a single company can provide the service more cost effectively than multiple companies, even if there are no economies of scale (think of infrastructure utilities).

A *natural monopoly* is often defined as an industry with increasing economies of scale for all realistic sizes. With this definition, only one company in an industry is able to survive in the long run. A more complete definition of a natural monopoly is an industry where a single company can provide services to customers at a lower cost than multiple companies. Competition might occur, but would be detrimental to society since prices would increase and money that could have been used for other purposes is required to duplicate existing infrastructure.

Imagine a market for retail electricity with four competitors. The ability of customers to choose among suppliers will keep these companies focused on being cost competitive and reliable. But competition in this case would require four separate electric distribution systems. Where a single power pole exists now, four would exist under competition. Where a green utility box on your lawn exists now, four would exist under competition. Each of these companies would have to charge enough to recover the costs of their investments, meaning higher charges for the distribution system. Even if competition results in certain efficiencies and benefits, this scenario would result in much higher electricity bills, a negative visual impact on the community, and additional adverse effects such as road closures for maintenance and repair work. The same is true for all other infrastructure utilities deemed natural monopolies.

To reiterate, infrastructure utilities are natural monopolies, but might encounter competitors if left purely to a free market system. Since society would be worse off in this situation, laws and regulations are passed that give a single public utility the exclusive right to provide a utility service to customers within a certain geographic area. Competition is prevented, and the advantages of competition are foregone.

Without competition, utilities basically have no economic incentive to keep costs low or to keep quality high. Therefore, utilities are subject to regulatory oversight when it comes to rates and quality of service. The goal of utility regulation is to keep as much of the benefits as possible that would otherwise occur in a competitive environment.

Competition in the context of monopolies refers to direct competition where another company offers essentially the same service. It does not include indirect competition from substitute services. For example, local telephone service is experiencing competition from computer-based voice calls made over the Internet. Industrial customers can choose to generate some of their own electricity. Customers can choose to install a propane tank rather than receive natural gas from the gas utility. Customers can install their own water wells and septic tanks. Although the threat of substitutes provides a bit of incentive for utilities to keep rates low and quality high, utilities still have tremendous monopoly power and the need for regulation as a natural monopoly remains.

1.3   UTILITY OWNERSHIP

The term "public utility" is potentially confusing. Although the word public is not intended to refer to ownership, many assume that public utilities are owned by investors and are publicly traded. Others assume that public utilities are owned by the public through government organizations. Both assumptions are wrong. A public utility is simply an organization that provides an essential service to the public, regardless of ownership.

Because this is a book about business, the concept of ownership is of critical importance. Owners, by definition, determine the vision, mission, and goals of their business. It is their company. What they say goes (subject to ethical, legal, and regulatory constraints). Owners of for-profit businesses are primarily interested in profit maximization. Government owners are primarily interested in political considerations, which often includes profits for the government coffers.

Customer-owners are primarily interested in good value, and are happy when excess revenue, the equivalent of profit, is achieved.

Although ownership structures may differ, each utility still provides an essential public service and its customers are still the final arbiter of performance. Customers will fight to keep rates low and quality high, regardless of ownership. Because it is best for society, most customers desire all utilities to behave as if they were businesses in a free market under full competition.

1.3.1 Privately-Owned Utilities

Privately-owned utilities are for-profit businesses that are owned by investors. They are registered corporations, and are commonly called *investor-owned utilities (IOUs)*. Investors purchase ownership in IOUs with the expectation that profits made by the IOU will provide a sufficient financial return.

When an IOU or any other corporation is formed, shares of *common stock* are issued to owners. Typically, each share of common stock has a proportional claim to company profits and proportional voting rights. If a single person owns all common stock, this person controls all company decisions and profits.

Common stock can be privately held or publicly traded on stock market exchanges. Companies in the US with privately held stock are not required to submit detailed financial data to the Securities and Exchange Commission (SEC), and are not constantly pressured by investors and analysts on financial plans and profit projections. Companies in the US with publicly-traded stock have to provide this information, which takes significant time and management attention. In return, the financial situation of publicly traded companies is transparent to investors, making it much easier to raise money through the sale of common stock or bonds. In financial jargon, publicly-traded companies have good access to *capital markets*.

Historically, most large IOUs in the US were publicly traded. This makes sense since utilities are extremely capital intensive and constantly have to raise money through capital markets to expand and replace infrastructure. Publicly-traded companies have found recent legislation, such as the Sarbanes-Oxley act of 2002, to be costly and burdensome. Since privately held companies do not need to meet these and other requirements, there has been a trend for publicly-traded companies to "go private." This involves an investor or private equity firm purchasing all common stock and then delisting from all stock market exchanges. When public utilities go private, the purchasing investors must show that they can provide sufficient capital for utility growth and operations.

It is also possible for privately-held utilities to "go public." A private utility goes public by issuing a *public offering* (the first public offer of a company is called its *initial public offering* or *IPO*). This is done by partnering with an investment bank, called the *underwriter*. The underwriter helps to set the initial offer price, market the company to institutional investors, and ensure that shares of stock are sold at the highest possible price.

IOUs tend to be large. In the US, investor-owned electric utilities are only about 7% of all electric utilities, but serve about 70% of all electric customers. The largest IOUs have millions of customers.

1.3.2 Government-Owned Utilities

Many utilities are owned and operated by government entities. Because government entities are owned by citizens of the general public, government-owned utilities are also called publicly-owned utilities. The reader should note the difference between a publicly-owned utility, which is part of government, and a publicly-traded utility, which is part of the private sector and has its stock traded on public exchanges.

Utilities can be owned by many different levels of government including towns, cities, counties, states, and the federal government. City-owned utilities are very common. They are often called *municipal utilities*, or simply *munis*. Munis typically operate as a city department (e.g., the Los Angeles Department of Water and Power), and are run by a political appointee of the mayor. County-owned and state-owned utilities exist but are less common. The US federal government also owns some utilities, but these are mostly limited to large hydroelectric projects, the Tennessee Valley Authority (TVA), and the Bonneville Power Administration (BPA).

A *public utility district* (PUD) is a specialized type of government-owned utility with geographic boundaries not limited to cities or counties. They are similar to school districts in this respect. PUDs are led by commissioners that are either elected or appointed. PUDs are similar to munis in many ways, but their service territories tend to be less urban. Political pressures in PUDs are often less than munis since they are focused strictly on providing utility service and they have self-contained financial systems.

Governments around the world have chosen to incorporate some of their public utilities into private businesses. Sometimes the shares of the utility are sold to the public, but it is more common for the government (at least initially) to remain the sole shareholder. In these cases, utilities are bound by all of the rules and regulations of other IOUs. But their board of directors is appointed by the government, and all dividend payments go to the government. For example, Canada has many *Crown Corporations*, where IOUs are wholly-owned by a provincial government.

Government-owned utilities like munis have certain cost advantages when compared to IOUs. First, they do not have to pay income taxes or property taxes. Second, they are able to borrow money more cheaply through the issuance of government bonds, which have tax-free interest. These cost efficiencies do not always result in lower rates. It is common for munis to set rates comparable to nearby IOUs, and transfer excess profits from the utility department to the government general fund.

It is not uncommon for cities to threaten to take over the privately-owned utility systems that serve them. Reasons for this are varied and can include the hope of lower rates, frustration with quality of service, or to otherwise exert pressure on the IOU to do something beneficial for the city. Sometimes the cities actually assume ownership, typically by purchasing the infrastructure under the terms of the utility franchise agreement. This process is called *municipalization*.

It is also not uncommon for cities to privatize utility systems. This could be initiated by a city in an effort to raise money, by voters who feel that an IOU could do a better job, or by an IOU looking to grow its service territory.

Although some government-owned utilities are very large, they tend to be small on average. For example, government-owned electric utilities comprise about 60% of all electric utilities in the US, but only serve about 15% of all electric customers.

1.3.3   Member-Owned Utilities

Cooperatives (*co-ops*), sometimes called *electric membership corporations* (*EMCs*), are a special form of public utility where residents in an area own their utility system. In the US, cooperatives are formed under state law and take the form of a nonprofit business. The ownership structure of cooperatives is somewhat different than IOUs since voting rights are assigned to members instead of voting shares. Each member of a cooperative has the same rights as every other member.

Since cooperatives are nonprofit enterprises, they strive to provide utility services at a cost equal to collected revenues. Any excess revenue is typically returned to members in the form of credits on the utility bill or as dividend payments.

Nearly all cooperatives are formed to provide utility services in rural areas. Due to low customer density, IOUs are often not interested in serving these areas. The cooperative option allows customers to take control of their own destiny. The formation of cooperatives is aided by the availability of low interest loans from a federal agency called the *Rural Utility Service*, or *RUS*. Prior to 1995, the RUS was called the *Rural Electrification Administration*, or *REA*.

By their very nature, cooperatives tend to be small in terms of customers. An average electric cooperative serves twenty-thousand members, with many having only a few hundred. Also due to their nature, cooperatives require more infrastructure per customer since customers live farther apart. For example, an average electric cooperative has seventeen customers per mile of distribution facilities while IOUs have an average of thirty-five per mile and municipals have an average of forty-seven per mile. Some cooperatives have grown larger and denser as metropolitan areas sprawl into formerly rural areas. But even the largest cooperatives are small by IOU standards.

The REA and RUS have been fantastically successful at providing rural utility service in the US by encouraging the formation of cooperatives. In 2009, about 12% of all US electric customers and about 5% of US telephone customers are served by cooperatives. At this point, rural utility service in the US is nearly universal and the number of cooperatives is gradually declining. As IOUs pursue growth opportunities, they are gradually purchasing cooperatives, while few new cooperatives are forming.

1.4   UTILITY REGULATION

Since utilities are natural monopolies, they require government oversight so that they do not exercise monopoly power to the detriment of customers and society. This oversight typically comes from legislation establishing regulatory bodies. In turn, these regulatory bodies establish regulations and monitor utilities for compliance. In the US, utility regulation typically occurs at both the state and federal level.

It is sometimes helpful to interpret regulation in terms of the implicit pact a utility makes with society. Society grants a company the right to be the exclusive provider of a utility service for all customers within a defined service territory, making the company a public utility. In return for this right, the utility agrees to several conditions. First, the utility agrees to serve all customers within the service territory in a non-discriminatory way. Second, the utility agrees to provide adequate service quality. Third, the utility agrees to make only prudent management decisions that can generally be considered good business practice. Fourth, the utility agrees to rate regulation.

The goals of regulators are to have utility services provided with adequate service levels for the lowest possible rates. In this role, they strive to balance the business interests of the utility with the public interests. It is in everyone's long term best interest to have a financially viable utility. It is also in the interest of a local community to have affordable and reliable utility services.

Regulators are sometimes perceived to be the defenders of customer rights against the potential abuses of profiteering utilities. This is true in the sense that regulators must ensure that utilities do not use their monopoly status to the detriment of customers and society. However, regulators must also ensure that utilities are treated fairly. When an issue arises, regulators have a role similar to courtroom judges. They are obliged to give each side a sufficient opportunity to present its case.

Regulators formally address an issue by opening up a *docket*, which is a calendar of events that may include deadlines and dates for registration, requests for information (RFIs), written testimony, written rebuttals, depositions, hearings, final briefs, and decisions. Regulatory decisions are not necessarily final, and can be appealed to the courts.

When a docket opens, utilities will present their own recommendations, evidence, and testimony (often with the support of external consultants and experts). A variety of entities may present opposing views, including ratepayer advocacy groups, municipalities, environmental groups, and others. Entities that involve themselves in regulatory hearings are called *interveners*. The commission staff also plays a key role, since staffers often have more experience and detailed utility knowledge than the commissioners they serve. It is the job of regulatory commissioners to weigh the evidence presented by the utility and interveners, and make a fair decision that balances the interest of all stakeholders.

1.4.1 State Regulation

In the US, most utility regulation is performed at the state level through a regulatory body. The most commonly used terms for a state utility regulatory body are *public service commission (PSC)*, *public utilities commission (PUC)*, and *corporation commission*, although many other terms are used. These regulatory bodies are led by commissioners, who are either political appointees of the state governor or directly elected by voters.

Although each of the fifty states has its own regulatory body, regulatory commissioners talk to one another. Communication happens through a variety of venues, but primarily through the *National Association of Regulatory Commissioners (NARUC)*. NARUC allows commissioners in all states to address emerging issues, trends, successes, and failures. NARUC also presents the interests of state governments to federal utility regulators. The activities of NARUC lead to regulatory trends when states address similar regulatory issues and choose to adopt similar regulatory approaches.

1.4.2 Federal Regulation

In the US, there has been an increasing trend towards federal agencies assuming broader responsibilities for utility regulation. Many aspects of local utility systems, especially with regard to infrastructure and rates, remain regulated primarily by states. However, there are many interstate, public health, and national security issues that require federal regulation, even for utilities with only local systems.

Utility regulation by the federal government comes from a variety of agencies. Some of these agencies only regulate utilities, while others regulate a number of different entities. The following summarizes the federal agencies most involved with utility regulation in the US.

Securities and Exchange Commission (SEC). The primary role of the SEC is to regulate the trading of financial securities such as corporate stocks and bonds. In this respect, the SEC regulates utilities in much the same manner as other businesses. However, the SEC has historically played an important role that goes beyond this generic function. In 1935, the *Public Utility Holding Company Act (PUHCA)* authorized the SEC to regulate the securities of utilities operating in multiple states. The effect of this regulation was to limit businesses, in most cases, from owning utilities in geographically separated areas. The Energy Policy Act of 2005, among many other things, repealed PUHCA. However, state regulatory commissions can still present a major obstacle when utilities attempt mergers and acquisitions that otherwise would have been prevented by PUHCA.

Environmental Protection Agency (EPA). The US EPA is a large organization responsible for protecting human health and the environment. It creates and enforces national environmental standards under a variety of laws. Many of these laws affect all businesses, including utilities. In addition, the EPA regulates several areas that have specific implications for certain types of public utilities. The first are air quality and emission standards, which have a heavy impact on power plants that burn fossil fuels. The second is drinking water quality standards, which impact water utilities. The last is the discharge of pollutants into US waters, which primarily impacts wastewater utilities and electric generation facilities using cooling water.

Federal Energy Regulatory Commission (FERC). FERC regulates the interstate transmission of electric power, natural gas, and oil. According to FERC, its primary regulatory responsibilities include: the transmission and sale of natural gas; the transmission of oil by pipeline; the transmission and wholesale sales of electric power; the licensing and inspection of hydroelectric projects; the siting and abandonment of interstate natural gas pipelines and storage facilities; the safety and reliability of liquefied natural gas terminals; the reliability of interstate electric transmission systems; the monitoring of energy markets, including the assessment of fines; the monitoring of environmental matters related to natural gas, hydroelectric, and major electricity policy initiatives; and the administration of accounting and financial reporting regulations and conduct of regulated utilities. Prior to 1997, the FERC was called the *Federal Power Commission (FPC)*.

Nuclear Regulatory Commission (NRC). The NRC regulates radioactive materials with the goal of ensuring public safety and protecting the environment. As such, the NRC regulates all non-military nuclear reactors including electric power producing reactors and others used for research and test purposes. The NRC issues licenses for new reactors, monitors existing reactors, and reissues licenses or oversees decommissioning when the license of an active plant expires. The NRC also regulates radioactive material, including the nuclear fuel used in power producing reactors. Last, the NRC regulates the handling, storage, and disposal of nuclear waste.

Federal Communications Commission (FCC). The FCC provides federal regulations related to interstate and international communications by radio, television, wire, satellite, and cable. The FCC is the primary regulator for long distance telephone companies and cellular telephone companies. In certain situations, it also provides targeted regulation for local telephone companies. Many utilities, due to their dispersed geographic nature, use private communications networks such as microwave, private radios in licensed wavelength bands, private radios in unlicensed wavelength bands, and others. These networks must comply with FCC regulations.

1.5   UTILITY RATES

Most readers will be familiar with utility bills. Every month, utilities track the usage of each customer and issue a bill based on this usage. As discussed earlier, utilities do not have full discretion in setting rates. Utilities can propose rates to regulators. When this happens, regulators will either accept the proposed rates or require changes. The ratemaking process is discussed in detail in Chapter 7, but it is worthwhile to now provide some basic background material on utility rates.

Due to their status as a natural monopoly, utilities driven by pure profit motives will set rates higher than would otherwise occur in a competitive business environment (monopolistic pricing is discussed further in Chapter 3). With respect to rates, the job of regulators is to simulate the effects of competition. Good regulation will result in rates similar to what would occur under competition. Ratemaking is done by examining the costs utilities incur to provide services to customers. The target revenue for the utility is set so that these costs can be recovered with a fair profit left over for owners. Rates are then designed so that the utility can achieve the target revenue.

Most utility retail rates are based primarily on the amount of retail consumption. Power bills charge for electric power that is used. Natural gas bills charge for gas that is used. Water bills charge for water that is used. Long distance telephone bills charge for calls that have been made. Most customers are comfortable with a consumption-based rate design since heavy users pay proportionally more than light users.

The problem is that consumption-based utility bills do not properly reflect a utility's cost to provide service. For utilities, high initial costs are required to build infrastructure and additional high costs are required to maintain this infrastructure once it is built. Consider an electric utility primarily supplied by hydroelectric power. It costs the utility almost nothing in incremental costs when customers use more power. If all customers stop consuming electricity, the utility will still have to pay for its dams, transmission system, and distribution systems. Customer bills are based on consumption, but the cost to serve these customers has a large fixed component.

The mismatch between cost-to-serve and consumption is often used to justify lower rates for commercial and industrial customers. These customers typically require proportionally less infrastructure compared to the utility services they consume. Therefore, it costs the serving utility less on a per-unit basis of consumption. Having lower rates for large commercial and industrial customers is in alignment with the regulatory goal of rates being based on cost of service.

Imagine a utility customer that rarely uses the utility service. For example, a family may have a vacation home that is only occupied for a few weeks per year. From the perspective of this customer, no utility services are consumed during most months and so utility bills for these months should be very small. From a utility perspective, most infrastructure costs are incurred regardless of whether the vacation home is occupied, and utility bills should reflect this cost. As a compromise, many utility bills will have a minimum monthly charge to insure that at least some of the infrastructure costs are covered. Other utility bills may have a monthly fixed charge that serves a similar purpose.

Consider a local telephone company. It costs the phone company virtually nothing when customers make local calls. The connection is made automatically and incremental energy costs for each call are negligible. To reflect this structure of nearly 100% fixed costs, local telephone bills are typically a fixed charge regardless of local call volume.

Engineers design utility systems to handle times of peak usage. Most of the time there is plenty of unused system capacity. Peak demand may only occur a few hours per year, but the system is built to accommodate these times. Reducing peak demand during these few hours allows the utility to build and maintain less infrastructure, reducing costs. As such, utility rates, especially for large users, often contain incentives to reduce peak usage. Rates may include peak demand charges, which require meters that record peak usage (or other measures that impact peak usage) during a billing period. Rates may also include escalating block charges, where usage becomes increasingly expensive as more is used.

Peak demand can also be reduced by smoothing out customer usage. Customers do not necessarily have to consume less. They just have to shift some usage from peak hours to off-peak hours. There are various rate approaches to incentivize this behavior. The most common is to have high rates during peak hours and low rates during off-peak hours (e.g., long distance telephone rates with unlimited free calls during weekends, when usage is low). These are generically referred to as *time-of-use rates*. Another approach is to offer large customers lower overall rates if they agree to temporarily shut down operations or reduce load if requested by the utility, called *interruptible rates* and *curtailable rates*.

As mentioned before, utility rates are designed so that utilities can recover their costs and make a fair profit. In certain situations, regulators do not allow certain costs to be recovered. Costs that result from sound management decisions are said to by *prudently incurred*. Good faith must be presumed by the regulators, which means that management decisions must be judged based on when they occurred, without the benefit of hindsight. In addition, regulators are not allowed to substitute their judgment for the judgment of management. With these considerations in mind, costs that regulators find to be imprudently incurred, often through hearings called *prudency reviews*, may not be eligible for full recovery.

In a competitive environment, good management decisions sometimes lead to bad financial outcomes. For example, a business may build a new factory to accommodate a projected increase in consumer demand. For many reasons, such as an unexpected downturn in the economy, the increase in demand may not occur and the business will not recover its investment in new factory. But what about a utility that expands its system capacity for a projected increase in demand that never occurs? Even though these costs were prudently incurred, many consumer advocates argue that regulators should not allow for cost recovery since the intent of regulation is to simulate competition. Utilities argue that they are required by regulation to meet peak demand, and would be subject to sanctions if they did not build a system with sufficient capacity. At present, the utility argument prevails with regulators and all prudently incurred utility costs are recoverable through rates.

1.6 SERVICE STANDARDS

Businesses in a competitive environment pay close attention to product quality and customer service. Customers will purchase more from businesses that offer high quality for the money and purchase less from businesses that offer low quality for the money. Similarly, when businesses compete, customers not receiving good customer service can take their business elsewhere. Utilities do not directly compete, and are often accused of having poor customer service. The reader may recall Lily Tomlin's parody of a telephone operator speaking to a customer, "Next time you complain about your phone service, why don't you try using two Dixie cups with a string. We don't care. We don't have to. We're the Phone Company!"

A monopoly driven purely by profit will provide a lower quality of service than it otherwise would in a competitive business environment. It is the job of regulators to be aware of this fact and to ensure that the service quality provided by utilities is adequate. Certain aspects of service standards may have well-defined metrics with specific targets set by regulation or legislation. These metrics will vary based on the type of utility service being offered, as discussed in Section 1.1. But there are certain aspects of service standards that apply to all public utilities. Each of these is now briefly discussed.

Obligation to Serve. Utilities are given the exclusive right to provide essential services within a specified service territory. From the utility perspective, this means no direct competition. From a customer perspective, this means no choice among suppliers. Since public utilities by definition provide an essential public service, they are obligated to connect and serve all customers within their service territory. Existing customers cannot be disconnected without cause, such as a failure to pay bills or equipment tampering. New customers must be connected and served, even if the utility feels that it is a money losing proposition.

Continuity of Service. Being essential, utility services are expected to be available the vast majority of the time. Therefore, utilities are obligated to have infrastructure designs, inspection programs, maintenance procedures, and operational practices that result in a high continuity of service. When service interruptions inevitably occur, utilities are expected to have the ability to quickly identify the root cause, make repairs, and restore service. Regulators understand that perfect reliability is infinitely costly. Therefore, it is also the responsibility of utilities to balance continuity of service with cost.

Quality of Service. Utility services are of little good if quality is so poor that the service is essentially useless. For example, water pressure, gas pressure and electric voltage need to be sufficiently high, but not too high. Telephone voice distortion must not be excessive. Each public utility has its own extensive list of issues related to quality of service, some of which have been discussed in Section 1.1. Some aspects of service quality must fall within a range, with extreme values being undesirable. Other aspects of service quality can keep getting better and better (e.g., the purity of drinking water). As with reliability, regulators understand that perfect service quality is infinitely costly. Therefore, it is the responsibility of utilities to balance quality of service with cost.

Safety. Safety is important in all aspects of society and business, but is particularly important for utilities. A big part of the utility business is constructing, maintaining, and operating utility infrastructure, which is inherently hazardous. As such, all utilities have a heavy emphasis on safety training, safety equipment, and safety procedures. Many utilities also list safety as a core value, sometimes requiring safety messages to be given before all meetings. Concern for safety goes beyond employees. Since utility infrastructure is geographically distributed throughout populated areas, public safety is also of critical importance. The types of public safety concerns are too numerous to list exhaustively, but include things such as electrocutions, explosions, fires, drowning, vehicles colliding into utility structures, people trespassing in hazardous utility areas, and so forth. Several visual images depicting utility safety concerns are shown in Figure 1.7.

Cost Efficiency. Since utility rates are set based on their cost to serve, there are no economic incentives to become more cost efficient. Therefore, utilities are required by regulators to make decisions that result in the lowest possible rates for customers. Since rates determine revenue, the obligation to pursue cost efficiency is often called the *minimum revenue requirement*. If there are two effective ways to address a problem, the minimum revenue requirement obliges the utility to choose the one with lower cost. Expensive investments can be made today, but only if they are cost effective in the long run compared to alternatives.

Customer Service. From a customer perspective, there is much more to a utility than the service they supply. Customers interact with the utilities in a variety of ways and, like dealing with other businesses, expect a certain level of professionalism and a minimum amount of unnecessary frustration. Bills should be correct. Phone calls should be answered within a reasonable amount of time.

Figure 1.7. The picture on the left shows utility crews performing maintenance on a high voltage transmission structure. For this type of activity, proper training, procedures, and equipment are critical for worker safety. The picture on the right shows a sinkhole that opened up after a break occurred in a water main. This is just one example of how utilities must be extremely conscientious about public safely as well as worker safety.

Appointments should be kept and be on time. Complaints and questions should be handled with due process and in a timely manner. Conversations should be courteous and professional. Regardless of whether regulators define specific metrics and targets for customer service, the utility obligation to provide adequate customer service remains. Poor customer service will inevitably lead to a large number of customer complaints to regulators and elected officials, eventually leading to regulatory action.

Nondiscriminatory. Public utilities are required by regulation to provide services in a nondiscriminatory manner; no customer is allowed to get especially good treatment and no customer is allowed to get especially bad treatment. Restaurants often post signs stating, "We reserve the right to refuse service to anyone." Utilities do not have this luxury. Federal equal protection laws prohibit discrimination based on race, religion, sex, age, and several other factors. But restaurants can refuse service to people who have poor hygiene, are generally obnoxious, or for a variety of other reasons. Utilities have to serve everyone the same. For example, utilities are not able to provide better infrastructure, maintenance, and restoration priority to wealthy neighborhoods because they pay higher utility bills. Such actions would be considered discriminatory against customers in poor neighborhoods.

Although regulators can impose sanctions on utilities with poor service quality on a case-by-case basis, many are choosing to set service quality targets with pre-defined penalties and rewards. A typical approach is to select one or more service quality metrics that can be accurately computed by the utility and have been historically tracked for several years or more. Annual targets are then set for a period of years. Utilities must pay penalties if performance targets are not met. In some cases, utilities may be given rewards if performance targets are exceeded. There is often a range of service quality outcomes, called a *dead band,*

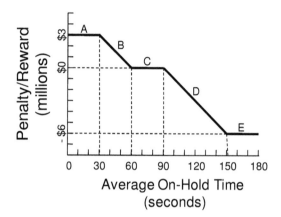

Figure 1.8. A performance-based rate example. This figure shows a possible PBR structure for average call center on-hold times. A dead band is set between sixty and ninety seconds, where no penalties or rewards are assessed (C). Penalties escalate for on-hold times greater than ninety seconds (D). Penalties are capped at $6 million for on-hold times greater than one hundred fifty seconds (E). Increasing rewards are earned for on-hold times less than sixty seconds (B). Rewards are capped at $3 million for on-hold times less than thirty seconds (E).

where neither penalties nor rewards are applicable. The use of financial penalties and/or rewards based on service quality is called performance-based regulation (PBR). An example of a possible PBR structure for a call center is shown in Figure 1.8, with penalties and rewards based on the average customer on-hold time until they first speak with a customer service representative.

1.7   DEREGULATION

Deregulation is a general term used to describe a relaxation of regulatory oversight, rules, and requirements of an industry. The extent of deregulation can be large for industries subject to competition, but is necessarily limited for utilities due to their monopoly status. In fact, so-called utility deregulation often results in more regulation, but of a different type. For this reason, many prefer the terms *industry restructuring* and *re-regulation.* Outside of the US, these regulatory changes are commonly called *liberalization.*

The goal of deregulation is to let free market forces, to the greatest extent possible, drive business behavior. Successful deregulation will result in lower prices, higher quality, greater innovation, greater customer choice, and improved customer satisfaction. It will also allow the free market to determine when, where, and by how much existing businesses expand their capacity and new businesses enter the market.

In the US, deregulation started in 1976 with the passage of the Railroad Revitalization and Regulatory Reform Act. For the railroads, as with other industries, the initial legislative action was just the beginning of a complex deregulation process taking many years. A summary of major deregulation milestones are:

US Industry Deregulation
1976 – Railways (Railroad Revitalization and Regulatory Reform Act)
1978 – Natural Gas (Natural Gas Policy Act)
1978 – Airlines (Airline Deregulation Act)
1980 – Interstate Trucking (Motor Carrier Act)
1982 – Interstate Busing (Bus Regulatory Reform Act)
1992 – Electric Utilities (National Energy Policy Act)
1996 – Telecom (Telecommunications Act)

Each of these acts is complicated in both content and industry impact. Detailed treatment is beyond the scope of this book, but it is appropriate to briefly discuss some of the acts that have had a large impact on utilities.

Prior to 1977, the wellhead price and transportation rates of natural gas were heavily regulated. This regulation led to very low gas prices and little incentive for energy companies to explore for new supplies. The oil embargo and resulting energy crisis of the mid-1970s highlighted the need for increased domestic energy supplies, leading to the Emergency Natural Gas Act in 1977 and the Natural Gas Policy Act in 1978. These acts gradually transitioned natural gas wellheads from regulated pricing to market-based pricing.

Prior to 1992, nearly all electric power was generated by regulated electric utilities; the cost of electric power generation was bundled together with all other costs of business and recovered through rates. The National Energy Policy Act of 1992 radically changed this business structure by allowing unregulated electric power producers to use utility transmission facilities to transport wholesale power to customers (called *open access*). This resulted in the construction of a large number of new power plants, called *merchant generation*, by business investors. It also resulted in an electricity market complete with spot prices, day-ahead prices, and futures. Utilities choosing to participate in the electric generation market were forced to *vertically unbundle*, where generation, transmission, and distribution are, at a minimum, operated as separate businesses.

Prior to 1982, all local and long distance telephone service was provided by the giant monopoly AT&T. A court settlement in 1982 forced AT&T to divest all of its local exchange operating companies. Local telephone service remained a monopoly, but long distance gradually became competitive with the success of companies like MCI and Sprint. The Telecommunications Act of 1996 further increased competition in several ways. First, it allowed the regional operating companies to compete in the long distance market. At the same time, it allowed long distance companies and cable television companies to offer local exchange service. Under the Telecommunications Act, the FCC oversees new entrants in the long distance market and state regulators oversee telephone competition in local markets.

Utility deregulation has primarily focused on production and wholesale rather than local service. This is reasonable since local utility systems would need to be duplicated to provide true local competition. Nevertheless, a form of competition exists for local telephone service, and many states have pursued similar goals for local gas service and local electric service. These efforts are commonly called *retail choice* and *retail competition*.

With retail choice, a customer is allowed to choose among a variety of service providers. These providers are responsible for purchasing natural gas or electricity from producers and reselling it to customers. These providers pay a fee for transmission and distribution services. With retail choice, traditional utilities are still responsible for all aspects of the utility infrastructure. The service providers only provide competition in terms of procurement and customer rate plans. A variety of studies have concluded that retail choice is often confusing to customers and may not result in significant benefits. Many customers tend not to choose a new service provider, and end up still being served by the local distribution company, who serves as the *provider of last resort*.

1.8   SUMMARY

Utilities are a special form of business in several ways. First, they provide services that are considered essential for the health, safety, and economic well-being of society. Second, they require extensive infrastructure that is not economical to duplicate. Based on these factors, utilities are given the exclusive right to provide service to customers in a specified service territory. In return, the utility agrees to regulatory oversight in terms of the rates they charge to customers and the service quality they provide to customers.

Voters in an area will typically decide whether they want utility services to be provided by the government or by private investors. Government-owned utilities are not for profit and have certain cost advantages, but tend to be influenced by political considerations. The behavior of IOUs is more predictable since they stay focused on profitability, but they are also subject to more regulatory oversight. Rural areas without utility service have historically formed member-owned cooperatives, which are not for profit, but resemble small investor-owned utilities in many other ways. Regardless of ownership, it is beneficial for all utilities to make strategic and operational decisions based on sound business principles.

Most utility oversight occurs through regulation rather than legislation. Governments will typically pass laws allowing the formation of regulatory bodies headed by elected or appointed commissioners. These commissioners are given broad powers to make decisions affecting utilities under their regulatory authority, such as setting rates and determining acceptable levels of service quality. In the US, extensive regulation occurs at both the federal and state levels.

In a free market with competition, customer choice will determine quality and price. Customers will choose a service that has higher quality for the same price or similar quality for a lower price. Without competition, a utility trying to maximize profits will tend to charge more and offer lower quality. As such, regulators require that utilities provide adequate levels of service in a way that minimizes cost. Rates are then set so that costs are recovered and, if applicable, a fair profit is made.

Regulation has its drawbacks, and recent efforts have been made to both simulate market forces and, more ambitiously, to deregulate certain aspects of certain utility industries. Market forces are simulated by performance-based ratemaking that financially rewards utilities for good performance and financially penalizes them for bad performance. Utility deregulation separates out aspects of a service where competition is practicable, and allows competition to occur. Aspects not suitable for competition remain regulated as usual, but now with the need to interface with the competitive parts of the industry.

A key regulatory requirement is for utility managers to make prudent business decisions, especially with regard to spending money. It is not enough to have good intentions. Utility managers must have both utility knowledge and business knowledge so that financial requirements can be balanced with infrastructure and operational requirements. In other words, managers are expected to run their utilities with the same level of competence as successful businesses in a competitive environment. New managers with business backgrounds from other industries need to learn the engineering aspects of the utility. New managers with engineering backgrounds from within the industry need to learn the business aspects of the utility.

1.9 STUDY QUESTIONS

1. What is meant by the term natural monopoly? What are some different ways that a natural monopoly can occur?
2. What is meant by the term public utility? Are public utilities completely free from competition? Explain.
3. Describe the three primary types of utility ownership. Is ownership structure likely to lead to different infrastructure and operational decisions? Explain.
4. Why are utilities regulated? What are some of the major goals of utility regulation?
5. What was the purpose of the 1935 Public Utility Holding Company Act? What was the impact, if any, of its repeal in 2005?
6. Explain how a typical utility bill for a residential customer is computed? What are some benefits and drawbacks of this approach?
7. Why do utilities care about peak demand? What are some things a utility can do in an attempt to reduce peak demand?

8. What is meant by the term minimum revenue requirement? How does this term relate to utility rates?
9. Name some of the major aspects of utility service standards. What is meant by obligation to serve?
10. What is the goal of utility deregulation? Describe a real example.

2
Accounting

Accounting is the language of business, and an understanding of basic accounting principles is critical for any person wishing to understand the business drivers of utilities today. This chapter presents a summary of accounting from an engineering and technical management perspective. Its purpose is not to turn the reader into an accountant. Therefore, detailed treatment of accounting mechanics, rules, nuances, and processes are avoided. The focus is to present the theory and principals underlying financial accounting, provide a sound basis for the interpretation of financial reports, and demonstrate an array of key accounting issues that are important for utility engineers to understand.

2.1 THE BASIC ACCOUNTING EQUATION

Although many perceive accounting to be arcane and esoteric, it can be boiled down to one simple equation: the assets of a company must be equal to the claims on these assets. This *basic accounting equation* is most commonly represented as the following:

Assets = Liabilities + Owner's Equity   (2.1)

An *asset* is something of value, and a company's assets are the sum of all things of value that a company owns. A *liability* is an obligation of the company, such as a loan or an unpaid bill. When the value of all liabilities is subtracted from the value of all assets, the residual value is left for the owners. This residual value is called *owner's equity*.

To illustrate the basic accounting equation in action, a simple example is now provided. Imagine a group of investor purchasing and running a small utility. These investors initially sell ten million shares of common stock at $10 per share. The sale of stock raises $100 million in cash, which is referred to as *paid-in capital*. The raised cash, an asset, corresponds to $100 million in owner's equity.

The stock transaction described above is reflected in the *balance sheet* of the company. The balance sheet can be thought of as an expanded version of the accounting equation, and describes all assets, liabilities, and owner's equity. The balance sheet of the company (all values in millions of dollars) after the sale of common stock is:

Assets		Liabilities		Owner's Equity	
100	Cash			100	Common Stock
100	**Assets**	**0**	**Liabilities**	**100**	**Owner's Equity**

Transaction: Issue 10 million shares of common stock at $10 per share.

The above table is organized into columns corresponding to assets, liabilities, and owner's equity. Each individual item in these categories is listed, starting from the top, in non-bold font. The total amount of each category is listed at the bottom in bold font.

After the first transaction of the company, the balance sheet confirms that assets are equal to liabilities plus owner's equity. In a more practical sense, the company now has (1) cash on hand, and (2) stock holders that expect a return on their investment. The company finds that it needs additional funds to start business operations. It therefore sells $50 million in bonds. The bond issuance raises another $50 million in cash, and results in a corresponding $50 million liability. The balance sheet is now:

Assets		Liabilities		Owner's Equity	
150	Cash	50	Bonds	100	Common Stock
150	**Assets**	**50**	**Liabilities**	**100**	**Owner's Equity**

Transaction: Issue $50 million in bonds.

The bond transaction causes the *cash account* to increase from $100 million to $150 million. To balance the increase in assets, a $50 million liability is recorded so that the total assets of $150 million equal the liabilities plus owner's equity.

The company now negotiates to buy a small utility system for $90 million and pays in cash. The effect is to reduce the cash account by $90 million and to add a new asset worth the same amount. Total assets have not changed, and total assets still equal liabilities plus owner's equity; the balance sheet still balances.

	Assets		Liabilities		Owner's Equity
60	Cash	50	Bonds	100	Common Stock
90	Utility system				
150	**Assets**	**50**	**Liabilities**	**100**	**Owner's Equity**

Transaction: Purchase a $90 million utility system.

The company is now a utility. In its first month, it provides utility services to its customers and then bills them $15 million for these services. The customers have not yet paid these bills, but their legal obligation to pay these bills is an asset to the company. This type of asset is typically recorded in a category called *accounts receivable*.

When a customer is obligated to pay for services rendered, the result is an increase in profits, also called *earnings*. Since the owners of the company have rights to these earnings, an increase in accounts receivable is balanced by an increase in an owner's equity account called *retained earnings*.

	Assets		Liabilities		Owner's Equity
60	Cash	50	Bonds	100	Common Stock
90	Utility system			15	Retained Earnings
15	Accounts receivable				
165	**Assets**	**50**	**Liabilities**	**115**	**Owner's Equity**

Transaction: Bill $15 million to customers.

Notice that the above balance sheet still balances. Assets are worth $165 million and the sum of liabilities and owner's equity is worth the same amount.

In the process of providing services to its customers, the utility incurs expenses of $5 million, which it owes to a variety of contractors and outsourcing companies. The utility has not yet paid its bills, but must record the obligation to pay these bills as a liability. Unpaid obligations are typically recorded in a category called *accounts payable*. An increase in accounts payable results in a decrease in earnings, which is recorded as a decrease in the retained earnings account.

	Assets		Liabilities		Owner's Equity
60	Cash	50	Bonds	100	Common Stock
90	Utility system	5	Accounts payable	10	Retained Earnings
15	Accounts receivable				
165	**Assets**	**55**	**Liabilities**	**110**	**Owner's Equity**

Transaction: Incur $5 million in expenses.

The company now pays the $5 million that it owes in bills. These payments come out of cash accounts, and are offset by a reduction in accounts payable. Since the company is transferring cash out of the company, total assets are lower. Since the company no longer has unpaid bills, total liabilities are also lower. The balance sheet reflects clearly the difference between a company with many unpaid bills and a company with few unpaid bills. The new balance sheet is:

	Assets		Liabilities		Owner's Equity
55	Cash	50	Bonds	100	Common Stock
90	Utility system	0	Accounts payable	10	Retained Earnings
15	Accounts receivable				
160	**Assets**	**50**	**Liabilities**	**110**	**Owner's Equity**

Transaction: Pay $5 million in unpaid bills.

Customers now pay $10 million of their unpaid bills. This is not the total amount owed, but has the effect of increasing cash by $10 million and reducing accounts receivable by the same amount. There is no net effect on total assets.

	Assets		Liabilities		Owner's Equity
65	Cash	50	Bonds	100	Common Stock
90	Utility system	0	Accounts payable	10	Retained Earnings
5	Accounts receivable				
160	**Assets**	**50**	**Liabilities**	**110**	**Owner's Equity**

Transaction: Customers pay $10 million of their bills.

The utility system will not live forever. To account for this, the value of the utility system on the balance sheet is reduced over time. This reduction in value is called *depreciation*. Depreciation is covered in more detail later, but for now the assumption is a utility system depreciation amount of $1 million. This reduces the *book value* of the utility system by $1 million, and is treated as an expense that lowers retained earnings. The updated balance sheet is:

	Assets		Liabilities		Owner's Equity
65	Cash	50	Bonds	100	Common Stock
89	Utility system	0	Accounts payable	9	Retained Earnings
5	Accounts receivable				
159	**Assets**	**50**	**Liabilities**	**109**	**Owner's Equity**

Transaction: Utility system depreciates by $1 million.

In addition to operating expenses and depreciation expenses, a utility is obligated to pay interest payments to bond holders. In this case, the utility pays $1 million in interest. These payments reduce the cash account and retained earnings accordingly. The interest payment does not affect the face value of the bonds, and therefore does not affect the amount of liabilities in the bond account. If the utility paid off part of the principle of the bonds, the cash account (an asset) and the bond account (a liability) would both be reduced.

The updated balance sheet after the $1 million in bond interest payments are made is:

Assets		Liabilities		Owner's Equity	
64	Cash	50	Bonds	100	Common Stock
89	Utility system	0	Accounts payable	8	Retained Earnings
5	Accounts receivable				
158	**Assets**	**50**	**Liabilities**	**108**	**Owner's Equity**

Transaction: Utility pays $1 million in interest to bond holders.

At this point, the utility has assets of $158 million. This is equal to the $50 million in liabilities plus the $108 million in owner's equity. Since the purchase of the utility system, the following observations can be made:

- Total assets have increased from $150 million to $158 million
- Total cash has increased from $60 million to $64 million
- Retained earnings have increased from zero to $8 million
- Liabilities have remained the same at $50 million
- No new common stock has been issued

From an accounting perspective, this utility is performing well. It is both profitable and generating cash. In order to transfer some of these profits to its owners, the utility now decides to distribute $2 million of retained earnings to common stock holders. This amount is taken from the cash account and is called a *dividend*. Since there are ten million shares of common stock, the dividend corresponds to twenty cents per share. The resulting balance sheet is:

Assets		Liabilities		Owner's Equity	
62	Cash	50	Bonds	100	Common Stock
89	Utility system	0	Accounts payable	6	Retained Earnings
5	Accounts receivable				
156	**Assets**	**50**	**Liabilities**	**106**	**Owner's Equity**

Transaction: Issue $2 million in dividends.

This example has been provided as an introduction to the basic accounting equation to show how the balance sheet corresponds to the basic accounting equation, and how financial transactions lead to changes in the balance sheet. There are a large number of possible financial transactions, with many being complicated and nuanced in how they affect the balance sheet. Some of these, those most important to the utility engineer, are addressed in the remainder of this chapter.

At this point, the reader should be comfortable with the simple utility example presented in this section, and the accounting terms used in the example. Those for whom the content of this section represents a new area of knowledge are encouraged to repeat the material if necessary. Good comprehension will increase the ease and value of remaining sections.

2.2 JOURNALS, LEDGERS, AND ACCOUNTS

In the previous section, financial transactions led to increases and decreases in account values. Accountants have specific terms for these increases and decreases. These terms are *debit*, which is abbreviated *dr.*, and *credit*, which is abbreviated *cr.*

A debit refers to either an increase in an asset account, a decrease in a liability account, or a decrease in an owner's equity account. Think of using a debit card to withdraw cash from an ATM. The cash in your pocket, an asset, increases. The meaning of debit is opposite for different sides of the accounting equation.

A credit is the opposite of a debit. It refers to either an increase in a liability account, an increase in an owner's equity account, or a decrease in an asset account. Think of using a credit card. Your credit card debt, a liability, increases. Like debit, the meaning of credit is opposite for different sides of the accounting equation. A summary of debits and credits is:

Debit (Dr.) – Increase in an asset account;
Decrease in a liability account;
Decrease in an owner's equity account.

Credit (Cr.) – Decrease in an asset account;
Increase in a liability account;
Increase in an owner's equity account.

Debits and credits typically come in pairs. Consider the following three common financial transactions: (1) an asset account increases and a liability account increases, (2) an asset increases and another asset decreases, and (3) a liability account increases and another liability account decreases. In each case, the transaction involves a debit and a credit of equal value. In the first case, the asset account is debited and the liability account is credited. In the second case, an asset account is debited and another asset account is credited. In the third case, a liability account is credited and another liability account is debited.

When a financial transactions results in a debit and a credit of equal amounts, it results in a *double entry* recorded in the financial system. This *double entry accounting* is performed in a financial transaction log called a *journal*. A journal showing the transactions described in Section 2.1 is shown in Table 2.1. In this journal, each transaction involves a debit (Dr.) and a credit (Cr.) of equal amount. The debit and credit totals at the bottom indicate that the transactions balance in accordance with the basic accounting equation. If these totals are not equal, there is a mistake in the journal entries.

Table 2.1. This table shows double entry accounting in a journal that corresponds to the financial transactions discussed in Section 2.1. Notice that each transaction involves a debit (Dr.) and a credit (Cr.) of equal amount. The debit and credit totals at the bottom indicate that the transactions balance in accordance with the basic accounting equation.

Line	Date	Account	Dr.	Cr.
1	Jan. 5	Cash	100	
2		Common Stock		100
3		*Issue common stock*		
4	Jan. 15	Cash	50	
5		Bonds		50
6		*Issue bonds*		
7	Jan. 31	Utility System	90	
8		Cash		90
9		*Purchase utility system*		
10	Feb. 28	Accounts Receivable	15	
11		Revenue (retained earnings)		15
12		*Bill customers for services*		
13	Feb. 28	Expense (retained earnings)	5	
14		Accounts payable		5
15		*Incur operating expenses*		
16	Feb. 12	Expense (retained earnings)	1	
17		Utility System		1
18		*Depreciation expense*		
19	Feb. 15	Expense (retained earnings)	1	
20		Cash		1
21		*Interest payment on bonds*		
22	March 5	Accounts payable	5	
23		Cash		5
24		*Pay operating expenses*		
25	March 10	Cash	10	
26		Accounts Receivable		10
27		*Customer payments*		
28	March 15	Expense (retained earnings)	2	
29		Cash		2
30		*Dividend payment*		
31				
32		**TOTAL (millions)**	**$279**	**$279**

Table 2.2. A summary of journal entries affecting the cash account. This is sometimes called a *T account* since the debit and credit columns were historically divided with a T shape (see bold borders). The first column references each transaction to its journal entry. In this case, the cash account start with a value of zero. After seven transactions, the final amount of the cash account is $62 million.

Journal Line	Description	Dr.	Cr.	Balance
	Beginning balance			**0**
1	Issue common stock	100		100
4	Issue bonds	50		150
8	Purchase utility system		90	60
20	Interest payment on bonds		1	59
23	Pay operating expenses		5	54
25	Customer payments	10		64
29	Dividend payment		2	62
	Ending balance (millions)			**$62**

Each journal entry is linked to an account. Upon being entered in the journal, an entry is *posted* to its associated account. The cash account associated with the journal entries of Table 2.1 are shown in Table 2.2. This account is sometime called a *T account* since the debit and credit columns were historically divided with a T shape (see bold borders).

An account will start with a beginning balance, and then sequentially list all journal transactions that have affected the account. Each transaction is referenced to its journal entry, allowing the entire transaction to be viewed if necessary. The ending balance is provided after the last transaction. In addition to consolidating all account activity in one place, the account clearly shows how the account balance is changing over time.

In Table 2.2, the cash account begins with a balance of zero. Several large debits increase the cash account to $150 million, followed by several credits that reduce the account to $54 million. At this point, customer payments increase the cash account, followed by the final entry, which takes $2 million out of the cash account for dividend payments.

Revenue and expense accounts are temporary. They are built over a period of time, typically one month, and are then closed. Revenue accounts are closed by reducing their value to zero and increasing retained earnings by the same amount. Expense accounts (which have negative balances in the owner's equity category) are closed by increasing their value to zero and decreasing retained earnings by the same amount. These transactions are called *closing entries* and are done in a process referred to as *closing*.

Consider a utility that earns $10 million in revenue and incurs $5 million in expenses over the course of a month. The closing entries are shown in Table 2.3. Notice that the revenue account is reduced to zero with a $10 million debit. Similarly, the expense account is increased from negative $5 million to zero with a $5 million credit. The net impact on retained earnings is a $5 million increase.

Table 2.3. This table shows closing entries for a revenue account of $10 million and an expense account of negative $5 million. The revenue account becomes zero with a $10 million debit (reduction) and the expense account becomes zero with a $5 million credit (increase). The retained earning account increases by a net value of $10 – $5 = $5 million.

Line	Date	Account	Dr.	Cr.
1	Jan. 31	Revenue	10	
2		Retained Earnings		10
3		*Closing entry*		
4	Jan. 31	Retained Earnings	5	
5		Expense		5
6		*Closing entry*		
7				
8		**TOTAL (millions)**	**$15**	**$15**

A utility can easily have thousands of accounts. Managing these accounts is the responsibility of accountants, and utility engineers will seldom find it necessary to address accounting issues at this level of detail. To facilitate higher level analyses, accounts are often grouped together and presented as combined information.

Related accounts can be grouped together in a *subsidiary ledger*. For example, a utility will have a separate account for the receivables of each customer, which may number in the millions. These accounts are typically grouped together in the *accounts receivable subsidiary ledger*. The sum of all account balances in the subsidiary ledger is reported in a *controlling account*. In this way, the total amount of unpaid customer bills is quickly known by viewing the accounts receivable controlling account.

All accounts in the utility constitute the *general ledger*. Since the general ledger reflects all financial transactions, the sum of all general ledger debits must equal the sum of all general ledger credits. When these sums are generated, it is called a *trial balance*. If the trial balance does not reflect an equal total for credits and debits, one or more accounting mistakes exist.

This section has only scratched the surface of accounting mechanics and processes, but it is appropriate at this point to end. Those interested in more in-depth treatment of the inner workings of accounting are encouraged to pursue textbooks, self-study workbooks, and courses. However, remember that mastering the details of accounting is essential for accountants but not for utility engineers.

2.3 ACCOUNTING PRINCIPLES

The primary purpose of financial accounting is to present information about a company's financial performance to external parties such as owners, lenders, potential owners, potential lenders, financial analysts, and so forth. Since financial performance is closely related to profitability, financial accounting is structured so that it can provide a fair and accurate representation of a company's profitability. To do this, there are some guiding principles that serve as the foundation of financial accounting. A few of these principles are important to utility engineers and are now presented.

Principle 1: Matching. In order to properly determine profit, revenues must be matched with the expenses used to generate the revenues. This means that revenues are recorded as soon as services are provided to customers, not when customers pay for these services. Similarly, expenses are recorded as soon as they are used to generate revenue, not when they are paid. Recording revenues and expenses when they occur is called *accrual accounting*.

Consider an electric utility that purchases fuel for power generation. This fuel has not been used to generate revenue and is therefore recorded as an asset in a fuel account. When this fuel is used to generate electricity that is sold to customers, the amount of fuel used is recorded as an expense, with a corresponding reduction in the fuel account. In this way, the fuel expense is matched with the revenue it helps to generate.

Principle 2: Historical Cost. Financial transactions record the costs at the time the transactions are made. These *historical costs* are generally not adjusted to reflect changes in market value. Therefore, assets on a balance sheet do not represent either (1) the cost of these assets if they were to be purchased today, or (2) the value of these assets if they were to be sold today.

Consider a utility that purchased a piece of property for $100 thousand thirty years ago. Today, the market value of this land is $1 million, but the value of the land on the balance sheet is still $100 thousand. The reverse effect is also possible. Consider a utility that purchases $10 million in steel pipe. If the price of steel goes down, the market value of this pipe could be $7 million but the value on the balance sheet is still $10 million.

There are a few special cases where assets values are periodically updated to reflect market value (e.g., publicly-traded investments such as stock holdings). However, the historical cost principle means that asset values shown on the balance sheet do not reflect the current market value of the asset.

Principle 3: Full Disclosure. Financial statements are supposed to present complete and clear financial information. The full disclosure principle means that financial statements are not supposed to present misleading or potentially confusing information. It also means that potentially important information is not supposed to be omitted, obfuscated, or hidden. Accountants are not supposed to "cook the books" or use "creative accounting" to meet earnings targets. These actions violate the full disclosure principle.

The full disclosure principle typically results in financial statements having a large number of *footnotes*. These footnotes discuss important financial issues that would not otherwise be evident through the financial statements alone. Footnotes are a critical part of accounting, and it is not possible to fully understand the financial situation of a company without closely examining the footnotes.

The issue of full disclosure can be illustrated through the Enron scandal. Enron, an energy company, had reported revenues of $111 billion in 2000 and was one of the darlings of Wall Street. It was soon discovered that much of this revenue was from companies mostly owned by Enron and controlled by Enron executives. These *special purpose entities* would borrow money from banks and use this money to purchase Enron services. Enron secured these loans, and was responsible for their repayment if the special purpose entities failed. This ultra-critical financial risk was not disclosed in Enron's financial footnotes. It is not clear whether Enron violated any specific accounting rules, even though its aggressive accounting practices clearly pushed the limits. However, it is clear that Enron was in violation of the principle of full disclosure.

Principle 4: Consistency. Accounting is complicated, and there are often multiple ways to record financial transactions while still conforming to all accounting rules. The principle of consistency requires that a utility take the same approach to similar financial transactions and to not change this approach over time. If a utility desires to make a change in accounting practices, the effect of this change on financial statements must be determined and reported (as per the principle of full disclosure).

For example, it is possible to record many utility maintenance activities as either an expense or as an increase in the value of an asset. Treating maintenance as an expense will reduce profits today. Treating maintenance as an increase in the value of an asset will reduce future profits as this increase in value is depreciated over time. The principle of consistency prevents a utility from expensing maintenance activities in years with high profits and then switching practices in years with low profits.

Principle 5: Conservatism. Conservatism refers to the criteria for recording revenues and expenses. Based on conservatism, revenue should be accrued when it is reasonably certain, but expenses should be accrued when they are reasonably possible. Due to the asymmetry of these criteria, profits based on accrual accounting are pessimistic.

Consider a $5 million lawsuit filed against the utility by a customer. Even if the utility feels that it will win the lawsuit, it must be conservative and assume that it will lose. Since the utility actions leading to the lawsuit have already occurred, the expense must be recorded as soon as the lawsuit is filed and is deemed to have merit. The accounting transaction would consist of a $5 million expense and the creation of a $5 million provisional account that is treated as a liability. The $5 million expense results in correspondingly lower profits. If the customer wins the lawsuit and the utility pays, the cash account and the provisional account are both reduced by $5 million. Profits are not affected since the "earnings hit" has already occurred due to the principle of conservatism. If the utility wins the lawsuit, the accounting transaction is *reversed*, which consists of a $5 million expense reversal entry and elimination of the provisional account.

Now consider a $5 million lawsuit filed by the utility against a customer. In this case, the principle of conservatism requires the assumption that the utility loses the lawsuit, even if it is confident in winning. Once the lawsuit is won, the utility can increase revenue by $5 million, increase accounts receivable by $5 million, and wait to get paid.

These accounting principles, along with several others, are the basis for a detailed set of accounting rules that address many specific accounting situations. For publicly traded companies in the US, the Securities and Exchange Commission (SEC) requires conformance to rules set by the Financial Accounting Standards Board (FASB, pronounced "fazbee"). This non-profit organization periodically adds new rules to address difficult accounting situations. For example, in 1986 the FASB issued the Statement of Financial Accounting Standard (SFAS) Number 90, titled *Regulated Enterprises–Accounting for Abandonments and Disallowances of Plant Costs*. This rule required many electric utilities to reduce the value of nuclear power plants on their balance sheet, leading to dramatic reductions in profits.

Financial accounting is governed by *generally accepted accounting principles* (GAAP, pronounced "gap"). GAAP in the United States (US GAAP) consists of the basic accounting principles, FASB rules and standards, and generally accepted industry practices. Federal law requires that all publicly traded companies have their accounting statements audited and certified by independent accountants to demonstrate that the statements are accurate and were prepared in accordance with US GAAP.

Many companies outside of the US conform to rules set by the International Accounting Standards Board (IASB). It appears that US GAAP and IASB reporting requirements will eventually converge so that separate financial statements are not needed for US and international markets. This has important implications for all publicly traded companies, but is of particular importance for regulated utilities. US GAAP currently allows for the use of regulatory assets and liabilities in a manner that is not supported by the IASB. Also, US GAAP uses the principal of historical cost for asset valuation while the IASB uses fair value. These differences present difficulties for utilities since it is generally very difficult and costly to assign a fair value to utility infrastructure.

There is a concern by some that the IASB standards are less transparent than US GAAP. In early 2009, SEC chair Mary Schapiro expressed this concern as follows, "American investors deserve and expect high standards of financial reporting, transparency and disclosure – along with a standard-setter that is free from political interference and that has the resources to be a strong watchdog. At this time, it is not apparent that the IASB meets those criteria, and I am not prepared to delegate standard-setting or oversight responsibility to the IASB."

Many regulated utilities are part of a group of companies owned by a parent company. For example, a parent company may own several regulated utilities and several non-regulated businesses. Each of the businesses owned by the parent company will have its own general ledger and will keep track of its own profits and losses. All of the controlling accounts for each company are then combined into *consolidated financial statements* to show the overall financial performance of the parent company.

2.3.1 Assets

In accounting, an asset is a valuable resource acquired at a measurable cost. An asset is not necessarily a tangible thing, and can include items such as owed money, patents, and pre-paid insurance. Assets that can be touched and felt are called *tangible assets* and others are called *intangible assets*. A breakdown of asset types is shown in Figure 2.1.

Tangible assets are "real stuff" such as infrastructure, vehicles, and buildings. Utility infrastructure assets that are used to generate and deliver utility services are typically called *utility plant*. A tangible asset is initially recorded in financial statements at its installed cost, which includes the cost of engineering, design, equipment purchase, installation, and commissioning.

Intangible assets have value, but are not physical items. Non-monetary intangible assets include items such as patents and trademarks, and are recorded at their purchase price. Because of this, patents and trademarks granted to a utility do not appear as an asset since there is no purchase price. Acquired patents and trademarks, in contrast, appear as assets with a value equal to their purchase price. This demonstrates an important point: not everything of value to a utility necessarily appears as an asset on financial statements.

Goodwill is a special type of non-monetary intangible asset. It refers to a premium paid above market value for an asset, typically during the acquisition of a company. Suppose a utility announces the intention to purchase another utility. Before the announcement, the target utility is valued by the stock market at $1 billion. The negotiated acquisition price is $1.2 billion, which is $200 million over market value. Accounting rules state that the target utility be recorded at the market value of $1 billion. The premium of $200 million is recorded as goodwill.

Goodwill assumes that the acquiring utility will be able to recover the paid premium in increased profits that were not achievable by the target company prior to the acquisition. Typical justifications for goodwill are employee reductions, increased efficiencies, increased market power, increased economies of scale, sharing of best practices, and other synergies. If at any time it is discovered that the expected increases in profits cannot be realized, the value of goodwill must be reduced, or *written down*. When the value of an asset is written down, it is said to be *impaired*.

Monetary assets consist of cash and short term investments such as stocks and bonds. The accounting value of a monetary asset is periodically updated to reflect market value. For example, a company may purchase one thousand shares of a random stock at $10 per share. This $10 thousand is spent as a financial investment, and has no other purpose. If the stock price increases to $12 per share, the asset value must be changed from $10 thousand to $12 thousand on the balance sheet, resulting in a $2 thousand increase in revenue.

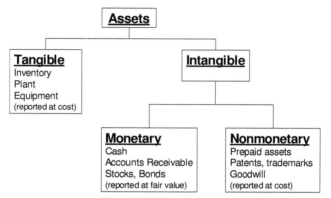

Figure 2.1. An asset is a thing of value that a utility owns. Assets can be further divided into tangible assets such as buildings and equipment, and intangible assets such as cash and patents. Since most assets are reported at their initial cost, financial statements may not be a good indication of actual asset value.

Another way to classify assets is based on the expected length of time until the asset is used up. Assets that are expected to be used up within one year are called *current assets*. Typical current assets include cash, cash equivalents, accounts receivable, inventory, and short-term investments. Assets that are not expected to be used up within one year are called *noncurrent assets* or *long-term assets*. Typical long-term assets include land, buildings, equipment, utility infrastructure, and long-term investments. Tangible long-term assets are also called *fixed assets*.

In summary, assets are things of value that the utility owns. These can be both physical items and nonphysical items. Typically, the value of assets in financial statements is based on the initial cost of the asset. Therefore, financial statements are not necessarily a good indication of true asset value.

2.3.2 Liabilities

Liabilities are financial obligations such as loans, unpaid bills, and unpaid taxes. Like assets, liabilities are grouped according to their timeframe. *Current liabilities* are expected to be paid off within one year. *Noncurrent* or *long-term liabilities* are not expected to be paid off within one year. Liabilities with an uncertain amount or timeframe are called *provisional*, such as the provisional liability set up for the lawsuit example discussed in the previous section.

Common long-term liabilities for utilities include long-term debt (e.g., bonds and bank loans), long-term leases, and employee pension obligations. These are of interest when examining long-term financial prospects. Long-term liabilities will eventually have to be paid, which leaves less available profits for shareholders.

Common current liabilities for utilities include unpaid wages, unpaid bills, unpaid interest, declared but unpaid dividends, and customer pre-payments for services. Current liabilities are of interest when examining short-term financial prospects. Within the next year, the utility will need enough cash to pay all of its current liabilities. Preferably, these will be paid with cash generated by operations. Just to be safe, investors typically like a utility to have cash and cash equivalents greater than current liabilities.

A utility is obligated to pay off a certain amount of its long-term debt over the next twelve months. This amount is treated as a current liability since it is a short-term financial obligation. Financial statements refer to this amount as the *current portion of long-term debt*.

In summary, liabilities are obligations that will eventually require a reduction in assets to fulfill. If these obligations must be met in the short-term, they are classified as current. If they do not have to be met in the short-term, they are classified as long-term.

2.3.3   Owner's Equity

The assets of a utility are everything valuable that it owns. If a utility were to cease operations, these assets would first be used to fulfill all outstanding liabilities. The remaining assets are left for the owners, and are therefore called owner's equity. Since individuals with ownership rights to a company are called shareholders, owner's equity is sometimes called *shareholder's equity*.

The basic accounting equation states that owner's equity is equal to assets minus liabilities. Owner's equity can further be divided into *paid-in capital* and *retained earnings*. Paid-in capital is the amount of money raised through the issuance of common stock. Retained earnings are everything else, and are equal to cumulative revenue minus cumulative expenses minus cumulative dividends.

Since paid-in capital is generated by selling common stock, it is sometimes called *common stock equity*. In financial statements, paid-in capital refers to the amount of money generated at the time of stock issuance, not market value. Consider a utility that sells one million shares of common stock for $10 per share, raising $10 million. This $10 million is recorded as paid-in capital. If the stock market price of these shares rises to $20 per share, the market value of the one million shares also doubles to $20 million. However, the paid-in capital value on the financial statements remains at the original value, consistent with the principle of historical cost.

Sometimes the term *equity* is used in place of the term *owner's equity*. For example, the financial ratio *return on equity* refers to earnings divided by owner's equity. However, many accountants define equity as any claim on assets, including liabilities. To avoid confusion, use of the term equity without clarification is discouraged.

2.3.4   Working Capital

The term *capital* classically refers to factors of production, such as land and equipment, which can be used by a society to generate wealth. In accounting, these factors of production are closely related to the assets of a company; an asset has value because it has the potential to generate revenue or to help in the generation of revenue. When a utility expenditure results in an asset that is listed in financial statements, the expenditure is said to be *capitalized*.

The term *financial capital* is used to describe money that is available for investing in a business. In the accounting world, the word capital is commonly used to refer to financial capital. An example is the term *venture capital*, which describes funds available for startup businesses.

Working capital is the net amount of financial capital tied up in daily business operations. Mathematically, working capital is equal to current assets minus current liabilities. Consider a utility that converts all of its current assets to cash. It now uses the money generated to pay off all current liabilities. The amount left is equal to the original amount of working capital. The equation for working capital is the following:

$$\text{Working Capital} = \text{Current Assets} - \text{Current Liabilities} \qquad (2.2)$$

The term *liquidity* refers to the ability of an asset to be easily and inexpensively bought and sold, or otherwise converted to cash, without affecting its value. When a less liquid asset is converted into a more liquid asset, it is said to be *liquidated*. The ability of a company to quickly and efficiently sell assets, if necessary, and pay off all of its debt obligations is called *liquidity*. Liquidity is desirable from an investor perspective since assets can easily be sold to pay off debt and avoid bankruptcy, if necessary.

Working capital is one measure of liquidity. If working capital is positive, the utility can use current assets to pay off current liabilities as they come due. Positive working capital guarantees that the utility will be able to pay its debts for at least a year. However, a utility has working capital to help generate revenue, not to pay off debts. Cannibalizing working capital to pay off debts is often a sign of financial trouble.

From an accounting perspective, smaller working capital is typically better. Money that is tied up in working capital could otherwise be used for purposes such as investments, the repurchase of stock, or higher dividend payments. Working capital should be enough to (1) satisfy liquidity expectations of investors, and (2) ensure efficient operations. But working capital should not be more than these factors require. Common ways to reduce working capital include having customers pay their bills as quickly as possible, paying bills as late as possible, encouraging customer pre-payment, collecting unpaid receivables, reducing inventory levels, and reducing cash reserves.

2.3.5   Capital Employed

It is of interest to consider the total amount of capital used to finance a utility. Accountants call this amount *capital employed,* and typically define it as total assets minus current liabilities:

Capital Employed = Assets – Current Liabilities   (2.3)

It is helpful to think of capital employed as all long-term investments that require a rate of return from the utility. To help clarify this point, the traditional definition of capital employed shown above can be transformed mathematically. Since assets are equal to noncurrent liabilities plus current liabilities plus owner's equity, capital employed is mathematically equal to the following:

Capital Employed = Noncurrent Liabilities + Owner's Equity   (2.4)

Long-term debt is typically the largest component of noncurrent liabilities. Long-term debt (such as bonds) is specifically sought after to provide financial capital. In this sense, it is similar to owner's equity, which also exists to provide financial capital. Taken together, owner's equity and long-term debt are called *total capitalization.*

2.3.6   Amortization and Depreciation

When a utility purchases an expensive piece of equipment, the principle of matching requires that the cost of this equipment be matched with the revenue that it helps to generate. Matching is accomplished by reducing the value of the asset over time. The process of reducing the value of an asset over its expected useful life is called *amortization.* When amortization is applied to a tangible asset, it is called *depreciation.*

Consider the purchase of an expensive construction vehicle for $1 million. If this full amount is incorrectly recorded as an expense, profits for the year of purchase would be $1 million lower. Consistently making this mistake for all similar purchases would result in lower profits for years with expensive equipment purchases and higher profits for years with minimal equipment purchases. This accounting outcome is not desirable; the equipment will provide value to the company for many years, and should not have its costs allocated to a single year.

Assume that the new construction vehicle is expected to last for twenty years. The matching principle requires that the total cost of the vehicle be distributed across each of these twenty years. Typically, this is done by allocating an equal amount to each year. For this example, $50 thousand of the $1 million is treated as a *depreciation expense* for a period of twenty years; each year the vehicle is depreciated by $50 thousand until the initial value has been depreciated.

Accounting records keep track of initial asset values, depreciation expenses for the current period, and accumulated depreciation expenses. The current asset value is equal to the initial value minus accumulated depreciation expenses, and is referred to as the *book value* of the asset. An example of several years of depreciation for the construction vehicle is now provided.

Year 1: Assets
           Construction Vehicle

Initial Cost	$ 1,000,000
Accumulated Depreciation:	$ 50,000
Book Value	$ 950,000

Year 2: Assets
           Construction Vehicle

Initial Cost	$ 1,000,000
Accumulated Depreciation:	$ 100,000
Book Value	$ 900,000

Depreciation continues until the book values reaches zero and the asset is *fully depreciated*. Even if the vehicle is still in service, no further depreciation expense is recorded. Therefore, a utility will seem more profitable if it has a large amount of infrastructure and assets in service that are fully depreciated.

A utility must do something when a physical asset reaches the end of its useful life. At a minimum, the asset must be removed, resulting in a *disposal cost*. Often, the asset can be sold for a *salvage value*. The depreciable amount of an asset is equal to its initial value plus its disposal cost minus its salvage value. Consider an asset with an initial cost of $50,000, a disposal cost of $10,000, and a salvage value of $5,000. The depreciable amount of this asset is equal to $50,000 + $10,000 − $5,000 = $55,000. If this asset is depreciated over ten years, the depreciation amount per year is $5,500. When the asset is actually retired, deviations from the assumed disposal cost and salvage value are recorded as gains or losses in the financial statements.

As mentioned previously, the most common method of depreciation is to record an equal amount of depreciation expense each year over the expected life of the asset. This approach is called *straight line depreciation* since the book value of the asset over time corresponds to a line with a negative slope. Other common methods include *declining balance depreciation*, which depreciates a constant percentage of book value each year, and *sum of digits depreciation*, which depreciates decreasingly smaller amounts over the life of the asset. Both of these alternate methods are called *accelerated depreciation*, since more depreciation occurs earlier when compared to straight line depreciation. It is also possible to use *units of production depreciation*, which depreciates based on some measure of usage (such as mileage for vehicles) rather than elapsed time. Detailed treatment of these alternate depreciation methods are beyond the scope of this book.

An asset may not last as long as its original depreciation schedule. For example, the construction vehicle may only last fifteen years. At this point, the asset still has a book value of $250,000 but is not able to be of further use to the utility. To account for this, the asset must be *written off* by reducing the book value of zero. This is done by recording a $250,000 depreciation expense.

A utility may also choose to invest in an asset in order to increase its value or to extend its useful life. Consider the construction vehicle again. At year fifteen, the utility decides to replace the engine for a cost of $100,000. The utility expects that this action will allow the vehicle to remain in service for another ten years. The construction vehicle remains on its current depreciation schedule, and the engine is recorded as a separate asset with a depreciation schedule of ten years. Maintenance work that is recorded as an asset and depreciated over time is commonly called *capital maintenance*.

Utilities have an incredibly large number of assets. Although it is useful to track depreciation expenses separately for expensive assets, it is problematic to do this for everything. Consider meters. A large utility may have millions of meters, an accounting nightmare if each one needs its own asset account with quarterly depreciation charges. Instead, most utilities use a technique called *group depreciation*, which calculates a single depreciation expense for all assets within a specific asset class.

A simple example of group depreciation treats all assets within a class as a single depreciable item. For example, a utility may have spent $30 million on meters in 2009. If these meters are expected on average to last 30 years, the account "2009 Meters" is depreciated by $1 million per year for 30 years, regardless of what actually happens to individual meters.

Utilities typically use a more sophisticated approach to group accounting that recognizes that not all assets within a group will be retired at the same time. Meters might on average last thirty years, but some will be retired much earlier and some will be retired much later. There are a group of curves that represent a range of retirement patterns that accountants use to determine the appropriate depreciation amounts for a group at each age. These curves, called *Iowa Curves*, were developed in 1931 at Iowa State University. Utilities determine average group retirement ages and appropriate Iowa Curves through *depreciation studies*, which are separate from the financial accounting function and are often performed by *depreciation engineers*.

2.3.7   Inventory

Inventory refers to assets that are used up in the process of producing goods and services that will be sold to customers. Classical categories of inventory include raw materials, work-in-progress, and finished goods. Inventory purchases start as raw materials, are converted into work-in-progress, and are further converted into finished goods. When finished goods are sold, the reduction in inventory determines the *cost of goods sold.*

For most utilities, a large amount of inventory will be used to build and repair infrastructure. An electric utility will have wire reels, wood poles, and electric meters. A gas utility will have pipes, valves, and gas meters. Other utilities will similarly have materials required to repair damaged equipment and to expand the existing system. These types of inventory items are the raw materials of a utility.

Utilities typically use raw materials to build infrastructure, not to make products that will be sold to customers. Therefore, work-in-progress is often classified as a fixed asset rather than part of inventory. This accounting treatment is especially important for construction projects that can take many months or years to complete. Consider a large utility project that will cost $60 million to complete and will take three years. After the first year, $15 million of inventory is spent on the project plus an additional $5 million in labor costs. To account for this, inventory is reduced by $15 million, a labor expense of $5 million is recorded, and the asset account *construction work in progress* is increased by $20 million. At the completion of the project, the $60 million that has been added to the construction work in progress account is transferred to the *utility plant in service* account.

Raw materials may physically remain the same but change in price. Consider the meter inventory for a utility. On June 1st, the utility purchases two hundred meters at $100 per meter, resulting in a meter inventory of $20,000. In June, the utility installs one hundred meters, cutting the meter inventory in half to $10,000. On July 1st, the utility replenishes its meter inventory by purchasing one hundred new meters. The price of the meters has gone up to $120 per meter, resulting in a purchase price of $12,000 and a total inventory of $22,000. The meter inventory now looks as follows:

Inventory: Meters

Date	Quantity	Unit Price	Total
6-1-08	100	$100	$10,000
7-1-08	100	$120	$12,000
Total	200	$110	$22,000

The bottom line represents the total number of meters in inventory, the weighted average purchase price, and the total value of meter inventory. At this point, assume that the utility installs fifty more meters, cutting inventory to 150 units. How should this be treated in inventory? Since all of the meters in inventory are identical, it is not possible to know whether each specific meter was purchased at $100 or $120. Should accountants use the earliest price, the latest price, or the average price?

All answers are correct. Utilities can choose how to account for inventory usage. Inventory accounting that always uses the earliest purchase price first is called *first-in, first-out (FIFO)*. Inventory accounting that always uses the latest purchase price first is called *last-in, first-out (LIFO)*. Inventory accounting that uses the average purchase price of all remaining items in inventory is called *weighted average*.

Assuming that prices rise over time, FIFO results in lower expenses, higher profits, and higher inventory values when compared to LIFO (the weighted average method is in between). Because of this, fair utility comparisons cannot be made without understanding their method of inventory accounting. According to the consistency principle, utilities must use the same method of inventory accounting from year to year. If a utility changes its method of inventory accounting, it must clearly state how the change affects its financial statements.

2.3.8 Income Taxes

A utility must pay income taxes on profits. Confusingly, profits computed in financial statements are not the same as profits computed for income tax purposes. Rules governing the preparation of financial statements are designed to reflect the financial situation of a company. Rules governing the preparation of tax statements are embedded in an exceedingly complex tax code that is far removed from basic accounting principles.

GAAP requires income tax expenses in financial statements to be based on pre-tax earnings within these financial statements, not the amount of taxes actually paid. Consider a utility subject to a federal income tax rate of 35% and a pre-tax financial statement income of $10 million. The corresponding income tax expense recorded in financial statements is required to be 0.35 x $10 million = $350,000. Over time, this amount of income tax will have to be paid to the government. This year, assume that the tax code results in a taxable income of $8 million and an actual tax bill of $280,000. The difference in financial statement tax expense and actual owed taxes is $70,000. This $70,000 will eventually have to be paid, but not this year. Financial accounting handles this tax situation with the following journal entry:

Account	Dr.	Cr.
Income tax expense	$350,000	
Income tax payable		$280,000
Deferred tax liability		$70,000

This transaction reduces retained earnings by $350,000 (through the first entry) and increases liabilities by the same amount (through the second and third entries). The *deferred tax liability* is the amount of taxes that have been deferred and will have to be paid sometime in the future.

Table 2.4. An example of deferred taxes where the financial depreciation of an asset is $1 million per year over ten years and the tax depreciation of an asset is $2 million per year over five years. In each of the first five years, the utility avoids paying $350 thousand in taxes. In years six through ten, the utility pays $350 thousand more in taxes.

Year	Depreciation Expense (Taxes)	Depreciation Expense (Financial)	Difference	Deferred Taxes for Year	Accumulated Deferred Taxes
1	2,000,000	1,000,000	1,000,000	350,000	350,000
2	2,000,000	1,000,000	1,000,000	350,000	700,000
3	2,000,000	1,000,000	1,000,000	350,000	1,050,000
4	2,000,000	1,000,000	1,000,000	350,000	1,400,000
5	2,000,000	1,000,000	1,000,000	350,000	1,750,000
6	0	1,000,000	-1,000,000	-350,000	1,400,000
7	0	1,000,000	-1,000,000	-350,000	1,050,000
8	0	1,000,000	-1,000,000	-350,000	700,000
9	0	1,000,000	-1,000,000	-350,000	350,000
10	0	1,000,000	-1,000,000	-350,000	0

The difference between tax accounting and financial accounting is important, especially when considering depreciation. Financial accounting typically depreciates an asset in equal amounts over its useful life. Tax accounting commonly allows certain assets to depreciate more quickly, which results in smaller tax bills in the near term but larger tax bills in the future.

Consider the purchase of a fleet of energy efficient vehicles. The fleet costs $10 million, and is expected to last ten years. Financial accounting depreciates the initial cost by $1 million per year over ten years. Assume that the tax code gives preferential treatment to energy efficient vehicles by allowing them to be depreciated over five years. Tax accounting will therefore depreciate the fleet by $2 million per year over five years. In the first five years, tax depreciation expenses exceed financial depreciation expenses by $1 million. Assuming a 35% tax rate, this results in a tax bill that is $350,000 less that the recorded tax expense in the financial statements (this is recorded as deferred taxes). From years six through ten, financial depreciation expenses exceed tax depreciation expenses by $1 million. This results in a tax bill that is $350,000 more than the recorded tax expense in the financial statements (reducing the deferred tax account). The details of this example are shown in Table 2.4.

Tax accounting is complicated, and it is not necessary for utility engineers to have anything other than a very basic knowledge of the tax code. Keeping this in mind, it is important to understand that favorable tax treatment generally *will not* increase accounting profits. Taking advantage of the tax code can shift tax payments from now into the future, which is good. However, the full tax expense must be recognized today so that it can be properly matched with revenues.

2.4   FINANCIAL STATEMENTS

Financial statements are used by a company to describe its financial position and financial activities. According to the International Accounting Standards Board, the objective of financial statements is to "provide information about the financial strength, performance, and changes in financial position of an enterprise that is useful to a wide range of users in making economic decisions."

When people speak of financial statements, they are typically referring to the *income statement*, the *balance sheet*, and the *statement of cash flows*. The income statement describes a utility's revenues and expenses over a specific period of time. The balance sheet describes a utility's assets, liabilities, and owner's equity at a specific point in time. The statement of cash flows describes changes in the cash account over a specific period of time.

One of the most important business skills for utility engineers is the ability to understand utility financial statements. To help develop this ability, this section analyzes the income statement, balance sheet, and statement of cash flows for a large investor-owned electric utility in the US. The specific analysis of this utility is not important. Rather, the goal is to establish a level of comfort and confidence when examining the financial statements of utilities in general.

2.4.1   Income Statement

An income statement, also called a *profit and loss statement*, describes the profitability of a utility over a specific period of time, such as a month, three months (quarter), or a year. It does this by presenting revenue and expenses in different categories so that income can be presented at different levels. Income statements start with company revenue, and successively subtract expense categories until there are no more expenses to subtract. This process generally results in operating income, earnings before interest and taxes, income before taxes, and net income (recall that earnings and income are synonymous). Definitions of these measures of profitability are now provided.

Operating Income – this is the basic profitability of a company's services. It is equal to operating revenue minus operating expenses. Operating revenue is revenue from sales of services and/or products during the normal course of business. Operating expense is equal to all of the costs required to produce these services and/or products, including overhead expenses. Operating income is a good measure of operational efficiency. This is because it excludes revenues and expenses that are not related to normal operations.

Earnings Before Interest and Taxes (EBIT) – EBIT (pronounced "eebit") is similar to operating income except it includes other sources of revenue and expenses not related to interest and taxes. This includes gains and/or losses on investments, gains and/or losses on the sale of equipment, and similar items. EBIT is a popular measure since it indicates the amount of profit that has been generated without factoring in how the business is financed (affecting interest payments) or how the business is taxed (affecting net profitability).

Earnings Before Interest, Taxes, Depreciation, and Amortization (EBITDA) – EBITDA (pronounced "ee bit duh") is similar to EBIT except it adds back the non-cash expenses of depreciation and amortization. EBITDA is an approximation of generated free cash flow that is available for capital investments, interest payments, taxes, and dividends.

Net Operating Profit After Taxes (NOPAT) – NOPAT (pronounced "no pat") represents the profits generated by a company that are available for distribution to all investors including owners (through dividends) and lenders (through interest payments). It is equal to EBIT minus taxes, and is also equal to net income plus interest payments.

Income Before Taxes (IBT) – IBT (pronounced "eye bee tee") considers all revenue and all expenses except for income taxes. IBT is a popular measure of overall profitability because it allows for the comparison of companies with different tax situations. For example, utilities in different countries may be taxed at different rates. IBT allows these companies to be more fairly compared.

Net Income – Net income is the "bottom line," and represents profits that remain after all expenses have been subtracted from all revenue. Net income should be viewed as utility profits that can either be retained or distributed to shareholders through dividend payments. Dividends are not subtracted from net income because they are treated as a distribution of net income to shareholders.

The relationship of revenue and income measures to each other is shown in Figure 2.2. All calculations start with operating revenue, which represents the "top line." Operating revenue minus operating expenses results in operating income. Operating income adjusted for non-operational revenue and expenses result in EBIT. EBIT can be adjusted in three separate ways to result in IBT, EBITDA, or NOPAT. IBT minus taxes results in net income, also called the "bottom line."

A sample income statement for a large US electric utility is shown in Table 2.5 (all values are stated in thousands of US dollars). Income is shown for three years so that the current year results (2002) can easily be compared to the prior two years.

The first part of the income statement is operating revenue. For this utility, operating revenue is divided into two categories: utility and diversified business. Utility revenue relates to regulated business operations. In this case, utility revenue corresponds to customer electricity bills. Diversified business relates to unregulated businesses that are owned by the utility. Total operating revenue is commonly referred to as the *top line*.

From 2000 to 2001, utility operating revenue increased dramatically from about $3.5 billion to $6.6 billion. This increase was due to a significant acquisition. Because of this, it is not clear whether utility revenue increased from 2000 to 2001. Utility revenue from the acquired company must be added to the 2000 numbers for meaningful comparisons to be made. Utility revenue increases by a bit less than 1% from 2001 to 2002. Without investigating further, it is not clear whether this growth is due to rate increases, new customers, permanent increases in electricity consumptions, or temporary increases due to weather (e.g., a hot summer leading to increased use of air conditioning).

From 2000 to 2001, diversified business revenue increased from about $223 million to $1.5 billion. Presumably, this large increase was also due to the acquisition; it is not clear whether individual unregulated businesses are growing or not. It is clear that diversified business revenue was about 6% of total operating revenue in 2000 and 19% in 2001. This represents a significant change in revenue mix. It will impact how executives organize and run the utility and how investors analyze and value the company. Diversified business revenue decreased by 12% from 2001 to 2002. Large revenue decreases like this are a potential concern for investors and warrant further investigation.

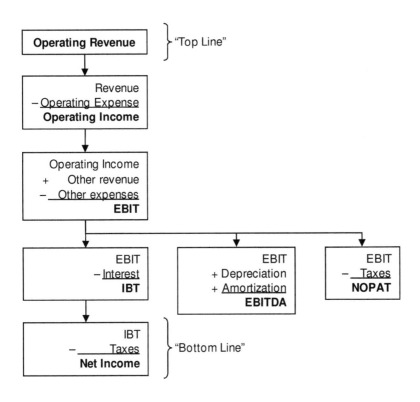

Figure 2.2. The relationship of revenue and income measures to each other. All calculations start with operating revenue, which represents the "top line." Operating revenue minus operating expenses results in operating revenue. Operating revenue adjusted for non-operational revenue and expenses result in EBIT. EBIT can be adjusted in three separate ways to result in IBT, EBITDA, or NOPAT. IBT minus taxes results in net income, also called the "bottom line."

Table 2.5. A sample income statement for a large US utility (all values in thousands of US dollars). Income is presented at different levels. Operating income is equal to operating revenue minus operating expenses. Adding all other income and subtracting interest charges results in income before taxes. Subtracting taxes and making all other adjustments results in net income. Net income can be retained or distributed to shareholders through dividends.

		2002	2001	2000
Operating Revenues	Utility	6,600,689	6,556,561	3,545,694
	Diversified business	1,344,431	1,528,819	223,228
	Total Operating Revenues	7,945,120	8,085,380	3,768,922
Operating Expenses	Utility			
	Fuel used in electric generation	1,614,879	1,559,998	682,627
	Purchased power	862,395	868,078	364,977
	Operation and maintenance	1,361,189	1,210,750	792,164
	Depreciation and amortization	820,279	1,067,073	735,353
	Taxes other than on income	386,254	379,830	162,268
	Diversified business			
	Cost of sales	1,433,626	1,422,890	81,376
	Impairment of long-lived assets	363,822	42,852	-
	Other	98,193	304,817	266,931
	Total Operating Expenses	6,940,637	6,856,288	3,085,696
Operating Income		**1,004,483**	**1,229,092**	**683,226**
Other Income	Interest income	14,526	22,481	18,353
	Impairment of investments	(25,011)	(164,183)	-
	Gain on sale of investment	-	-	200,000
	Other, net	33,804	(28,439)	15,423
	Total Other Income (Expense)	23,319	(170,141)	233,776
Interest Charges	Net interest charges	641,574	689,694	261,570
	Borrowed construction funds	(8,133)	(16,801)	(18,992)
	Total Interest Charges, Net	633,441	672,893	242,578
Income Before Taxes (IBT)		394,361	386,058	674,424
Income Tax Expense (Benefit)		(157,808)	(154,338)	196,502
Income from Continuing Operations		552,169	540,396	477,922
Discontinued Operations, Net of Tax		(23,783)	1,214	439
Net Income		**528,386**	**541,610**	**478,361**

Table 2.6. A breakdown of operating income into utility operations and diversified business operations (all values in thousands of US dollars). In all years, utility operations are profitable and diversified business operations are not profitable. From 2001 to 2002, utility profits are increasing modestly and diversified business losses are increasing dramatically.

	2002	2001	2000
Utility			
Revenue	6,600,689	6,556,561	3,545,694
Expense	5,044,996	5,085,729	2,737,389
Operating income	1,555,693	1,470,832	808,305
Diversified business			
Revenue	1,344,431	1,528,819	223,228
Expense	1,895,641	1,770,559	348,307
Operating income	(551,210)	(241,740)	(125,079)

Total operating revenue decreased about 2% from 2001 to 2002. Since most investors are looking for *top line growth*, executives at this utility will most likely be looking for ways to (1) explain why the decrease in operating revenue was not their fault, and (2) identify ways to increase operating revenue in future years.

The next section in the income statement is operating expenses. These are the expenses that were required to generate the operating revenues, and are similarly divided into utility operating expenses and diversified business operating expenses.

Utility operating expenses are divided into several large categories that are listed separately on the income statement. This includes the cost of fuel used in electricity generation and the cost of purchased power. The other major categories are operations and maintenance (O&M), depreciation/amortization, and taxes other than income taxes (e.g., property taxes). From 2001 to 2002, O&M increased by about 12%, raising questions that should be examined further. At the same time, depreciation/amortization expenses decrease by 23%. It is possible that this represents an accounting shift where expenses that were previously treated as assets are now treated as O&M expenses. Questions like these can be answered by examining income statement footnotes.

Operating income declines significantly from 2001 to 2002. This is not good news. It is not clear from the income statement whether this decline is from utility or diversified operations. To explore further, the operating income is computed separately for utility and diversified operations in Table 2.6. In all years, utility operations are profitable. Additional good news is that utility operating income increases by 5.8% from 2001 to 2002. In contrast, diversified business operations are losing substantial amounts in each of the three years, with losses more than doubling from 2001 to 2002. Top management is sure to be under intense investor pressure to stop these losses.

A utility may have income from sources other than operations. A common source of other income is interest earned on investments. Another is through the sale of an asset at a price above book value. For example, if the book value of a piece of property is $8 million, and the property is sold for $10 million, a *gain on sale of investment* of $2 million will be recorded in *other income* (if an investment is sold at a price lower than book value, the difference is recorded as a *loss on sale of investment*).

Sometimes the value of assets must be lowered for reasons other than depreciation or amortization. This is called *writing down* or *impairing* the asset. For example, goodwill associated with an acquisition must be written down if it becomes apparent that this goodwill cannot be justified through increased profits. For this company, impairment expenses amounted to $164 million in 2001 and $25 million in 2002. These are not small amounts, and correspond to 13% and 2.5% of operating revenue for 2001 and 2002, respectively.

Since utilities borrow a lot of money to fund operations, interest payments are always a large expense. This utility spent $690 million on interest payment in 2001 and $642 million in 2002. The reduction of 7% in interest payments is of potential significance, and could be due to refinancing at lower interest rates, lower debt, or a combination.

When utilities borrow money for a specific construction project, interest payments on this borrowed money can typically be capitalized and included as part of the asset value. Consider a project that costs $10 million to design, construct, and commission. Interest on loans to finance this specific project amounted to an additional $1 million. When completed, the financial records assign a value of $11 million for this project. The $1 million in interest payments are included in net interest charges on the income statement, but are subtracted out in the *borrowed construction funds* line. In 2002, this utility paid over $8 million in interest charges for construction loans.

In 2000, the net interest charges of $262 million for this utility amounted to 38% of operating revenue. This means that for every dollar of profit earned through operations, thirty-eight cents were required to be paid in interest payments. In 2001, this percentage rose to 56%. In 2002, this percentage rose again to 64%. Spending this amount of operating income on interest payments is a concern. A higher interest payment means a higher risk that the utility will not be able to meet its interest payments in the future, increasing the likelihood of bankruptcy. It also means less profit for shareholders.

IBT is equal to operating income plus other income, minus net interest charges. This is basically the profits of a company before income taxes. For this utility, IBT dramatically dropped after 2000, even as revenue significantly increased. The result is a stunning drop in IBT when measured as a percentage of revenue. In 2000, IBT is about 18% of revenue. In 2001 and 2002, IBT dropped below 5% of revenue.

The effect of income taxes is shown just below IBT. In 2000, the accounting expense for income taxes was nearly $200 million. In both 2001 and 2002, there is an income tax accounting benefit of more than $150 million. For this utility, the benefits in 2001 and 2002 are due to primarily to deferred federal income taxes.

IBT minus income tax expense is equal to *income from continuing operations*. This represents the net profitability of the component of business operations that will continue into the future. For this utility, income from continuing operations has increased slightly from year to year. However, income tax effects have a dramatic impact on this number, with income tax benefits propping up profits in 2001 and 2002.

When a utility closes down a part of its business, the net income related to these activities is typically reported separate on the income statement from items related to continuing operations. Often, these revenues and expenses are categorized as *discontinued operations*. In 2000 and 2001, discontinued operations were slightly profitable. In 2002, discontinued operations lost nearly $24 million. If there are no discontinued operations, income from continuing operations is equal to net income.

Table 2.7. Typical per share earnings ratios (values in US dollars). Earnings per share is equal to net income divided by the number of outstanding shares. Typically, a utility will have committed shares that have not been issued. If all of these committed shares are added to the divisor, the resulting ratio is called diluted earnings per share. Since dividends are only issued to outstanding shares, dividends per common share does not consider dilution.

	2002	**2001**	**2000**
Common Shares Outstanding	217,247	204,683	157,169
Basic Earnings per Common Share			
Income from continuing operations	2.54	2.64	3.04
Discontinued operations, net of tax	(0.11)	0.01	-
Net income	2.43	2.65	3.04
Diluted Earnings per Common Share			
Income from continuing operations	2.53	2.63	3.03
Discontinued operations, net of tax	(0.11)	0.01	-
Net income	2.42	2.64	3.03
Dividends per Common Share	2.20	2.14	2.08

The sum of income from continuing operations and income from discontinued operations results in net income. Net income, as stated previously, represents the "bottom line" profits that can be retained by the company or distributed to shareholders through dividends. Net income for this utility increased in 2001 and then decreased slightly in 2002.

It is common for income statements to state earnings on a per share basis. This is done by dividing net income by the number of issued common shares of stock, and is called *basic earnings per common share*. Utilities often have committed additional shares of stock that are not yet issued. For example, a utility may grant *stock options* to employees that allow for the future purchase of shares at a stated price. When new shares are issued, the current shares are said to be *diluted* since less profit is available for each share. When earnings are divided by common shares outstanding plus committed shares, the ratio is called *diluted earnings per common share*.

Per share earnings information is shown in Table 2.7. Both basic and diluted earnings per share have been decreasing since 2000, with the dilution impact being minimal. This trend is a problem for the utility. Investors and analysts focus heavily on earnings per share, with a particular emphasis on growth in earnings per share. Typically, a utility will publicly state its expected earnings per share for each quarter in the upcoming year. Financial analysts follow the company closely and try to predict whether the utility will meet its quarterly earnings targets. When a fiscal quarter ends, the utility issues a quarterly earnings report and will present financial results to the investor community. At this point, the utility will either meet its earnings per share target, exceed its target by a number of cents per share, or miss its target by a number of cents per share.

Meeting quarterly and annual earnings per share targets are some of the highest priorities for utility executives. Utility engineers will help align their actions with these priorities if they consistently ask themselves questions such as "How will this decision impact earnings per share?" and "What approach is best from an earnings per share perspective?"

Focusing on quarterly earnings targets requires a short term perspective that must be balanced through long term life cycle management of utility infrastructure. These perspectives are not necessarily in conflict. A long term infrastructure plan should be translated into an annual budget with quarterly earnings targets. Meeting earnings targets must then be done while accomplishing the corresponding goals of the infrastructure plan.

Dividends are the way profits are distributed to shareholders. As such, the amount of declared dividends per share is closely watched by both financial analysts and shareholders. Utilities are under intense pressure to consistently increase dividends per share. In contrast, a reduction in dividends (or a complete dividend omission) is viewed negatively and is typically only done if absolutely necessary.

For this utility, dividends per common share have consistently increased, even as net income declined over the same period. However, the percentage of net income that is distributed as dividends increased from 68% in 2000 to 81% in 2001 to 91% in 2002. In the long term, net income must increase to support continued dividend growth. In the short term, the message to shareholders is that strong dividend payments is a top priority.

In summary, an income statement presents the revenue and expenses of a utility over a specific period of time. This section has presented income statement concepts by analyzing the actual income statement for a large US utility. The point is not to understand the profits and losses of this specific utility *per se*. Rather, the intent is to increase the confidence and comfort level of the reader when reviewing utility income statements in general.

2.4.2 Balance Sheet

This chapter began by introducing the basic accounting equation: assets = liabilities + owner's equity. To demonstrate the basic accounting equation in action, a simple balance sheet was developed from scratch through a set of financial transactions. This section continues the balance sheet topic by examining the actual balance sheet of the same utility used for the income statement analysis.

A balance sheet shows the financial status of a utility at a specific point in time. Information is typically categorized into assets, liabilities, and owner's equity. Consider Table 2.8, which summarizes the utility balance sheet at a high level. Monetary values are shown both in thousands of US dollars and as a percentage of total assets for the year under consideration.

Table 2.8. Balance sheet summary for a large US utility (values in thousands of US dollars). In accordance with the basic accounting equation, total assets are equal to total liabilities plus total owner's equity. Monetary values are also expressed as a percentage of total assets.

	2002	2001	2002	2001
ASSETS				
Total utility plant, net	10,656,234	10,521,767	49.9%	50.4%
Total current assets	2,856,206	2,897,800	13.4%	13.9%
Total other assets	7,840,264	7,471,134	36.7%	35.8%
Total Assets	21,352,704	20,890,701	100.0%	100.0%
LIABILITIES				
Total noncurrent liabilities	9,747,293	8,618,960	45.6%	41.3%
Total current liabilities	2,734,308	3,438,823	12.8%	16.5%
Total other liabilities	2,101,263	2,736,554	9.8%	13.1%
Total Liabilities	14,582,864	14,794,337	68.3%	70.8%
OWNER'S EQUITY				
Common stock	4,589,782	3,960,928	21.5%	19.0%
Retained earnings	2,087,227	2,042,605	9.8%	9.8%
Preferred stock in subsidiaries	92,831	92,831	0.4%	0.4%
Total Owner's Equity	6,769,840	6,096,364	31.7%	29.2%
LIABILITIES + OWNER'S EQUITY	21,352,704	20,890,701	100.0%	100.0%

This utility has assets with book value amounting to $21 billion. This amount grew by slightly more than 2% from 2001 to 2002, showing that asset investment rates are higher than depreciation rates. About half of the utility's asset value is from total utility plant.

The utility has liabilities amounting to 68% of asset value. This is considered a high number, but the percentage decreased slightly from 2001 to 2002. The largest portion of liabilities is noncurrent, which likely consists of long-term debt. For this utility, noncurrent liabilities are approximately equal to the book value of total utility plant; it has fully funded its infrastructure with long-term debt.

The utility has owner's equity amounting to 32% of asset value, which increased 2.5% from 2001 to 2002. About one third of owner's equity is from the $2 billion in retained earnings, which has not changed significantly from 2001 to 2002. The increase in owner's equity is due to the issuance of about $630 million in common stock. This stock issuance is not particularly surprising since the debt level of the utility is so high. However, it is not clear from this high level view of the balance sheet why the utility needed the additional cash. This question is best answered by examining the statement of cash flows, which is the topic of the next section.

A more detailed statement of balance sheet assets is shown in Table 2.9. Major asset categories are utility plant, current assets, and other assets. These correspond to the high level categories used in Table 2.8.

Table 2.9. A more detailed statement of balance sheet assets for a US utility (values in thousands of US dollars). Major asset categories are utility plant, current assets, and other assets. The largest line items within these categories are net utility plant in service and goodwill.

	2002	2001	2002	2001
ASSETS				
Utility Plant				
Utility plant in service	20,152,787	19,176,021	94.4%	91.8%
Accumulated depreciation	(10,480,880)	(9,936,514)	-49.1%	-47.6%
Utility plant in service, net	9,671,907	9,239,507	45.3%	44.2%
Held for future use	15,109	15,380	0.1%	0.1%
Construction work in progress	752,336	1,004,011	3.5%	4.8%
Nuclear fuel, net of amortization	216,882	262,869	1.0%	1.3%
Total Utility Plant, Net	10,656,234	10,521,767	49.9%	50.4%
Current Assets				
Cash and cash equivalents	61,358	53,708	0.3%	0.3%
Accounts receivable	737,369	779,286	3.5%	3.7%
Unbilled accounts receivable	225,011	199,593	1.1%	1.0%
Inventory	875,485	871,643	4.1%	4.2%
Deferred fuel cost	183,518	146,652	0.9%	0.7%
Assets of discontinued operations	490,429	552,458	2.3%	2.6%
Other current assets	283,036	294,460	1.3%	1.4%
Total Current Assets	2,856,206	2,897,800	13.4%	13.9%
Other Assets				
Regulatory assets	393,215	463,837	1.8%	2.2%
Nuclear decommissioning trust funds	796,844	822,821	3.7%	3.9%
Diversified business property, net	1,884,271	1,072,123	8.8%	5.1%
Miscellaneous investments	463,776	441,932	2.2%	2.1%
Goodwill	3,719,327	3,656,970	17.4%	17.5%
Prepaid pension costs	60,169	487,551	0.3%	2.3%
Other assets and deferred debits	522,662	525,900	2.4%	2.5%
Total Other Assets	7,840,264	7,471,134	36.7%	35.8%
Total Assets	**21,352,704**	**20,890,701**	**100.0%**	**100.0%**

The utility has spent over $20 billion in infrastructure, as shown on the line "utility plant in service." This amount increased by about $1 billion from 2001 to 2002. About half of the initial cost of the utility plant in service has been depreciated, indicating that on average, equipment is about half as old as its expected useful life.

The utility has about $15 million in assets that are "held for future use." Typically, this consists of pre-purchased land that the utility intends to utilize sometime in the future. It could also include facilities that are not presently used for utility services, but have definite plans for use in the future.

About $752 million is for "construction work in progress (CWIP)." This represents the accumulated costs of construction projects that will later be transferred to "utility plant in service" when the construction is completed. The construction work in progress declined by about 25% from 2001 to 2002. This could be due to either a reduction in construction activity or the completion of one or more large construction projects. In regulatory accounting, interest on loans for CWIP is sometimes allowed to be capitalized into an account called *allowance for funds used during construction* (AFUDC).

An asset specific to electric utilities is nuclear fuel. In most cases, storable fuel such as coal is treated as inventory. Since nuclear fuel rods are not physically consumed when used, they are treated as a fixed asset and amortized based on their expended heat content. This utility has about $217 million left in unused nuclear fuel.

As previously discussed, current asset are expected to be liquidated within the next year. Perhaps the most important current asset is "cash and cash equivalents," which include all very safe investments that can be easily converted into cash (e.g., bank accounts, treasury bills, money market funds). The utility has about $61.4 million in cash, which is up from $53.7 million in 2001. Compare this to accounts receivable (unpaid customer bills) and unbilled accounts receivable, which together total almost $1 billion. Clearly, this utility relies on customers paying their bills for most of its ongoing cash needs.

Inventory levels are $875 million, which is up slightly from 2001. Inventory amounts to 4.3% of undepreciated utility plant in service and 9% of net utility plant in service. These values can easily be compared to other utilities to determine whether they represent high, typical, or low inventory levels.

Deferred fuel cost is another asset specific to electric utilities. Some utilities are able to recover actual fuel costs on a deferred basis. For example, rates may assume an average fuel cost of five cents per kilowatt-hour. At the end of the year, actual average fuel cost may be six cents per kilowatt hour. This utility is able to recover the difference (in this case $183 million) in the upcoming year, resulting in a current asset on the balance sheet.

Regulatory assets are incurred costs that have been approved by regulatory agencies for recovery over a period of years. Instead of being treated as an expense in a single year, these costs are booked as assets and are depreciated over time. This utility has about $393 million in regulatory assets, which is significantly lower than 2001.

A nuclear decommissioning trust fund has been set up for this utility to pay for future cleanup of nuclear power plants when they are eventually shut down. Each year, a portion of revenue is placed into the trust fund and is invested in a financial portfolio. This utility has drawn about $26 million from this fund over the past year.

The last major asset item on the balance sheet is goodwill. For this utility, goodwill amounts to $3.7 billion, and represents payments over market value for acquired companies. Recall that goodwill must be impaired (reduced) if the utility does not believe that it can be justified by expected future profits. For this utility, goodwill increased from 2001 to 2002, showing that no significant impairment occurred.

A more detailed statement of balance sheet liabilities is shown in Table 2.10. Major liability categories are noncurrent, current, and other. These correspond to the high level categories used in Table 2.8.

Table 2.10. A more detailed statement of balance sheet liabilities for a US utility (values in thousands of US dollars). Major liability categories are noncurrent, current, and other. The largest line item by far is long-term debt, which increased by more than $1 billion from 2001 to 2002.

	2002	2001	2002	2001
LIABILITIES				
Noncurrent Liabilities				
Long-term debt	9,747,293	8,618,960	45.6%	41.3%
Total Noncurrent Liabilities	9,747,293	8,618,960	45.6%	41.3%
Current Liabilities				
Current portion of long-term debt	275,397	688,052	1.3%	3.3%
Accounts payable	756,287	760,116	3.5%	3.6%
Interest accrued	220,400	211,731	1.0%	1.0%
Dividends declared	132,232	117,857	0.6%	0.6%
Short-term obligations	694,850	942,314	3.3%	4.5%
Customer deposits	158,214	151,968	0.7%	0.7%
Liabilities of discontinued operations	124,767	162,917	0.6%	0.8%
Other current liabilities	372,161	403,868	1.7%	1.9%
Total Current Liabilities	2,734,308	3,438,823	12.8%	16.5%
Other Liabilities				
Accumulated deferred income taxes	932,813	1,408,155	4.4%	6.7%
Accum. deferred investment tax credi	206,221	224,688	1.0%	1.1%
Regulatory liabilities	119,766	291,789	0.6%	1.4%
Other liabilities and deferred credits	842,463	811,922	3.9%	3.9%
Total Other Liabilities	2,101,263	2,736,554	9.8%	13.1%
Total Liabilities	**14,582,864**	**14,794,337**	**68.3%**	**70.8%**

Noncurrent liabilities for this utility are entirely from long-term debt, which is typical. For utilities, long-term debt is normally raised by selling bonds. This utility has outstanding bonds with a total original issue value of $9.7 billion. Eventually, all of these bonds will have to be repurchased at their original sale value, which is why this amount appears as a noncurrent liability.

From 2001 to 2002, long-term debt increased by about 13%, amounting to more than $1 billion dollars. Recall that this amount is about equal to the increase in utility plant in service (see Table 2.9). It appears that this utility is raising cash to fund the expansion of its utility plant.

Eight line items are listed under current liabilities, with the following six in need of descriptions. "Current portion of long term debt" is the amount of debt principle that will be paid off in the next year. "Accounts payable" represents bills that the utility has yet to pay. "Interest accrued" are unpaid interest payments that the utility must issue to bond and preferred stock holders. "Dividends declared" are similar and will be paid to common stock holder. "Short-term obligations" are typically short-term loans that need to be repaid within the next year. "Customer deposits" consist of payments by customers that will eventually have to be returned.

Table 2.11. A more detailed statement of owner's equity for a US utility (values in thousands of US dollars). Major categories are retained earnings, preferred stock equity, and common stock equity. From 2001 to 2002, this utility experienced a decrease in total retained earnings and an increase in total common stock equity (through an issuance of new common stock).

	2002	2001	2002	2001
OWNER'S EQUITY				
Retained Earnings				
Basic retained earnings	2,087,227	2,042,605	9.8%	9.8%
Other comprehensive losses	(237,762)	(32,180)	-1.1%	-0.2%
Total Retained Earnings	1,849,465	2,010,425	8.7%	9.6%
Preferred Stock Equity				
Preferred stock of subsidiaries	92,831	92,831	0.4%	0.4%
Total Preferred Stock	92,831	92,831	0.4%	0.4%
Common Stock Equity				
Common stock	4,950,558	4,121,194	23.2%	19.7%
Unearned restricted shares	(21,454)	(13,701)	-0.1%	-0.1%
Unearned ESOP shares	(101,560)	(114,385)	-0.5%	-0.5%
Total Common Stock Equity	4,827,544	3,993,108	22.6%	19.1%
Total Owner's Equity	**6,769,840**	**6,096,364**	**31.7%**	**29.2%**

There is one large change that occurs in current liabilities. The current portion of long term debt was reduced dramatically (by more than half) from 2001 to 2002. It appears that a lot of long-term debt was coming due in 2001. When the utility raised the $1 billion in new long-term debt, it also refinanced existing debt, pushing payments out into the future.

"Other liabilities" are primarily related to deferred expenses that will have to be paid sometime in the future. This included "deferred income taxes" and "deferred investment tax credits." These deferred amounts have been reduced significantly from 2001 to 2002. "Regulatory liabilities" are similar to regulatory assets, but represent amounts that are expected to be refunded to customers in future periods through rate reductions. Regulatory liabilities have also reduced significantly from 2001 to 2002.

The last expanded section of the balance sheet, owner's equity, is shown in Table 2.11. Major categories are retained earnings, preferred stock equity, and common stock equity. These correspond to the high level categories used in Table 2.8.

The utility separates retained earnings into "basic retained earnings" and "other comprehensive losses," which consists of one-time non-recurring expenses that do not directly relate to normal business operations. Basic retained earnings stayed essentially the same from 2001 to 2002. But "other comprehensive losses" increased by more than $200 million. A detailed explanation of this expense will be found in the footnotes (not included in this book).

Preferred stock is listed as owner's equity, but typically resembles bonds rather than common stock. When preferred stock is issued, the amount of money raised is recorded on the balance sheet in a manner similar to common stock. Most preferred stock shares have a fixed dividend amount that must be paid before any common stock dividends are paid. Preferred stock typically guarantees than any missed dividend payments must be paid with similar priority. Last, most preferred stock shares have a date in which the utility will repurchase the share at the issuing price. These features amount to a bond-like instrument that does not result in bankruptcy if it is not paid in a given year.

The utility has about $93 million in outstanding preferred stock, which is unchanged from 2001 to 2002. This is a very small amount when compared to total owner's equity and long-term debt. For this utility, preferred stock has not been the "preferred" way to finance its operations.

This utility has raised almost $5 billion through the issuance of common stock. This amount includes more than $800 million raised from 2001 to 2002. Some of the shares of stock are being held by the utility so that they can be given to employees in the future. Since these shares have not yet been earned by the employees, their value is subtracted from common stock equity on the balance sheet. For example, a utility may grant an employee $30 thousand in restricted stock, corresponding to a specific number of shares. A contract states that the restricted stock will vest over a three year period. After each of the first three years, one third of the shares are transferred to the employee and $10 thousand is removed from the unearned shares account. The same process is followed for employee stock ownership plans (ESOP). The utility has a value of unearned shares that, if given to employees, will reduce total common stock equity by about 2.5%.

In summary, a balance sheet shows the assets, liabilities, and owner's equity of a utility for a specific point in time. This section has presented balance sheet concepts by analyzing the actual statement for a large US utility. Like the income statement, the intent is not to understand the balance sheet details of this specific utility. Rather, the intent is to increase the confidence and comfort level of the reader when reviewing utility balance sheets in general.

2.4.3   Statement of Cash Flows

At this point, the reader has probably noticed some limitations with income statements and balance sheets. The three major concerns are (1) identical businesses can have very different income statements and balance sheets if different accounting practices are used, (2) income statements and balance sheets can be misleading, and (3) income statements and balance sheets are relatively easy to manipulate.

Income statements can vary widely depending upon a number of factors including (1) classification of spending as assets or expense, (2) treatment of inventory, (3) provisional accounts for anticipated future costs, and (4) impairment of assets. Consider a utility that is trying to meet quarterly earnings targets. If it needs additional profit, it can look for ways to capitalize some items that otherwise would have been an expense. The cost of these items will then be depreciated over time instead of immediately, raising short-term profits. Another utility may be having a great quarter. To protect itself against future bad quarters, it may contribute some of its earnings to an asset account (such as a bonus pool) that can be drawn from in the future when profits are low. These types of "income smoothing" practices are not supposed to occur, but unfortunately are not uncommon.

The problems described for income statements have a corresponding impact on balance sheets. Balance sheets have an additional difficulty: values shown on the balance sheet do not reflect the fair value of assets. Consider a utility that has gradually issued common stock over the last century. Today, the stock is trading at $50 per share. If the average issue price of all outstanding stock is $25, the value of common stock on the balance sheet will only be half that of market value. Similarly, a utility may have a large number of in-service assets that are fully depreciated. These assets show zero value on the balance sheet, but have real value to the utility.

Because of the potential difficulties with income statements and balance sheets, investors and analysts pay close attention to the statement of cash flows. Income is subjective, but cash is objective. Identical utilities could have different balance sheets, but cash flow would be the same. Most importantly, the statement of cash flows is very difficult to manipulate.

Since large utilities have thousands of cash transactions every day, cash account records are of limited use in understanding cash flows. For a utility with a million customers, each month would have roughly one million transactions for customer payments alone. Instead, high level cash flow information is calculated from income statement and balance sheet information.

An important concept in cash analysis is that of *free cash flow*. This is typically defined as the amount of cash generated by a company that is available for distribution to owners as dividends. In the long run, net income must be equal to free cash flow. This makes sense since net income is supposed to represent profits that can be given to owners. In some years, net income may be higher than free cash flow, such as when a lot of cash is used to make large asset purchases. In other years, net income may be lower than free cash flow, such as when a lot of money is borrowed but not spent. Over the long run, these variations must equal out.

Because they are equal in the long run, net income is a good first approximation of free cash flow. To make this approximation better, several adjustments must be made. First, all non-cash expenses must be added back. Non-cash expenses, by definition, reduce net income but do not reduce cash. Examples include depreciation expenses, amortization expenses, and asset impairment expenses.

Next, changes in working capital must be considered. If current assets decrease, cash is freed up. For example, a reduction in accounts receivable by $100 million or a reduction in inventory levels by $100 million will free up this amount in cash. Similarly, increases in current liabilities free up cash. An increase in accounts payable by $50 million will free up this much in cash (cash conscious utilities try to pay their bill at the last possible moment).

Generating free cash flow by reductions in working capital is not sustainable. A utility can only reduce its inventory so much without hurting business operations. A utility can only be so effective at getting customers to pay their bills early. A utility can only defer bill payments so much without incurring late fees. Regardless, these issues can have a large impact on cash flow and must be considered.

The last adjustment to net income that must be made to determine free cash flow is to subtract all capital expenditures. When capital expenditures are made, cash is spent but net income is not immediately reduced. To compensate for this, the value of all capital expenditures must be subtracted.

To summarize, free cash flow is equal to net income plus non-cash expenses plus reductions in working capital minus capital expenditures. This corresponds to the following equation:

	Net Income
+	Non-Cash Expenses
+	Decrease in Working Capital
−	Capital Expenditures
=	Free Cash Flow

Although compact, the above equation is easily calculated from information on the income statement and balance sheet. The equation becomes somewhat clearer when depreciation and amortization are listed as a separate item and when working capital is separated into current liabilities and current assets. The expanded equation for free cash flow is:

	Net Income
+	Depreciation & Amortization
+	Other Non-Cash Expenses
+	Increase in Current Liabilities
+	Decrease in Current Assets
−	Capital Expenditures
=	Free Cash Flow

The free cash flow of a utility is recorded in the statement of cash flows, which is regarded by many financial analysts as the most important financial statement. This is partly due to its factual nature. Income statements and balance sheets can vary based on accounting practices, but cash is cash. In 1987, the FASB mandated that firms provide cash flow statements. In 1994, the IASB mandated that firms provide cash flow statements.

A high level statement of cash flows corresponding to the income statement and balance sheet of the prior two sections is shown in Table 2.12. Cash from operations represents net income plus non-cash expenses plus reductions in working capital. Cash from investing represents capital expenditures. Cash from operations minus cash from investing is equal to free cash flow.

Table 2.12. A sample statement of cash flows for a large US utility (all values in thousands of US dollars). Cash flow is typically categorized as cash generated through operating activities, cash used in investing activities, and cash generated and/or used through financing activities. In 2002, this utility increased its cash by more than $7.6 million, a reversal of the prior year when cash decreased by more than $46 million.

	2002	2001	2000
Cash from operations	1,597,911	1,422,750	852,878
Cash from investing	(2,211,834)	(1,655,785)	(4,434,363)
Free Cash Flow	(613,923)	(233,035)	(3,581,485)
Cash from financing	621,521	187,687	3,602,017
Cash from discontinued operations	52	(843)	525
Net Increase (Decrease) in Cash	7,650	(46,191)	21,057
Cash at Beginning of Year	53,708	99,899	78,842
Cash at End of Year	61,358	53,708	99,899
Supplemental Disclosures			
Interest (net of amount capitalized)	630,935	588,127	244,224
Income taxes (net of refunds)	219,278	127,427	367,665

The utility experienced negative free cash flow from 2000 to 2002. This is not necessarily bad, as long as investments will repay themselves through future profits. For example, a significant expansion of infrastructure will require a lot of cash today, and will hopefully result in increased revenue over time.

To compensate for negative free cash flow, the utility resorts to financing. In 2000, the utility had about $3.6 billion in negative free cash flow and about $3.6 billion in financing. In 2001, the utility had about $233 million in negative free cash flow and about $188 million in financing. This shortfall caused a reduction in cash. In 2002, the utility had about $614 million in negative free cash flow and about $621 million in financing, slightly increasing the cash account.

Table 2.12 provides several supplemental disclosures related to cash. The first is the amount of interest paid to lenders. Some analysts prefer to think of free cash flow as the amount of cash generated by a company that is available for distribution to all investors, not just owners. To them, free cash flow is equal to free cash flow as defined in Equation 2.5 plus interest payments:

$$
\begin{array}{rl}
& \text{Net Income} \\
+ & \text{Non-Cash Expenses} \\
+ & \text{Interest Payments} \\
+ & \text{Decrease in Working Capital} \\
- & \underline{\text{Capital Expenditures}} \\
= & \text{Free Cash Flow} \qquad \text{(alternative definition)} \qquad (2.5)
\end{array}
$$

The other supplemental piece of information in Table 2.12 is the amount of income taxes that are actually paid. In 2002, the income tax payment of about $220 million is far different than the almost $158 million income tax *benefit* shown on the income statement. As discussed in the section on income taxes, this discrepancy is due to the differences in financial accounting and tax accounting.

An expanded version of the statement of cash flows is shown in Table 2.13. The organization of the statement is the same, but details are provided for operating activities, investing activities, and financing activities.

Operating activities begins with net income as reported on the income statement. Non-cash expenses are then added back. This includes depreciation and amortization, impairment of long-lived assets, and deferred income taxes. These values do not correspond directly to the high level income statement since they mix values from the utility business, diversified businesses, and discontinued operations. All of these non-cash items are large in comparison to net income, and demonstrate the significant difference between net income and free cash flow.

Reductions in working capital are further broken down. These values, like the non-cash expenses, do not correspond directly to balance sheet values since they include the effects of discontinued operations. In 2002, working capital decreased by about $15 million. Though small, this represents a dramatic change from 2000 and 2001. In these years, working capital increased by $79 million and $271 million, respectively, for a total cash reduction of $350 million.

Investment activity details correspond closely to balance sheet items. In most years, investment activities are dominated by property additions to both utility and diversified operations. Consider 2002, where investment levels are similar for both categories. This is a potential concern since the utility business has much larger revenue than diversified operations. In addition, diversified operations have not been profitable in the last three years.

The largest investment item is 2000 acquisitions, amounting to $3.4 billion. This value does not include any cash that came with the acquisition. For example, if a company is purchased for $600 million and this company has $100 million in cash, the net acquisition price is recorded as $600 − $100 = $500 million.

Investment activities also include cash generated when equipment or other assets are sold. For example, when an old vehicle is sold, the amount of the sale is included in "net proceeds from sale of assets." Typically this category is minor, but becomes significant if a utility starts to sell off assets to satisfy cash needs (sometimes referred to as *cannibalizing*).

As mentioned previously, from 2000 to 2002 the cash generated from operating activities was lower than cash spent in investing activities. The cash deficiency is addressed through financing activities. Of particular interest for this utility is its debt history. In 2000, the utility paid off about $710 million in long-term debt and issued a similar amount of $783 million. This type of debt rollover is normal. However, the acquisition was funded by about $3.8 billion in short-term debt. In 2001, this short-term debt was converted to long-term debt. In 2002, additional short-term debt was converted to long-term debt.

Table 2.13. A sample statement of cash flows for a large US utility (all values in thousands of US dollars). Cash flow is typically categorized as cash generated through operating activities, cash used in investing activities, and cash generated and/or used through financing activities. In 2002, this utility ended up with a net cash increase of more than $7.6 million, a reversal of the prior year when cash decreased by more than $46 million.

	2002	2001	2000
OPERATING ACTIVITIES			
Net income	528,386	541,610	478,361
Depreciation and amortization	1,099,128	1,266,162	846,984
Impairment of long-lived assets	388,833	208,983	-
Deferred income taxes	(402,040)	(367,330)	(93,379)
Reductions in working capital			
Net (increase) decrease in accounts receivable	(45,172)	182,514	(34,754)
Net (increase) decrease in inventories	(48,785)	(298,733)	15,931
Net (increase) decrease in other current assets	(39,141)	(20,797)	57,141
Net increase (decrease) in accounts payable	57,387	(162,940)	229,117
Net increase (decrease) in other current liabilities	56,356	123,297	(148,813)
Other	34,509	(94,806)	(197,725)
Net decrease (increase) in working capital	15,154	(271,465)	(79,103)
Other	(31,550)	44,790	(299,985)
Cash from Operating Activities	**1,597,911**	**1,422,750**	**852,878**
Cash from Discontinued Operations	**52**	**(843)**	**525**
INVESTING ACTIVITIES			
Gross utility property additions	(1,174,220)	(1,177,727)	(853,584)
Diversified business property additions	(934,910)	(349,713)	(157,510)
Nuclear fuel additions	(80,573)	(115,663)	(59,752)
Acquisitions, net of cash	-	-	(3,441,775)
Net proceeds from sale of assets	42,825	53,010	200,000
Net contributions to nuclear decommissioning trust	(18,502)	(50,649)	(32,391)
Investments in non-utility activities	(27,030)	(15,043)	(89,351)
Other	(19,424)	-	-
Cash Used in Investing Activities	**(2,211,834)**	**(1,655,785)**	**(4,434,363)**
FINANCING ACTIVITIES			
Issuance of common stock, net	687,000	488,290	-
Issuance of long-term debt, net	1,797,691	4,564,243	783,052
Net increase (decrease) in short-term debt	(247,464)	(4,018,062)	3,782,071
Net increase (decrease) in cash provided by bad checks	79	(45,372)	115,337
Retirement of long-term debt	(1,157,286)	(322,207)	(710,373)
Dividends paid on common stock	(479,981)	(432,078)	(368,004)
Other	21,482	(47,127)	(66)
Cash from Financing Activities	**621,521**	**187,687**	**3,602,017**
CASH FLOW SUMMARY			
Net Increase (Decrease) in Cash	7,650	(46,191)	21,057
Cash at Beginning of Year	53,708	99,899	78,842
Cash at End of Year	61,358	53,708	99,899

In 2001, total debt increased by about $224 million. This amount is equal to the issuance of new long-term debt plus the retirement of new long-term debt plus changes in short term debt. In addition, the utility issued about $488 million in common stock. In 2002, total debt increased by about $393 million, and $687 million in common stock was issued. So, in just two years, this utility financed an incremental $1.8 billion in cash. Over the same two years, the utility issued $912 million in dividends.

Consider 2001, when the utility issued $488 million in new common stock and paid out $432 million in dividends. In effect, the utility was given a large amount of money by owners and then gave this money right back to owners as dividends (on which the owners are taxed!). In 2002, the utility issued $687 million in common stock and paid out $480 million in dividends. Utilities are under intense pressure to keep dividend levels high, but financing dividends through common stock issuance is a potential concern.

The cash flow summary shows the net increase or decrease in the cash account each year. This includes the starting cash balance, the change, and the ending cash balance. Presumably there is an appropriate amount of cash for a utility to have on hand. It should be enough to ensure the payment of all upcoming bills, but not too much, since idle cash could otherwise be invested in money-making activities. From 2000 to 2002, the cash balance for this utility has oscillated widely, from a high of nearly $100 million to a low of about $54 million. It is not clear whether the appropriate level of cash is at the low range, middle, or high range of these oscillations.

In summary, a statement of cash flows presents the cash received and used by a utility over a specific period of time in terms of cash from operations, cash from investing, and cash from financing. Like the income statement and balance sheet, this section has presented cash flow concepts by analyzing the actual income statement for a large US utility. The point is not to understand the cash position of this specific utility *per se*. Rather, the intent is to increase the confidence and comfort level of the reader when reviewing utility statements of cash flow in general.

2.5   OTHER TYPES OF ACCOUNTING

The primary topic of this chapter is financial accounting, which attempts to fairly represent the profitability of a company by matching revenues to expenses. For publicly traded companies in the US, financial accounting statements must be filed regularly with the Securities and Exchange Commission (SEC).

There are other types of accounting. Some, like tax accounting and regulatory accounting, are required by government agencies. Others, like management accounting, are for internal use and are not disclosed outside of the utility. In the past, utilities kept a separate "set of books" for each type of accounting. Each transaction would be separately recorded in the financial books, the tax books, the regulatory books, and any other accounting books being maintained. Today, sophisticated accounting software systems keep track of all accounting books simultaneously, eliminating the need for multiple transaction entries and ensuring consistency.

The rest of this section provides brief summaries of different types of accounting, other than financial accounting, commonly performed within utilities. Summaries are given along with comparisons to financial accounting.

2.5.1 Tax Accounting

Companies must pay income taxes in countries where operations are performed. In the US, these income taxes are based on a set of tax accounting principles, enacted by federal law, that vary substantially from GAAP. The financial accounting method of the utility is typically used as a reference point, but allowed expenses can vary significantly.

A large part of the US tax code is designed to encourage certain types of behavior. This is typically done through a combination of tax deferrals and tax credits. Tax deferrals, as discussed previously, involve accelerated depreciation schedules. For example, the government may choose to allow electric vehicles to be depreciated over five years even though they have a longer expected life. Tax credits are a permanent reduction in tax liability, and are therefore more valuable than tax deferrals. For example, the government may choose to allow a one thousand dollar tax credit for each electric vehicle purchased. If the utility purchases an electric vehicle, its total tax liability for the year is reduced by one thousand dollars.

In most other countries, the profit for tax purposes is largely based on income before taxes as dictated by GAAP. This difference makes it tricky to compare US utilities with non-US utilities (beyond the differences in US GAAP and International GAAP). In the US, utilities typically desire to make financial income as high as possible and taxable income as low as possible. If this can be achieved, it shows high profitability to investors and simultaneously results in a small tax bill. Outside of the US, utilities will often desire to make financial income as low as possible since a high financial income will result in a high tax bill.

2.5.2 Regulatory Accounting

The accounting needs for regulated utilities are often different than the needs of typical businesses. This is primarily due to the ratemaking process which may allow today's rates to pay for historical costs and/or future costs. For traditional business, customers pay directly for products and services. For a regulated utility, customers reimburse utilities for satisfying regulatory obligations. It is therefore appropriate in some cases to match customer payments with the costs of satisfying regulatory requirements rather than the direct cost of the utility service being provided.

Consider a utility subject to periodic damage from catastrophic events like hurricanes, earthquakes, or ice storms. Regulators may allow a portion of rates to be set aside to pay for such events. In this situation, revenue for this portion is realized but not allowed to increase earnings. Instead, a *storm reserve asset* is created that is drawn from when a catastrophic event occurs.

Utilities often have regulatory assurance that costs incurred today can be recovered through rates in the future. Consider a utility that is allowed to recover actual fuel costs. Rates today only cover an anticipated average fuel cost. If fuel costs exceed this average value, the utility is allowed to create an asset equal to the amount that will be recovered in the future.

Rate-related issues of regulated companies are addressed in FASB Standard 71. Where this rule applies, financial accounting is in alignment with regulatory accounting needs. However, regulated utilities must comply with the financial reporting requirements of federal, state and local regulatory agencies, which typically go beyond the scope of Standard 71. For example, a telephone company must report financial data to the Federal Communications Commission (FCC) and state public service commissions. Similarly, electric and gas utilities must report financial data to the Federal Energy Reliability Commission (FERC) and state public service commissions.

The accounting requirements of federal and state agencies are not always consistent with financial accounting. Financial accounting is primarily used to provide investors and analysts with profit, loss, and cash flow information. Regulatory accounting is primarily used for rate making and utility comparisons.

Federal and state regulators require data useful for setting rates and tracking utility performance. The SEC, on the other hand, requires data which comply with GAAP for the protection of investors, creditors and other users of general purpose financial statements. The inconsistency of data requirements necessitates the application of separate accounting procedures and the maintenance of separate accounting records to satisfy both regulatory and financial reporting demands. Because of these differences, utilities must maintain regulatory accounting records that are different and separate from financial accounting records.

Typically, regulatory accounting uses a uniform system of accounts so that utilities can be directly compared. For example all major electric utilities in the US must submit financial data to the chief accountant of FERC using the uniform system of accounts in FERC Form 1. For example, Account 400 is titled "Electric Operating Revenue" and has specific sub-accounts such as: 440 Residential Sales; 442 Commercial and Industrial Sales; and 444 Public Street and Highway Lighting. These account numbers make it straightforward to compare utilities with each other and with overall industry statistics.

For regulated utilities, regulatory accounting is at least as important as financial accounting. However, the principles of financial accounting are the same across all industries, while regulatory accounting requirements can vary widely.

It is typically sufficient for utilities engineers to be familiar with financial accounting with the understanding that regulatory accounting is somewhat different.

2.5.3 Project Accounting

Many activities within utilities are self-contained projects with a specific scope, start date, and finish date. Many projects are small, but others can cost hundreds of millions of dollars. It is important to have good financial information about ongoing projects so that important questions can be answered. Is the project financially on schedule? Have contractors been paid? Are equipment purchases costing more or less than estimates? Are labor charges more or less than estimates? Project managers need to examine these questions regularly, requiring project accounting systems to be updated more frequently than typical financial accounting systems.

Project accounting is of particular importance for regulated utilities since construction work in progress is sometimes allowed to be capitalized. The amount spent on a project as calculated through project accounting appears as an asset on the balance sheet. Consider a two year construction project with an estimated cost of $50 million per year. After the first year, project accounting records show that $63 million has been spent, resulting in a $63 million increase to the construction work in progress account. A project accounting summary for this example is shown in Table 2.14.

The project accounting summary breaks down costs into five categories. Each of these categories has an associated cost estimate, an actual accrued cost, and an estimate of percent completion. Engineering for this project is 100% complete and ended up with actual costs equal to the estimate. Equipment costs are running $7 million over budget with 60% of equipment (based on the original cost estimate) having been purchased. Labor is running $2 million under budget, with 60% of labor complete. Project management costs are on budget, and commissioning costs have not started to be incurred.

Table 2.14. A high level report from a project accounting system. This project is 58% complete, and has exceeded its budget by $5 million. Equipment costs are running $7 million over budget. Labor is running $2 million under budget. This type of information can be used to update project cost and schedule estimates.

	Estimated Total Cost	% Complete	Budgeted Cost	Accrued Cost	Variance
Engineering	$10	100%	$10	$10	---
Equipment	$30	60%	$18	$25	$7
Labor	$40	60%	$24	$22	($2)
Project Management	$10	60%	$6	$6	---
Commissioning	$10	0%	$0	$0	---
Total	**$100**	**58%**	**$58**	**$63**	**$5**

Based on project accounting reports, the project manager is able to update estimates for costs and completion dates. In this case, the project is ahead of schedule, having completed 58% of the project in the first year. The project is also running over budget, with the potential opportunity to stretch out the schedule in an attempt to reduce costs.

A key part of project accounting systems is cost information. Cost data is needed both for estimates and for tracking actual incurred costs. This cost information is typically based on cost accounting.

2.5.4   Cost Accounting

The purpose of cost accounting is to determine the required funds for activities such as building infrastructure and performing maintenance tasks. Cost accounting information is used for (1) the prediction of costs for work to be performed in the future, and (2) an assessment of the actual costs for completed work.

Cost accounting is relatively simple for outsourced work. An outsourced construction project will include a quote from the contractor that forms the basis for the internal cost estimate. At the end of the project, payments to the contractor constitute the project cost. A big benefit of outsourced work is cost transparency. A utility knows exactly how much the outsourced work costs.

Cost accounting is more complicated for work performed with internal resources. Consider identical construction projects performed by similar utilities. The calculated cost of these projects can differ widely, based on accounting treatment of a variety of factors. For example:

Potential Difficulties in Cost Accounting
- **Material Costs** – Materials for a project can be purchased new or can be taken from inventory. If taken from inventory, material costs can vary depending upon the inventory accounting method (e.g., FIFO, LIFO, or average cost).
- **Labor Costs** – Most utility craft labor is performed on an hourly basis. From this perspective, it is simple to calculate direct labor costs such as salary and benefits. However, these workers cannot perform their job function without other costs such as tools, warehouses, training, shift supervisors, and so forth. There is a wide range of practices with regard to what is included and what is not included in labor costs, and these variations can lead to large differences in cost accounting.
- **Construction Vehicles and Equipment** – Construction vehicles and equipment represent large expenses for most utilities. Once purchased, the initial price is a fixed cost whether it gets used or not. Therefore, some utilities do not allocate the capital cost of vehicles and equipment to projects. Other utilities allocate vehicle and equipment costs, typical-

ly by an hourly rate that can vary widely for different utilities. Variations in the treatment of construction vehicles and equipment can lead to large differences in cost accounting.

- **Overhead** – Overhead costs are not directly related to utility operations. This includes items such as CEO pay, income taxes, interest payments, accounting, insurance, headquarter facilities, website, and advertising. These costs are necessary and significant. Some utilities do not allocate overhead costs to work activities. Other utilities allocate overhead cost in a wide variety of ways. Variations in the treatment of overhead costs and equipment can lead to large differences in cost accounting.

Consider again two identical construction projects performed by two similar utilities. For cost accounting purposes, Utility A only considers costs directly attributable to the project. This includes $25 million for materials, $20 million for direct labor, and $5 million for the direct operational cost of vehicles. Utility B aggressively allocates as many indirect costs to the project as possible, including $10 million for indirect labor, $20 million for overhead, and $20 million for the capital cost of vehicles and equipment. It is extremely difficult to know whether these allocations represent accurate project costs. This is why Utility A does not include them. However, these costs are real and significant, which is why Utility B tries to include them. The results are shown in Table 2.15. Cost accounting results in a project cost of $50 million for Utility A and $100 million for Utility B.

Many utilities have sophisticated cost accounting systems that track the cost of all actual work and use this data to compute *unit costs* for a variety of activities. For example, these systems can compute the average cost per linear foot of trenching, the average cost to install a vault, the average cost to relocate a pole, and so forth. Because these unit costs are based on real data, there is a common misconception that unit costs represent true costs.

Table 2.15. Variations in cost accounting practices can result in large differences. In this example, Utility A only considers costs directly attributable to the project, resulting in a project cost of $50 million. Utility B aggressively allocates as many indirect costs to the project as possible, resulting in a project cost of $100 million.

Cost Category	Utility A	Utility B
Material Cost	25	25
Direct Labor (salary + benefits)	20	20
Indirect Labor (tools, training, etc.)	0	10
Allocated Overhead	0	20
Vehicles & Equipment (capital expense)	0	20
Vehicles & Equipment (operations expense)	5	5
Total	$50	$100

Consider again Table 2.15. Assume that both utilities have the exact same cost accounting systems and processes. Because of this, direct project costs for both utilities are identical. However, since cost accounting rules are different, the cost estimate for Utility B is twice that of Utility A, even though each utility has exactly the same cost structure.

It is generally safe to use unit cost information (as computed by cost accounting systems) to compare internal projects. For example, unit cost information can be used to select the most cost effective project from among several options. Unit cost information can also be used to determine whether costs for similar activities are increasing over time. However, it is generally not appropriate to use unit cost information for comparisons with other utilities or with contractors. If one utility has unit costs higher than another, it does not necessarily follow that the utility has higher actual costs. If the unit cost for a utility is lower than the quoted cost of a contractor, it does not necessarily follow that the utility is more cost effective.

In summary, cost accounting is a critical part of utility operations. It allows future costs to be estimated, and allows actual costs to be captured. However, different utilities have a wide range of cost accounting practices. These differences can lead to substantial unit cost differences for essentially the same work. When using cost accounting and unit costs, it is important to understand what this information represents and what is does not represent.

2.5.5   More on Management Accounting

Project accounting and cost accounting are types of management accounting. Financial accounting, tax accounting, and regulatory accounting follow strict rules and are intended for external stakeholders. Management accounting has no set rules and is intended for internal use. It uses existing accounting data to develop reports, understand business results, enable performance management, identify opportunities for profit increases, and so forth. When used properly, management accounting allows for better-informed business decisions.

In addition to project accounting and cost accounting, there are several other common management accounting practices. These include contribution margins and activity-based costing.

Contribution margins compute the marginal profit of an activity by only considering variable costs. Any cost that will not increase if the activity is slightly increased is considered a fixed cost, and is excluded from the contribution margin analysis. Consider a utility with a fixed infrastructure, such as a gas or electric utility. With the infrastructure in place, the only variable cost is the cost of purchased wholesale energy. Therefore, the contribution margin is equal to sales minus purchases energy. Assume that retail sales are $100 million and energy purchases are $40 million. The contribution margin analysis is:

Contribution Margin Analysis	
Revenue (millions)	
Retail sales	$100
Total Revenue	$100
Variable Costs (millions)	
Wholesale Energy	$40
Total Variable Cost	$40
Contribution Margin (millions)	$60 (60%)

This contribution margin analysis shows that every additional dollar in sales will result in sixty cents of additional profit. This will continue as long as (1) the existing infrastructure can support the additional sales, and (2) the cost of wholesale energy does not increase. This analysis also shows that the utility can sell incrementally retail energy at up to a 40% (the difference between 100% and the 60% contribution margin) discount and still make money.

Activity-based costing (ABC) is somewhat the opposite of contribution margins. It attempts to capture all costs, both variable and fixed, to determine the true profit margin of a product or service. To do this, ABC breaks down fixed costs into activities. It then examines the percentage of time that each activity is spent on each product and service. Consider a utility with three operating companies and a centralized accounting group. The ABC process will examine the accounting group and determine the percentage of time spent on each operating company and allocate accounting costs accordingly.

In theory, ABC has the ability to identify whether a product or service is truly profitable. Consider a utility value added service that shows a contribution margin of 30%. The utility has an overall corporate overhead cost of 20%. A traditional analysis would subtract the overhead cost from the contribution margin and show a profit margin of 10%. However, an ABC analysis shows that overhead functions are spending twice as much time for this service as for typical services. Therefore, the ABC analysis would subtract the activity-based overhead cost, in this case 40%, from the contribution margin. The ABC analysis now shows that due to very high overhead requirements, the value added service actually *loses* money in the amount of 30% – 40% = 10% of revenues.

There are many types of management accounting approaches in addition to those already described. Managers should be encouraged to design customized management accounting reports that will help them better understand financial performance and make better informed decisions. Utilities have a wealth of data in their financial accounting systems. Management accounting leverages the value of this data for uses beyond financial reporting.

2.6   SUMMARY

Financial accounting consists of a few concepts and a large amount of vocabulary. For the level of understanding required for utility engineers, nothing is particularly difficult or complicated, especially when compared to typical engineering topics. However, accounting topics are voluminous and nuanced. The best way to develop a good comfort level is to start thinking about financial accounting issues as they relate to daily work activities. How will my activities affect the balance sheet? What is the contribution margin of my part of the company? How is overhead allocated my project? Will my decision increase earnings per share? Thinking about these questions will gradually balance the engineering mindset (i.e., What are the engineering implications?) with an accounting mindset (i.e., What are the profit implications?).

Readers are also encouraged to study their company's financial statements. Going through financial statements will naturally raise questions in context. When these questions are investigated and answered, insight into the utility's finances is gained along with general accounting knowledge. Gaining familiarity with financial statements will gradually balance the engineering and accounting mindsets with an investor mindset (i.e., What is the financial position of the utility?).

Learning the basics of accounting is well worth the effort. Accounting is a profession with undergraduate degrees, graduate degrees, professional certifications, and so forth. There is no need for utility engineers to have accounting knowledge equivalent to accounting professionals. However, it is important to know enough to effectively interact with and utilize the accounting staff when necessary. This chapter has provided the accounting material necessary for this level of understanding. The terms and concepts introduced in this chapter are also foundational for the following chapters. Careful reading will magnify the clarity and usefulness of the rest of this book

2.7   STUDY QUESTIONS

1. What is the basic accounting equation? Explain in words the meaning of this equation.
2. Briefly explain the accounting principles of matching and conservatism.
3. What are the three basic financial statements? What is the purpose of each of these statements?
4. What is meant by capitalizing an expense? What is the impact of capitalizing an expense on current profits and future profits?
5. What is goodwill? How can the amount of goodwill on a balance sheet change?
6. What are the three methods of inventory accounting? How will each of these affect earnings and the balance sheet?

7. What is meant by free cash flow? What is the relationship of free cash flow to net income?
8. Explain some ways in which accounting practices can impact the income statement. Can accounting practices impact the statement of cash flows? Explain.
9. What are some differences between financial accounting, regulatory accounting, and tax accounting?
10. If a utility has a robust cost accounting system, do unit costs derived from this system represent true costs? Explain.

3
Economics

The subject of *engineering economics* typically deals with the time value of money and net present value calculations. In the business world, these topics are included in the subject of finance. Business economics (or simply economics) may occasionally use these tools, but addresses topics very different from traditional engineering economics.

Economists commonly say that economics is about the allocation of scarce resources. In a world of plenty, people can have anything they want and as much as they want. Utility services are not infinite. Service levels cannot be perfect. Safety cannot be perfect. Capacity cannot be infinite. Each would require infinite resources, which any utility engineer can verify is far from reality. How then, should questions related to price, quality, and quantity of service be answered? There are several possibilities, but most countries use economic theory as a basis for answering these questions.

Economics can broadly be classified into *microeconomics* and *macroeconomics*. Microeconomics addresses the allocation of scarce products and services in local markets. It includes topics such as supply, demand, prices, and monopolies. Macroeconomics addresses national and international economies as a whole, including topics such as business cycles, unemployment, inflation, national savings, lending rates, monetary policy, and international trade. Most of this chapter focuses on microeconomics. Knowledge of macroeconomics is a good supplement to microeconomics, but is not essential to understand the basic business drivers of utilities.

In a competitive market, decisions made by producers and consumers determine the amounts of products and services produced and their corresponding market prices. Since the price of utility services is typically set by regulators, the dynamics of a pure competitive market do not apply. However, most regulators attempt to set rules and rates in an attempt to capture the efficiencies of a competitive market. For this reason, a basic understanding of microeconomics is an important part of understanding the business of utilities.

3.1   SUPPLY AND DEMAND

In a market economy, consumers choose the type and amount of products and services they purchase. Typical factors that impact purchasing decisions are quality and price. For the rest of this chapter, "products and services" will be replaced with the term "services."

If a service is nearly free, customers are typically willing to purchase a large amount. If the service has a low price, customers are willing to purchase slightly less. As prices continue to increase, customers will almost always be willing to purchase increasingly less of the service. If a service is infinitely expensive, customers will not be willing to purchase any amount.

The combination of price levels for a service and the corresponding quantity desired by customers is called a *demand curve*. A typical demand curve is shown in Figure 3.1. For each price level, there is a corresponding quantity demanded by customers. The high price of P_0 corresponds to the low quantity of Q_0. The low price of P_1 corresponds to the high quantity of Q_1.

Consider a utility service with fixed rates. For these rates, customers will consume a certain quantity of the service. If rates increase, customers will consume less of the service. If rates decrease, customers will consume more of the service.

If small changes in price lead to large changes in the quantity demanded by customers, the demand is called *elastic*. If large changes in price only result in small changes in the quantity demanded by customers, demand is called *inelastic*. Mathematically, *elasticity of demand* is equal to the percent change in quantity demanded divided by the percent change in price. The equation for elasticity of demand, E_D, is the following:

$$E_D = \frac{\Delta Q_D}{Q_D} \cdot \frac{P}{\Delta P} \tag{3.1}$$

where E_D is elasticity of demand; Q_D is quantity demanded; ΔQ_D is change in quantity demanded; P is price; and ΔP is change in price.

Elasticity of demand is demonstrated in Figure 3.2. A horizontal demand curve represents a perfectly elastic demand. A price just below this curve corresponds to infinite demand. A price just above the curve corresponds to no demand whatsoever. A vertical demand curve represents a perfectly inelastic demand. The quantity of the service demanded is the same regardless of price. Demand is classified as elastic if elasticity of demand is below negative one and is classified as inelastic if it is between zero and negative one.

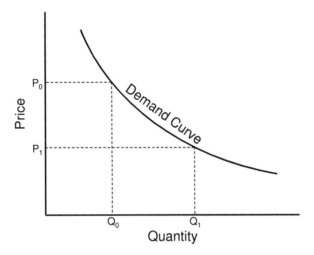

Figure 3.1. A typical demand curve showing the relationship between the price of a service and the quantity of the service demanded by customers at that price. For the high price of P_0, customers demand the low quantity of Q_0. For the low price of P_1, customers demand the high quantity of Q_1.

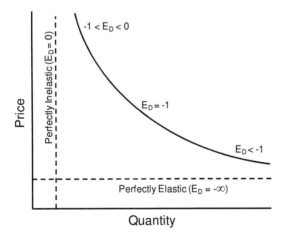

Figure 3.2. Elasticity of demand, E_D, is equal to the percent change in quantity demanded divided by the percent change in price. A horizontal demand curve represents a perfectly elastic demand; a price just below this curve corresponds to infinite demand. A price just above the curve corresponds to no demand whatsoever. A vertical demand curve represents a perfectly inelastic demand; the quantity of the service demanded is the same regardless of price.

Elasticity of demand determines the impact of price changes on revenue. If demand is inelastic, a small price increase will increase revenue. Consider a unit price of utility service of P with an elasticity of demand of 50%. The quantity demanded at this price is Q, resulting in utility revenue of P x Q. If price is increased by 2%, the quantity demanded will only decrease by half of this percentage since the elasticity of demand is 50%. The new revenue is 1.02 x P x 0.99 x Q = 1.01 x P x Q. A 2% increase in price results in about a 1% increase in revenue.

If demand is elastic, a small price increase will decrease revenue. Consider a unit price of utility service of P with an elasticity of demand of 200%. Like the previous example, the quantity demanded at this price is Q, resulting in utility revenue of P x Q. If price is increased by 2%, the quantity demanded will decrease by twice this percentage since the elasticity of demand is 200%. The new revenue is 1.02 x P x 0.96 x Q = 0.98 x P x Q. A 2% increase in price results in about a 2% decrease in revenue.

But customer demand is irrelevant unless companies are willing to offer the demanded service. If the price of a service is zero, for-profit companies will not be willing to supply any of the service. If the price is the service is small, companies will supply a small amount of the service. As prices continue to increase, companies will continue to supply more of the service if it results in increasingly higher profits.

The combination of price levels for a service and the corresponding quantity supplied by companies is called a *supply curve*. A typical supply curve is shown in Figure 3.3. For each price level, there is a corresponding quantity supplied by companies. The low price of P_0 corresponds to the low quantity of Q_0. The high price of P_1 corresponds to the high quantity of Q_1.

Consider a utility value added service that is not subject to regulation. For a given price, utilities will desire to provide a certain quantity of the service. If the price is higher, utilities will desire to supply more of the service. If the price is lower, utilities will desire to supply less of the service.

If small changes in price lead to large changes in the quantity supplied by companies, supply is called *elastic*. If large changes in price only result in small changes in the quantity supplied by companies, supply is called *inelastic*. Mathematically, *elasticity of supply* is equal to the percent change in quantity supplied divided by the percent change in price. The equation for elasticity of supply, E_S, is the following:

$$E_S = \frac{\Delta Q_S}{Q_S} \cdot \frac{P}{\Delta P} \tag{3.2}$$

where E_S is elasticity of supply, Q_S is quantity supplied, ΔQ_S is change in quantity supplied, P is price, and ΔP is change in price.

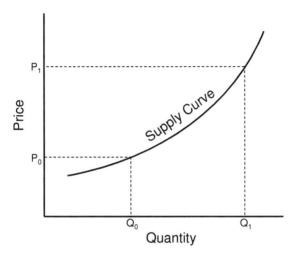

Figure 3.3. A typical supply curve showing the relationship between the price of a service and the quantity of the service supplied by companies at that price. For the low price of P_0, companies supply the low quantity of Q_0. For the high price of P_1, companies supply the high quantity of Q_1.

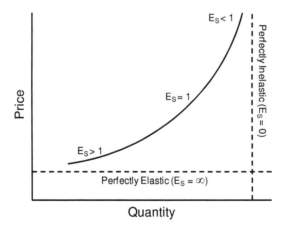

Figure 3.4. Elasticity of supply, E_S, is equal to the percent change in quantity supplied divided by the percent change in price. A horizontal supply curve represents a perfectly elastic supply. A price just above this curve corresponds to infinite supply. A price just below the curve corresponds to no supply whatsoever. A vertical supply curve represents a perfectly inelastic supply. The quantity of the service supplied is the same regardless of price. Supply is classified as elastic if the elasticity of supply is greater than one and is classified as inelastic if it is less than one.

Elasticity of supply is demonstrated in Figure 3.4. When supply is elastic, small changes in market price lead to large changes in production volume by suppliers. A horizontal supply curve represents a perfectly elastic supply. A price just above this curve corresponds to infinite supply. A price just below the curve corresponds to no supply whatsoever. A vertical supply curve represents a perfectly inelastic supply. The quantity of the service supplied is the same regardless of price. Supply is classified as elastic if elasticity of supply is greater than one and is classified as inelastic if it is less than one.

In summary, a demand curve describes how much of a service customers want at a certain price and a supply curve describes how much companies want to provide at a certain price. Taken together, supply curves and demand curves allow market dynamics and pricing to be examined. This interaction is discussed in the next section.

3.2 MARKET PRICING

The supply curve and demand curve for a service can easily be shown on the same graph. The intersection of these two curves occurs at the price where the amount of the service demanded by customers is exactly equal to the amount of the service supplied by companies. This price is called the *market equilibrium price*. When supply is equal to demand, price will equal the market equilibrium price and the market is said, as one might expect, to be in *equilibrium*. Consumers can purchase the amount of service they desire without encountering shortages. Suppliers can sell all of the services they produce without the fear of unsold production. The point of market equilibrium for a supply curve and a demand curve is shown in Figure 3.5.

What happens if prices are higher than the market equilibrium price? Consider Figure 3.6, where prices start out at the high level of P_H. At this price, companies are willing to supply more of the service than customers are willing to buy. The companies are forced to lower prices to entice customers to buy more. When prices are lowered, customers demand more of the service and the companies are willing to supply less of the service. This results in motion along the supply and demand curves as indicated by the arrows. Price will continue to decline until the quantity supplied is equal to quantity demanded. This is shown as the market equilibrium price, P_E.

The logic is similar if prices are too low. If the price is below the market equilibrium price, customers will demand more of a service than companies are willing to supply. The companies will raise prices to increase profits. When prices are raised, customers demand less of the service and the companies are willing to supply more of the service. This results in motion along the supply and demand curves towards market equilibrium. Price will continue to increase until it reaches the market equilibrium price where quantity supplied is equal to quantity demanded.

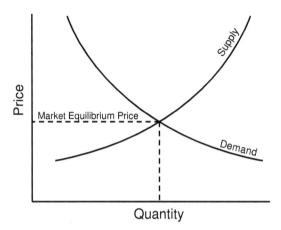

Figure 3.5. Market equilibrium occurs at a price where the quantity of a service supplied by companies exactly matches the quantity demanded by customers. Graphically, this occurs where the supply curve and the demand curve intersect. The price where the intersection occurs is called the market equilibrium price.

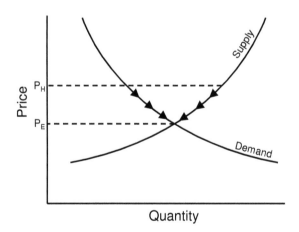

Figure 3.6. Market forces drive price towards the market equilibrium price, P_E. In this figure, prices start out at the high level of P_H. At this price, companies are willing to supply more of the service than customers are willing to buy. The companies are forced to lower prices to entice customers to buy more. When prices are lowered, customers demand more of the service and the companies are willing to supply less of the service. This results in motion along the supply and demand curves as indicated by the arrows. Price will continue to decline until the quantity supplied is equal to quantity demanded.

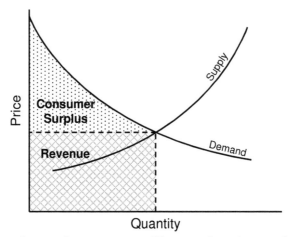

Figure 3.7. The total amount that customers are willing to pay for a given quantity of services is equal to the area under the demand curve. The actual amount paid is equal to price times quantity and corresponds to company revenue. The difference between the total amount that customers are willing to pay and the amount that they actually pay is called consumer surplus.

As its name suggests, a market in equilibrium tends to stay in equilibrium. If prices are increased above the equilibrium point, an excess of supply will force prices to come down. If prices are decreased below the equilibrium point, an excess of demand will force prices to come up.

Competitive market pricing is generally a good deal for consumers. In fact, the extent to which this is a good deal can be quantified. Consider the demand curve shown in Figure 3.7. There are a small number of customers that are willing to pay a very high price for the service being considered. This corresponds to the very left part of the demand curve. These customers are willing to pay more than the market equilibrium price for a certain amount of services, but only pay the market equilibrium price. The difference in the amount that a customer is willing to pay and the amount they actually pay is called consumer surplus. At market equilibrium, total consumer surplus is equal to the area between the demand curve and the market clearing price.

The supply and demand curves that result in consumer surplus can change. Consider an increase in fuel costs, which will directly increase the cost of a utility to provide services to its customers. To supply the same quantity of services, the utility now requires an increase in rates to offset the increase in fuel prices. The effect is to shift the supply curve up. Consider Figure 3.8, which shows the original supply curve, S_0, and the shifted supply curve, S_1. Initially, the supply curve S_0 intersects the demand curve at a price of P_0 and a quantity of Q_0. After the fuel price increase, the new supply curve S_1 intersects the demand curve at a price of P_1 and a quantity of Q_1. The price has increased. Because of the price increase, the quantity demanded by customers decreases to Q_1. An upward shift in the supply curve increases prices and decreases consumption.

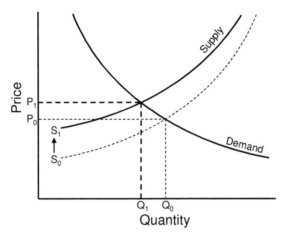

Figure 3.8. An upward shift in the supply curve causes an increase in price and a decrease in quantity demanded by customers. In this case, the supply curve shifts up from S_0 to S_1. Initially, the supply curve S_0 intersects the demand curve at a price of P_0 and a quantity of Q_0. After the shift, the new supply curve S_1 intersects the demand curve at a price of P_1 and a quantity of Q_1. The price has increased. Because of the price increase, the quantity demanded by customers decreases to Q_1. An upward shift in the supply curve increases prices and decreases consumption.

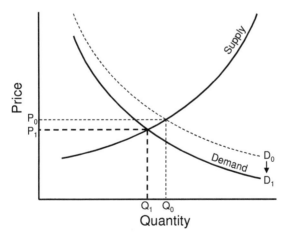

Figure 3.9. A downward shift in the demand curve causes a decrease in price and a decrease in quantity supplied. In this case, the demand curve shifts down from D_0 to D_1. Initially, the demand curve D_0 intersects the supply curve at a price of P_0 and a quantity of Q_0. After the shift, the new demand curve D_1 intersects the supply curve at a price of P_1 and a quantity of Q_1. A downward shift in the demand curve decreases prices and decreases consumption.

The supply curve is shifted down when the cost to produce services is reduced. For example, fuel costs could go down, new technologies could improve operational efficiencies, automated processes could reduce workforce costs, and so forth. The effect of a downward shift in the supply curve is the opposite of an upward shift. Consider Figure 3.8, but assume that the supply curve starts at S_1 and shifts down to S_0. The price of the service lowers from P_1 to P_0 and the quantity of the service demanded by customers at this lower price increases from Q_1 to Q_0.

Like supply curves, demand curves can also shift. A *shift in demand* should not be confused with a *shift in quantity demanded*. A shift in quantity demanded is due to a change in price. A shift in demand is due to a change in a need or desire for the service that is separate from price. Common causes for shifts in demand are changes in quality, new uses for the service, and the availability of substitutes.

As quality improves, demand increases. For example, many facilities historically used steam engines and internal combustion engines for their industrial processes. As the reliability of electricity supply improved, these facilities converted to the use of electric motors, increasing the demand for electricity.

Demand for a service will also increase when new uses are found. This is especially true for utility services. Demand for electricity increased with the invention of electrical appliances such as electric ovens, electric water heaters, radio, television, air conditioning, and computers. Demand for telephone bandwidth increased with the inventions of Telex teleprinters, fax machines, dial-up modems, and broadband modems. Demand for natural gas increased with the invention of gas-based clothes dryers, gas-based air conditioning, combined heat and power generation, and combined-cycle power plants.

Although there is typically little direct competition for utility services, viable substitutes for these services are sometimes available. Consider regulated telephone service. At one time, there was essentially no cost-effective substitute for flexible long distance voice communication. Today, people can use cellular telephones, satellite telephones, and internet-based telephones.

On an appliance level, electricity and natural gas can often be substituted for each other, although there are often high switching costs (such as converting from an electric clothes dryer to a natural gas drier). As technology advances, it will be increasingly cost effective to substitute utility service with self-generated electricity through photovoltaic solar panels, small wind turbines, and natural gas powered microturbines.

Consider a federal law requiring the use of compact fluorescent light bulbs instead of the more energy-intensive incandescent bulbs. After installing the new bulbs, consumers will require less electricity, shifting down the demand curve. Consider Figure 3.9, which shows the original demand curve, D_0, and the shifted demand curve, D_1. Initially, the demand curve D_0 intersects the supply curve at a price of P_0 and a quantity of Q_0. After the light bulb law, the new demand curve D_1 intersects the supply curve at a price of P_1 and a quantity of Q_1.

The price has decreased. A downward shift in the demand curve decreases prices and decreases consumption.

The effect of an upward shift in the demand curve is the opposite of a downward shift. Consider Figure 3.9, but assume that the demand curve starts at D_1 and shifts up to D_0. The price of the service increases from P_1 to P_0 and the quantity of the service supplied with this higher demand increases from Q_1 to Q_0.

In summary, competitive markets are driven by supply and demand. For a given price, companies are willing to supply a certain amount of a service. For the same price, customers demand a certain amount. If supply exceeds demand, companies will lower prices until supply equals demand. If demand exceeds supply, companies will raise prices until supply equals demand. When supply is equal to demand, the market is in equilibrium.

3.3   PRODUCER SURPLUS

Market dynamics are driven by competition. In the previous sections, mismatches in supply and demand result in price and quantity shifts until market equilibrium is reached. These shifts are described in terms of the overall market, with all suppliers lumped into a single group. To better understand market pricing, it is necessary to look at pricing shifts in terms of a single company.

Consider a company that is producing a service. Due to fixed costs, the average total cost to produce the first few units is high. As more units are produced, fixed costs are spread over more units, resulting in a declining unit cost. The incremental cost to produce additional units is called the *marginal cost* of the service. Typically, the marginal cost of making additional units starts off decreasing, further reducing average cost. The combination of lower fixed costs per unit and declining marginal costs results in *economies of scale*. That is, the unit cost of a company, up to a certain point, tends to decrease as production volume increases.

Beyond a certain production volume, marginal cost will start to increase. Eventually, increasing marginal cost will lead to increasing average cost. This phenomenon is called *diseconomies of scale*. An example of an increasing marginal cost curve is shown in Figure 3.10.

At low quantities, the price a company can charge for its service will exceed its marginal cost (due to demand exceeding supply). Graphically, this occurs when the demand curve is above the marginal cost curve. When this is the case, the company can produce an additional unit of service and charge customers more than its incremental cost. Consider a utility producing 10 million units of a service. At this quantity, the marginal cost is $10 to produce an additional unit, and customers are willing to pay $11 for this additional unit.

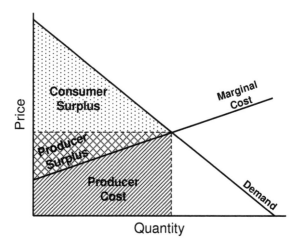

Figure 3.10. The quantity of a service supplied by companies will increase until the price of the service is equal to the marginal cost of the service. This occurs where the marginal cost curve intersects the demand curve. The profit made by companies, called producer surplus by economists, is equal to total revenue minus producer cost.

In a competitive market, suppliers will always choose to produce additional units when the current market price is higher than the marginal cost of producing the next unit; a unit can be sold for more than it costs to be produced, resulting in additional profits. Since there are more units to buy in the marketplace, prices fall. Production continues to increase and prices continue to fall until marginal cost is equal to market price. A company or group of companies may wish to keep prices artificially high, which will be discussed further in the next section. However, price fixing and other agreements between companies that allow prices to exceed marginal cost are illegal, harmful to customers, and typically prosecuted under antitrust legislation.

The quantity of a service supplied by companies will increase until price is equal to marginal cost. This occurs where the marginal cost curve intersects the demand curve. The total cost to produce services is equal to the area underneath the marginal cost curve. The total revenue is equal to price times quantity. The profit made by companies, called *producer surplus* by economists, is equal to total revenue minus producer cost. Producer cost and producer surplus are shown graphically in Figure 3.10.

As discussed in the previous section, the amount that consumers would be willing to pay for a certain quantity of service is equal to the area under the demand curve. This amount minus revenue is equal to consumer surplus. In a competitive market, the equilibrium price is always equal to marginal cost. In a competitive market, consumer surplus, producer surplus, and producer cost will have relationships similar to those shown in Figure 3.10.

3.4   MONOPOLISTIC PRICING

Sometimes market competition does not happen. An example is an industry with large fixed costs to begin production and a relatively small incremental cost to produce each unit of service. As production levels increase, average cost per unit decreases, resulting in *economies of scale*. Eventually average cost per unit will go up, but this may occur at production levels that will never be reached.

What will happen when a group of companies are competing in an industry with decreasing average costs? The company that is producing the most will have a cost advantage, and will be able to sell its service at prices that are not profitable for the smaller companies. The largest company will be the most profitable, allowing it to grow further and reduce prices further. Eventually, smaller companies are not able to survive at current market prices and go out of business. The remaining company is the sole provider of the services and is called a *monopoly*. The industry is said to be a *natural monopoly* since it leads to a single provider in the long run and this single provider can supply quantities demanded at costs lower than would otherwise be possible with competition.

Regulated services provided by public utilities are typically considered natural monopolies. Providing utility services requires large initial investments in infrastructure and large ongoing investments for operating and maintaining the infrastructure. Providing incremental services through this infrastructure is relatively inexpensive. Once a public utility is operational and serving many customers, it would be extremely difficult for a competitor to become established due to the massive investment required.

Utility infrastructure typically requires the use of public land and easements of private land. This is another justification for not having competition to provide utility services. Imagine having five competing electric companies, all with their own poles, towers, and lines. Imagine having five natural gas pipelines running under the streets. These scenarios are expensive, logistically difficult, and potentially ugly. Similar conclusions can be reached for water, local telephone, cable television, roads, railroad tracks, and other societal infrastructure.

The problem with monopolies is their ability to set prices. In a competitive market, competition leads to the market equilibrium price where supply equals demand. In contrast, an unregulated monopoly can set any price it chooses and customers have no say in the matter.

What price will unregulated monopolies choose to set? Consider a utility that is able to sell a quantity of Q_0 at a service price of P_0. If the utility increases its price to P_1, the quantity demanded at this price decreases to Q_1. The change in revenue from the price increase is equal to the new revenue, $P_1 \times Q_1$, minus the old revenue, $P_0 \times Q_0$. This change in revenue can then be compared to the cost savings realized through the decrease in quantity. The utility will implement the price increase if the change in revenue minus the change in cost is positive. The difference between marginal revenue and marginal cost corresponds to increased profits.

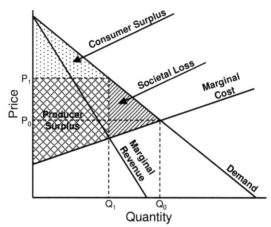

Figure 3.11. This figure compares competitive market pricing to monopolistic pricing. Competitive market pricing occurs where the marginal cost curve intersects the demand curve (price of P_0, quantity of Q_0). Monopolistic pricing occurs where the marginal cost curve intersects the marginal revenue curve (price of P_1, quantity of Q_1). Monopolistic pricing results in higher prices, lower quantities, and a reduction in consumer surplus.

Using this logic, monopolies will continue increasing price until the marginal cost reduction of supplying less services is equal to marginal revenue. At this point, higher prices will decrease profits since the marginal savings of supplying fewer services is greater than marginal revenue.

A comparison of competitive market pricing and monopolistic pricing is shown in Figure 3.11. Competitive market pricing occurs where the marginal cost curve intersects the demand curve. In this case, competitive market pricing corresponds to a price of P_0 and a quantity of Q_0, just as in Figure 3.10. Consumer surplus is equal to the area between the demand curve and the P_0 price level.

Monopolistic pricing occurs where the marginal cost curve intersects the marginal revenue curve. In this case, monopolistic pricing corresponds to a price of P_1 and a quantity of Q_1. The monopoly chooses a higher price and a lower quantity of sales than otherwise would occur in a competitive market. Producer surplus has increased, and consumer surplus has decreased (compare to Figure 3.10). More concerning is the impact to societal benefit. Under competitive market pricing, societal benefit (i.e., consumer surplus plus producer surplus) is maximized. Monopolistic pricing has reduced consumer surplus by reducing consumer surplus more than the corresponding increase in producer surplus. This reduction in societal benefit is called *societal loss*. Societal loss is equal to the area bounded by the demand curve, the marginal cost curve, the quantity at competitive market pricing, and the quantity at monopolistic pricing. In Figure 3.11, the societal loss due to monopolistic pricing is almost equal to the remaining consumer surplus.

When a monopoly sets a price higher than would otherwise occur in a competitive environment, it foregoes sales to customers that will not pay the higher price, but would purchase some services at a lower price. If possible, the monopoly will try to capture some of this business through *price discrimination*, which attempts to charge higher prices to those willing to pay, and lower prices to others. For practical reasons, price discrimination is typically implemented by setting a standard price that is very high and discounting based on willingness to pay. For example, adults in their working years typically have more purchasing power than children or retirees. Therefore, many businesses set a high standard price for services and then offer discounts for children and senior citizens. Other common examples of price discrimination are student discounts, financial aid packages, and coupons.

For price discrimination to work, a company must be able to set a price higher than their marginal cost. Effectively, this implies that price discrimination can only be accomplished by companies that have a certain amount of monopolistic pricing ability. In addition, effective price discrimination must prevent customers with a high willingness to pay from accessing lower prices. For example, coupons will not be effective from a price discrimination perspective if they are easily accessible to all customers, not just those who would not make the purchase at the undiscounted price.

Figure 3.12 shows the effect of price discrimination on consumer surplus. In this example, customers have been segmented into three price tiers. The first price tier targets customers willing to pay a high price for services, and corresponds to the monopolistic price that occurs when marginal cost is equal to marginal revenue. To capture more business, two lower price tiers are offered. The second price tier targets customers willing to pay a moderate price for services. The third price tier uses the markets equilibrium price. The effect of this tiered pricing strategy, assuming that it can be effectively implemented, is to both increase consumer surplus and societal benefit when compared to the single Tier 1 monopoly price. However, consumer surplus is lower than would result from a single Tier 3 price, which would occur under competitive conditions.

It is not uncommon for utilities to use tiered rates. A typical approach is to charge customers a low rate for a certain amount of a service used in a month. Beyond this amount, an intermediate rate is charged until another threshold quantity is used. Beyond this threshold, a high rate is used. The effect is to charge higher average rates to customers using large quantities of the utility service. If heavy users are willing to pay higher rates, this tiered pricing approach will reduce customer surplus when compared to a single low rate.

Consider a utility that sells to one hundred thousand residential customers: sixty thousand are light users, thirty thousand are moderate users, and ten thousand are heavy users. Initially, this utility sells to everyone at $1 per unit, regardless of the amount consumed. The utility then switches to a tiered rate structure. Monthly charges for a customer are $1 for the first 100 units, $1.50 for the next 100 units, and $2.00 for each additional unit beyond 200. Consumption and revenue for this scenario is shown in Table 3.1.

Table 3.1. Comparison of a single-tier rate scheme and a three-tier rate scheme. Under the three-tier rate scheme, higher rates are charged after 100 units per month are consumed and even higher rates are charged after 200 units per month are consumed. The impact of the three-tier rate scheme is to slightly reduce consumption, but significantly increase revenue from moderate and heavy users.

Customer Class	#	$ per unit	Units per Month	Units	$
Single-Tier Rates					
Light User	60,000	1.00	100	6,000,000	6,000,000
Moderate User	30,000	1.00	200	6,000,000	6,000,000
Heavy User	10,000	1.00	400	4,000,000	4,000,000
Total	**100,000**	**1.00**	**160**	**16,000,000**	**16,000,000**
Three-Tier Rates					
Light User	60,000	1.00	100	6,000,000	6,000,000
Moderate User	30,000	1.00	100	3,000,000	3,000,000
		1.50	75	2,250,000	3,375,000
Heavy User	10,000	1.00	100	1,000,000	1,000,000
		1.50	100	1,000,000	1,500,000
		2.00	150	1,500,000	3,000,000
Total	**100,000**	**1.21**	**147.5**	**14,750,000**	**17,875,000**

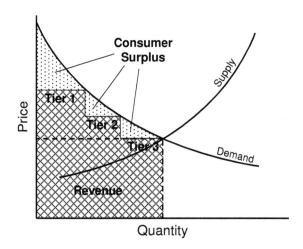

Figure 3.12. Companies with monopolistic pricing power can increase sales through price discrimination. In this example, monopolistic pricing corresponds to Tier 1, where marginal cost is equal to marginal revenue. To capture more business, two lower price tiers are offered. The second price tier targets customers willing to pay a moderate price for services. The third price tier uses the market equilibrium price, where price is equal to marginal cost. The effect of this tiered pricing strategy is to both increase consumer surplus and societal benefit when compared to the single Tier 1 monopoly price. However, consumer surplus is lower than would result from a single Tier 3 price, which would occur under competitive conditions.

The single-tier rate scenario shown in Table 3.1 results in 16 million units sold and $16 million in utility revenue. With the three-tier rate structure, consumption goes slightly down for moderate and heavy users (this is typically part of the intent of tiered rates). However, revenue increases to nearly $18 million and the average price per unit increases by 21%.

A conservative estimate of consumer surplus that is converted to utility revenue is equal to price difference multiplied by volume. In this case, the average price difference is $0.21 per unit and customers are consuming 14.75 units. In this case, customers are paying $3.1 million more per month than they would under the single-tier rates. This incremental revenue comes from moderate and heavy users. Under the three-tier rate scheme, consumer surplus for moderate and heavy users is reduced; they are not getting as "good a deal" as they otherwise would get from a single-tier rate scheme.

From one perspective, natural monopolies are good, since a single company can provide services at a lower cost than multiple companies. From another perspective, natural monopolies are bad, since monopolistic pricing leads to a reduction in societal benefits. Public utilities are natural monopolies, and are subject to these competing tensions.

The most common solution to utility pricing is to subject public utilities to regulation. Through the regulatory process, utility service prices are set in a manner that is similar to prices that otherwise would occur in a competitive environment. The subject of utility regulation and utility ratemaking is addressed in detail in Chapter 7, which will rely heavily upon the economic foundations presented in this chapter.

3.5   BUSINESS CYCLES

Economies have their ups and downs. These ups and downs are part of business cycles, which exist but are not fully understood. This section provides a brief overview of business cycles to familiarize the reader with some of the more important terms and concepts.

The examination of business cycles requires a measure of the size of an economy. The most common measure is *gross domestic product* (GDP). GDP measures the market value of the total output of an economy. Total output includes all final goods and services, but excludes intermediate goods and services. For the most part, a society consumes what it produces and GDP is approximately equal to consumption. The final GDP value must adjust for investment and net exports as follows:

$$\text{GDP} = \text{consumption} + \text{investment} + \text{exports} - \text{imports} \qquad (3.3)$$

When something is produced, it must be consumed, saved, or exported. When something is imported, it does not have to be produced domestically.

Some definitions of GDP separate consumption into private consumption and government consumption. This is to emphasize the potential role of government policy on the overall economy.

Another common measure of economic activity is *gross national product* (GNP). The US used GNP to measure its national accounts until 1992, when it switched to GDP. GNP considers net foreign income whereas GDP considers net exports. The GDP of a country measures all goods and services produced within the borders of the country, regardless of the nationalities of the people producing these goods and services. The GNP of a country measures all goods and services produced by citizens of the country, regardless of where they live.

2007 GDP values for various regions of the world (as calculated by the International Monetary Fund) are shown in Table 3.2. The total world GDP is about $55 trillion. In 2007 the world population was about 6.7 billion, resulting in a global GDP average of about $8200 per person. The country with the largest economy by far is the United States (although the total GDP of the European Union is slightly larger). The US accounts for about 25% of global GDP. In 2007, the US population was about 300 million, resulting in a US GDP of about $44,000 per person.

A modern economy needs supporting infrastructure. Without sufficient infrastructure, economic growth will be either restricted or prevented. Utility infrastructure and economic growth tend to go hand in hand, and it is important for utility engineers to have a basic understanding of business cycles that lead to oscillating GDP growth and contraction.

Recall that the consumption of an economy is roughly equal to the production of an economy. When people consume more, they must also produce more, causing the economy to grow. When people stop consuming, they must also produce less, causing the economy to shrink. These facts link the economy to psychological factors such as *consumer confidence*. If people feel that their economic prospects are weakening, they tend to consume less and cause the economy to weaken. This is both a self-fulfilling prophecy (I believe that the economy will weaken and therefore cause it to weaken) and a vicious cycle (a weakening economy results in lower consumer confidence that further weakens the economy). The author Aldous Huxley addresses this psychological factor in his book *Brave New World*, where people are encouraged to buy new stuff instead of fixing old stuff. From youth, citizens in Huxley's world are trained to believe that, "Ending is better than mending. The more stitches, the less riches."

When people refer to the business cycle, they are typically referring to the *fixed investment cycle*. When consumer confidence is high, companies increase their output capacity by investing in buildings, factories, equipment, and other fixed assets. Investors are happy to lend these businesses money at low interest rates since the economy is good and the risk of default is low. This is characteristic of an economy that is in the *expansion phase* of the business cycle (also called a *boom*). During a boom, easy access to financial capital and low interest rates will result in increased risk taking (the government controls short-term

Table 3.2. Gross Domestic Product (GDP) statistics for 2007 as compiled by the International Monetary Fund. The total world GDP is about $55 trillion. The Unites States has the largest country GDP by far, although the European Union in total is slightly larger.

Rank	Region	GDP ($B)	%	Rank	Region	GDP ($B)	%
	World	54,585	100%	44	Israel	164	0.30%
	European Union	16,906	31.0%	45	Chile	164	0.30%
1	United States	13,808	25.3%	46	Singapore	161	0.30%
2	Japan	4,382	8.0%	47	Philippines	144	0.26%
3	Germany	3,321	6.1%	48	Pakistan	144	0.26%
4	China	3,280	6.0%	49	Ukraine	142	0.26%
5	United Kingdom	2,804	5.1%	50	Hungary	138	0.25%
6	France	2,594	4.8%	51	Algeria	134	0.25%
7	Italy	2,105	3.9%	52	New Zealand	129	0.24%
8	Spain	1,440	2.6%	53	Egypt	128	0.23%
9	Canada	1,436	2.6%	54	Kuwait	112	0.20%
10	Brazil	1,314	2.4%	55	Peru	107	0.20%
11	Russia	1,290	2.4%	56	Kazakhstan	105	0.19%
12	India	1,101	2.0%	57	Morocco	75	0.14%
13	Mexico	1,023	1.9%	58	Slovakia	75	0.14%
14	South Korea	970	1.8%	59	Bangladesh	74	0.13%
15	Australia	909	1.7%	60	Qatar	73	0.13%
16	Netherlands	777	1.4%	61	Vietnam	71	0.13%
17	Turkey	659	1.2%	62	Libya	70	0.13%
18	Sweden	455	0.83%	63	Angola	61	0.11%
19	Belgium	453	0.83%	64	Croatia	51	0.09%
20	Indonesia	433	0.79%	65	Luxembourg	50	0.09%
21	Switzerland	427	0.78%	66	Sudan	46	0.08%
22	Poland	422	0.77%	67	Slovenia	46	0.08%
23	Norway	389	0.71%	68	Ecuador	46	0.08%
24	Taiwan	383	0.70%	69	Belarus	45	0.08%
25	Saudi Arabia	382	0.70%	70	Dominican Rep.	41	0.08%
26	Austria	371	0.68%	71	Oman	40	0.07%
27	Greece	314	0.57%	72	Serbia	40	0.07%
28	Denmark	312	0.57%	73	Bulgaria	40	0.07%
29	Iran	285	0.52%	74	Syria	39	0.07%
30	South Africa	283	0.52%	75	Lithuania	39	0.07%
31	Ireland	261	0.48%	76	Tunisia	35	0.06%
32	Argentina	260	0.48%	77	Guatemala	34	0.06%
33	Finland	246	0.45%	78	Sri Lanka	32	0.06%
34	Thailand	245	0.45%	79	Azerbaijan	31	0.06%
35	Venezuela	228	0.42%	80	Latvia	27	0.05%
36	Portugal	223	0.41%	81	Kenya	27	0.05%
37	Hong Kong	207	0.38%	82	Costa Rica	26	0.05%
38	Colombia	203	0.37%	83	Turkmenistan	26	0.05%
39	UAE	191	0.35%	84	Lebanon	25	0.05%
40	Malaysia	187	0.34%	85	Uruguay	23	0.04%
41	Czech Republic	175	0.32%	86	Uzbekistan	22	0.04%
42	Nigeria	167	0.31%	87	Yemen	22	0.04%
43	Romania	166	0.30%	88	Cyprus	21	0.04%

interest rates and will sometimes raise them in an attempt to prevent the economy from "overheating"). New and existing businesses will begin to invest in new ways that may turn out not to be profitable.

At a certain point, some businesses will overbuild, leading to overcapacity. When this happens, the businesses will lower prices in an attempt to cover their fixed costs. These price wars tend to result in lower corporate profits and bankruptcies. At the same time, some of the new risky business ventures may start to fail. These *malinvestments* eventually lead to financial difficulties and additional bankruptcies.

At this point, lenders will increase interest rates due to higher risk exposure (again, the Government controls short-term rates and will try to keep them low). Bankruptcies will lead to liquidation and abandonment or reallocation of capital investments. An increase in bankruptcy filings will also cause banks to start calling in loans (leading to bank runs prior to federal deposit insurance), tightening the credit market. Consumers will naturally be concerned about the economy, and start to spend less. Businesses will start laying people off due to lower consumer demand, causing unemployment to rise and wages to fall. This is characteristic of an economy that is in the *contraction phase* of the business cycle (also called a *bust*).

A *recession* occurs when economic contraction leads to negative GDP growth for more than two consecutive quarters. A severe recession that lasts a long time is called a *depression*.

Eventually, the trough of the contraction will occur and businesses will begin to produce positive financial results. This *economic recovery* gradually leads to lower interest rates, increased capital investment, lower unemployment, and higher wages. Consumer confidence returns and the business cycle returns to its expansion phase. It is estimated that this fixed investment cycle is typically between seven and eleven years long.

The *infrastructural investment cycle* is similar to the fixed investment cycle, but is more closely related to construction activity. It has a longer cycle time of fifteen to twenty-five years. The infrastructural investment cycle begins with low vacancy rates, rising rents, and increasing land prices. Builders respond by constructing more houses (i.e., increased *housing starts*). This construction activity helps the overall economy, and also requires supporting infrastructure such as electricity, gas, telephone, water, and sewer.

Overbuilding will occur when housing starts exceed new home sales, causing vacancy rates to increase. Builders will begin to reduce construction activity and lower prices. Real estate owners will start to lower rents. Eventually, housing prices and rents will stabilize, construction activity will increase, and the infrastructure investment cycle will begin a new expansion phase. Utilities are keenly interested in the infrastructure investment cycle since it is a strong driver of infrastructure expansion needs.

There are a host of other terms that are commonly used when the greater economy and business cycles are discussed. Some of the more important of these terms are now described.

Consumer Price Index (CPI). The CPI is a cost of living index, representing the average price of consumer goods and services purchased by households. The percent change in the CPI is typically used as a measure of inflation. The CPI is computed by determining the price of a bundle of goods, and weighting these prices based on observed consumption patterns. The US CPI has been found to slightly overestimate inflation for two reasons. First, weights are only adjusted once every ten years, and are therefore not able to account for rapid adoption of new product or rapid changes in product quality. Second, the CPI is not able to account for product substitution, where consumers shift spending from more expensive products to less expensive products.

GDP Deflator. The GDP deflator is a general price index that reflects the set of all goods that were produced domestically, weighted by the market value of the total consumption of each good. For example, if more natural gas is purchased in a specific year, it is weighted higher than in years where less is purchased. GDP deflators are typically normalized to a specific year. For example, a table could define the GDP deflator for 2005 as 100%; by definition, 2005 prices are 100% of 2005 prices. Deflator values for other years compare price levels to 2005 prices. For example, 2004 might have a GDP deflator of 97%, indicating lower prices than in 2005. Similarly, 2006 might have a GDP deflator of 105%, indicating that prices overall rose by 5% from 2005 to 2006.

Inflation. Inflation means that prices are increasing. Inflation occurs when price indicators like the CPI and GDP Deflator are increasing in value.

Deflation. Deflation is the opposite of inflation, and occurs when price indicators like the CPI and GDP Deflator are declining in value.

Stagflation. Stagflation occurs when inflation occurs at the same time the economy is shrinking or not growing. In the typical business cycle, busts are accompanied by deflation, which eventually leads to economic recovery. When stagflation has occurred, it has proven difficult and expensive to fix.

Unemployment Rate. The unemployment rate is the number of unemployed people actively seeking employment divided by the total workforce (employed people plus active job seekers). It only includes people who have looked for a job in the last four weeks, and therefore does not include discouraged unemployed people who may have given up looking. In the US, unemployment is measured by the Bureau of Labor Statistics from a monthly survey of sixty thousand households.

Natural Unemployment Rate. Economists argue that very low levels of unemployment are bad for the economy. When there are not enough people looking for jobs, there tends to be upward pressure on wages as businesses compete for workers. Businesses will compensate for higher payroll costs by raising prices, resulting in accelerated inflation. The best a nation can do is to achieve a level of unemployment that will not begin accelerated inflation. This level is called the *natural unemployment rate*. It is also called the *non-accelerating inflation rate of unemployment* (NAIRU). In the US, the natural rate of unemployment is typically estimated to be between 5% and 6%.

Money Supply. The money supply loosely refers to the amount of liquidity in an economy. It is measured in levels, M0 through M3, ranging from most liquid to least liquid. M0 is the narrowest definition of money supply, and includes physical currency circulating in the economy. M1 also includes checking accounts. M2 adds savings accounts and non-institutional money market accounts. M3 is the broadest measure, and further adds institutional money market accounts and various other large liquid assets. M2 is the economic indicator most closely associated with inflation.

Consumer Confidence Index (CCI). The US CCI is a normalized measure of consumer views about the current and future state of the economy. CCI is based on a monthly survey of five thousand US households. Each household is asked to respond positive, neutral, or negative to the following five questions: (1) current business conditions, (2) business conditions for the next six months, (3) current employment conditions, (4) employment conditions for the next six months, and (5) total family income for the next six months. Positive responses are divided by total responses. This ratio is normalized based on 1985 results being equal to a CCI of 100.

Business cycles are part of the broader topic of macroeconomics. The discussion of business cycles in this section has focused on a single economy, emphasizing the US. A full discussion of macroeconomics would involve the global economy and address topics such as exchange rates, foreign investment, and international trade. It would also discuss a range of economic shocks that arguably lead to macroeconomic cycles (e.g., new technologies, monetary shocks, changing energy prices, wars, natural disasters). These topics are interesting to many, but are not closely related to the business of utilities and are therefore beyond the scope of this book.

3.6   SUMMARY

In a competitive market, societal benefit is maximized when the price of a service is equal to the marginal cost of providing this service. This price is reached naturally through market forces. If the market price is too high, an excess in supply will cause prices to drop and production to be scaled back. If the price is too low, a shortage in supply will cause prices to increase and production to decrease. Supply is equal to demand at the market equilibrium price.

The market forces that drive prices to market equilibrium require competition. If a company is subject to competition, it will continue to increase supply until the marginal cost of the increase is equal to the market price of the service. Competition results in a price equal to marginal cost, which corresponds to the market equilibrium price.

The regulated services of public utilities are natural monopolies and, therefore, are not subject to competitive forces. An unregulated monopoly can set prices at any level it wishes. To maximize profits, a monopoly will continue to increase price until the marginal cost of the increase is equal to the market revenue of the increase. In most cases, monopolistic pricing results in higher prices and lower quantities. Worse, monopolistic pricing increases producer surplus at the expense of overall societal benefits. Since public utilities are natural monopolies, they are typically regulated to avoid monopolistic pricing.

3.7   STUDY QUESTIONS

1. The field of economics is about the allocation of what?
2. Explain the difference between microeconomics and macroeconomics.
3. Why does the supply curve slope upwards?
4. Describe some products or services with a high elasticity of demand and separately describe some products or services with a low elasticity of demand.
5. What is meant by the market equilibrium price? What happens if prices are higher or lower than the market equilibrium price?
6. What is meant by producer surplus, consumer surplus, and societal benefit?
7. What are some methods that companies use in an attempt to transfer consumer surplus to producer surplus?
8. What are some potentially good and bad things about a natural monopoly?
9. How does monopolistic pricing differ from competitive market pricing? What is the impact of monopolistic pricing on producer surplus and consumer surplus?
10. From an economic perspective, what should the role of utility regulation strive to achieve?

4
Finance

Finance is the science of money and markets. It dictates how to value a stream of uncertain cash flows. It shows how to determine the stock price of a company. It defines the optimal tradeoff between financial risk and financial return. It tells companies how much long-term debt should be issued. It allows portfolios of investments to be managed in a rigorous and systematic manner.

As a verb, finance means raising money for an investment. *Corporate finance* deals with raising money for a company, typically through stocks and bonds. Because investors purchase these stocks and bonds, a key goal of finance is to understand investor expectations and behavior. Utility engineers should understand the goals of corporate executives. Understanding the goals of corporate executives requires the understanding of investors, potential investors, and financial analysts. This can only happen by gaining a basic understanding of finance and its application to utilities.

This chapter presents a summary of finance from an engineering and technical management perspective. Its purpose is not to turn the reader into a financial wizard. Therefore, detailed treatment of historical context, derivations, and nuances are generally avoided. The focus is to present the theory and principals underlying corporate finance, and to demonstrate an array of key financial issues that are important for utility engineers to understand.

As mentioned in the previous chapter, the subject of "engineering economics" typically addresses the time value of money, valuing streams of future cash flows, and related topics. In the business world, these topics are part of the subject of finance, not economics. No aspect of finance can be properly comprehended without a full understanding of the time value of money, and this topic is addressed first.

4.1   TIME VALUE OF MONEY

An amount of money received today is more valuable than the same amount of money received sometime in the future. Consider $100 received today. Assume that this amount could be put into a savings account that offers 5% interest per year. After one year, the $100 has turned into $105. In the terminology of finance, the *present value* of $100 is has a *future value* of $105 in one year. After two years, the amount increases by an additional 5% to $110.50. The increase in the first year is $5.00 and the increase in the second year is $5.50. Left alone, the amount of increase will grow, resulting in an exponential increase in account value. This exponential increase is caused by *compound interest* – interest being earned on all previous interest payments.

4.1.1   Compound Interest

The future value of cash after n periods of interest, C_n, can be computed by knowing the initial amount of cash, C_0, and the interest rate per period, r. The equation for future value is:

$$C_n = C_0 \cdot (1+r)^n \quad ; \text{future value} \tag{4.1}$$

Using the same logic, an amount of money received in the future is less valuable than the same amount of money received today. Consider $105 promised one year from today. Since an investment of $100 today can result in the same amount in one year, $105 in one year has a present value of $100. The equation for present value is derived by rearranging the terms in the equation for future value:

$$C_0 = C_n \cdot (1+r)^{-n} \quad ; \text{present value} \tag{4.2}$$

Future value and present value are shown graphically in Figure 4.1 on a *cash flow diagram*. The horizontal axis corresponds to elapsed time with each tick mark representing an interest-earning period. Vertical arrows pointing up represent cash flows. In this case, present value, C_0, corresponds to the short arrow at $t = 0$. Future value corresponds to the tall arrow at $t = n$.

Because of compounding, future value will vary depending upon the frequency with which interest is calculated. Consider an annual interest rate of 12%. If interest is calculated once per year, $100 today has a future value of $112 in one year, corresponding to $r = 12\%$ and $n = 1$. If interest is calculated twice per year, $r = 6\%$ and $n = 2$. Therefore, $100 is worth $106 after six months and $112.36 after twelve months.

Figure 4.1. This cash flow diagram shows a graphical representation of present value, C_0, and future value, C_n. The horizontal axis corresponds to elapsed time with each tick mark representing an interest-earning period. Present value corresponds to $t = 0$ and future value corresponds to $t = n$.

Future values for different compounding frequencies are shown in Table 4.1. As expected, a higher compounding frequency leads to a higher future value. However, the amount of increase reduces quickly. Going from one to two compounding periods is worth \$0.36. Going from 12 (once per month) to 365 (once per day) is worth \$0.07. Going from once per day to once per microsecond is not even worth a single penny.

For an interest rate r, it is straightforward to mathematically derive future value based on continuous compounding. This is done by taking the limit of Equation 4.1 as n approaches infinity and with the interest rate per period equal to r ÷ n. The result is the following equation:

$$C_1 = C_0 \cdot e^r \quad ; \text{continuous compounding} \tag{4.3}$$

The above equation was used to compute the future value for the last row in Table 4.1. With a single compounding period, an annual interest rate of 12% also has an *effective annual interest rate* of 12%. With continuous compounding, an annual interest rate of 12% has an *effective annual interest rate* of 12.75%.

Table 4.1. This table shows future values for different compounding frequencies. Higher compounding frequencies leads to higher future values. However, the amount of increase reduces quickly, with little value possible beyond daily compounding.

Present Value	Annual Interest Rate	Compounding Periods Per Year	Future Value in 1 Year
\$100.00	12%	1	\$112.00
\$100.00	12%	2	\$112.36
\$100.00	12%	3	\$112.49
\$100.00	12%	4	\$112.55
\$100.00	12%	6	\$112.62
\$100.00	12%	12	\$112.68
\$100.00	12%	365	\$112.75
\$100.00	12%	Infinite	\$112.75

4.1.2 Net Present Value

It is common to consider a number of cash flow events for a single analysis. For example, a project may consist of a number of costs requiring cash outflows at various times. Once completed, the project may also result in revenue from customer payments at various times. The sum of all positive present values minus the sum of all negative present values is called *net present value*, or *NPV*. When cash inflows are from revenue and cash outflows are from expenses, NPV is equal to the following:

$$NPV = \sum_{n=0}^{\infty} (R_n - E_n)(1+r)^{-n} \tag{4.4}$$

NPV	=	Net Present Value
R_n	=	Revenue in year n
E_n	=	Expense in year n
r	=	Interest rate

NPV calculations are commonly used for three purposes. First, NPV can be used to determine whether a project is worth pursuing, from a purely economic perspective. Second, NPV can be used to identify a solution to a problem with the least rate impact to customers. Third, NPV can be used to determine the value of a company.

It is always desirable to identify ways to spend money now so that more money can be saved and/or earned in the future. The proper way to determine whether this type of proposal should proceed is through an NPV analysis. First, the present value of all costs should be determined including, but not limited to, initial costs and recurring costs. Next, the present value of all savings should be determined including, but not limited to, incremental revenue, reduced operational costs, and avoided investments. NPV is equal to the sum of all savings minus the sum of all costs. If the calculation is correct, positive NPV projects should be pursued and negative NPV projects should not be pursued. Economic analyses of projects are addressed in detail in Chapter 8.

Sometimes utilities must spend money so that adequate levels of service can be provided to customers. New customers must be served, growing demand must not overstress the system, reliability must be adequate, safety requirements must be maintained, unacceptable risks must be mitigated, and so forth. When an issue arises that must be addressed, utilities should identify the solution that keeps service rates as low as possible. In theory, this will be the project with the least *life cycle cost*, which corresponds to the smallest NPV of costs. The NPV calculation process is the same as in the previous paragraph, but the NPV of costs will typically be much higher than the NPV of benefits. Utilities must spend money to fix the problem, but do so for the smallest possible NPV.

NPV is also the methodology that financial analysts use to determine the inherent worth of companies. In short, the value of a company is equal to the NPV of expected future free cash flows. An analyst who believes that the NPV of expected future free cash flows is greater than current market value will recommend buying stock in the company (i.e., the stock is underpriced). A sell recommendation results from the NPV of expected future free cash flows being less than current market value (i.e., the stock is overpriced). Company valuation using NPV is addressed in detail in Section 4.2.

There are several types of cash flow streams that commonly arise in an NPV analysis. For this reason, it is convenient to provide a quick method of NPV calculation. The first type of cash flow stream is a *perpetuity*, which corresponds to a constant amount of cash at the beginning of each year, starting in Year 1. An example of a perpetuity is $1000 paid to the holder each January 1st, this year and every year into the future. The NPV of a perpetuity with an interest rate of r is equal to:

$$NPV = \frac{C_1}{r} \qquad ; NPV \text{ of a perpetuity} \qquad (4.5)$$

For example, the NPV of a $10,000 perpetuity with an interest rate of 10% is equal to $10,000 \div 0.1 = \$100,000$. This means that $100,000 in the bank today earning interest at 10% per year is equal in value to a $10,000 annual payment starting next year and continuing forever.

The second type of common cash flow stream is a *growing perpetuity*, which provides payments that increase by a fixed percentage each year. The payment is C_1 in Year 1. With a growth rate of g, the payment is $C_1 \times (1 + g)$ in Year 2 and $C_1 \times (1 + g)^{n-1}$ in Year n. NPV of a growing perpetuity with an interest rate of r is equal to:

$$NPV = \frac{C_1}{r - g} \qquad ; NPV \text{ of a growing perpetuity} \qquad (4.6)$$

Consider payments that start at $10,000 in Year 1, and grow by 5% per year (the interest rate is still 10%). The NPV is equal to $10,000 \div (0.10 - 0.05) = \$10,000 \div 0.05 = \$200,000$. In this case, a growth rate of 5% doubles the value of the annuity when compared to a zero growth rate. An amount of $200,000 in the bank today earning 10% interest is equal in value to a $10,000 annual payment starting next year and growing by 5% each year.

The perpetuity equations are simple to remember and convenient to use. However, cash flows are typically not expected to last forever. These *annuities* have formulae similar to perpetuities but with an additional term that essentially subtracts out the perpetuity values in distant years. The NPV of an annuity with an interest rate of r that pays out for n periods is equal to:

$$NPV = \frac{C_1}{r} \cdot \left(1 - \frac{1}{(1+r)^n}\right) \qquad ; \text{NPV of an annuity} \qquad (4.7)$$

For example, the NPV of a $10,000 annuity with an interest rate of 10% and a payout of 20 years is equal to $85,136. If the annuity pays out forever (i.e., a perpetuity), the NPV is equal to $100,000. In this case, about 85% of the value of the perpetuity is paid out in the first twenty years.

A *growing annuity* is similar to a growing perpetuity except it only pays out after n periods. The NPV of a growing annuity with an interest rate of r that grows at a rate of g and pays out for n periods is equal to:

$$NPV = \frac{C_1}{(r-g)} \cdot \left(1 - \left(\frac{1+g}{1+r}\right)^n\right) \qquad ; \text{NPV of a growing annuity} \qquad (4.8)$$

For example, the NPV of a $10,000 growing annuity with an interest rate of 10%, a growth rate of 5%, and a payout of 20 years is equal to $121,121. If the growing annuity pays out forever (i.e., a growing perpetuity), the NPV is equal to $200,000. In this case, only 61% of the value of the growing perpetuity is paid out in the first twenty years, a lower percentage than the constant annuity/perpetuity case. Without growth, most of the NPV of cash flows occur in the early years.

Cash flow diagrams and their associated NPV equations for perpetuities and annuities are shown in Figure 4.2. These are the most commonly used cash flow streams, and can be combined to represent many cash flow scenarios. Spreadsheet applications can be used to easily compute NPV for any arbitrary stream of cash.

4.1.3   Annualized Cost

Often, competing projects have different expected lifetimes, and comparing their economic attractiveness is problematic from an NPV perspective. Consider the purchase of a wood structure with a 30 year life costing $500 versus a steel structure with a 60 year life costing $1000, with no recurring costs for either. Even though the steel structure is expected to last twice as long as the wooden pole, the wooden pole has a present value that is half the price of the steel pole. The situation becomes more complicated when recurring costs are considered, since projects with longer lifetimes will appear worse due to the increased number of recurring charges.

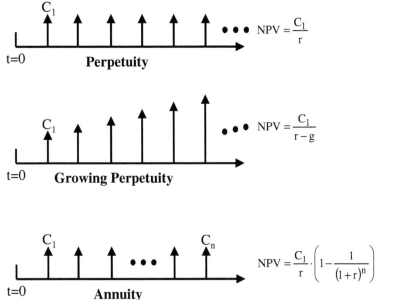

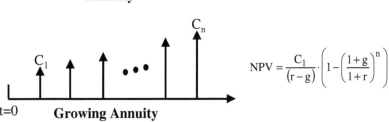

Figure 4.2. This figure shows cash flow diagrams and corresponding equations for perpetuities and annuities. Perpetuity payments start on time period 1 and continue forever. Annuity payments start on time period 1 and continue until time period n. In these equations, t is the time period, C_t is a cash payment in time period t, r is the interest rate, and g is the growth rate.

Projects with different lifetimes can be compared by converting their NPV into equal annual payments spread over the lifetime of the project. These payments, referred to as *annualized costs* or *levelized costs*, can be thought of as an annuity with duration equal to the expected life of the project. For example, the $500 wood structure can be financed over 30 years at an interest rate of 15% for $76.15 per year (assuming no financing charges), corresponding to an annualized cost of $76.15 per year. Similarly, the $1000 steel structure can be financed over 60 years at an interest rate of 15% for $150.03 per year. From this analysis, it is clear that the wooden structure costs much less per year than the steel pole considering the time value of money.

The annualized cost of a project is computed by calculating its net present value and then converting this result into an annuity over the life of the project. If the life of a project is n years and the interest rate is r, the annualized cost is computed as follows:

$$\text{Annualized Cost} = \text{NPV} \cdot \frac{r(1+r)^n}{(1+r)^n - 1} \tag{4.9}$$

Often a project will consist of an initial cost and a recurring annual cost. Assume that the $500 wood structure with a 30 year life has an annual maintenance cost of $50. Obviously, a stream of cash payments of $50 per year for 30 years is the same thing as a $50 annuity for 30 years. Therefore, the annualized cost can be calculated by adding the annual recurring cost to the annualized initial cost as follows:

$$\text{Annualized Cost} = \text{Recurring Cost} + \text{Initial Cost} \cdot \frac{r(1+r)^n}{(1+r)^n - 1} \tag{4.10}$$

This equation assumes that recurring costs do not increase in value over time. Typically, recurring costs can be expected to grow at least at the rate of inflation. If a price escalation is assumed for recurring costs, the author recommends computing the NPV of recurring costs using the growing annuity equation. This value is then added to the initial cost of the project and the sum is converted to an annualized cost.

4.1.4 Discount Rates

The previous sections presented the time value of money using interest rates. Although the use of interest rates adds clarity to the subject, it is often necessary to analyze the time value of money without an obvious associated interest rate. For example, there is no interest rate directly associated with stocks, even though investors expect a return on purchased stock. There is no interest rate directly associated with proposed infrastructure projects. There is no interest rate directly associated with proposed maintenance projects.

The generic term for the "r" value used in time value of money calculations is *discount rate*. It is possible that a discount rate corresponds to an interest rate, but commonly it is a number agreed upon so that the time value of money can be considered in cash flow analysis. Factors that are typically considered when choosing a discount rate are inflation, time frame, corporate risk exposure, and project risk exposure.

Inflation refers to price increases and the resulting loss of purchasing power of a fixed amount of cash. For example, if the price of a bundle of goods today is P, and expected inflation rate is r_i, then the expected price of the bundle of goods in future years is as follows:

Year	Price
0	P
1	$P \times (1 + r_i)$
2	$P \times (1 + r_i)^2$
3	$P \times (1 + r_i)^3$
n	$P \times (1 + r_i)^n$

Inflation is typically viewed as an increase in all monetary measures including wages, utility rates, and prices. In a microeconomic sense, inflation does not matter. People earn more money, but prices have increased by the same amount. Long-term contracts are typically not able to adjust themselves based on actual inflation, but these contracts consider the expected levels of inflation when they are valued.

The stated value of an interest rate or a discount rate is called the *nominal interest rate*. For example, the nominal value for interest on a bank deposit might be 5%. If money is deposited in the bank, the amount of money in the account will increase by 5% per year, but the purchasing power of the money will increase by less than 5% due to inflation. Consider an inflation rate of 2%. In one year, a deposit of $100 will have grown to $105. However, a typical bundle of goods will have grown in price from $100 to $102. Therefore, purchasing power has only grown by the nominal interest rate minus the inflation rate. This increase in purchasing power is called the *real interest rate*:

$$\text{Nominal Rate} = \text{Real Rate} + \text{Inflation Rate} \tag{4.11}$$

For the most part, inflation is a measure of the size of the money supply, and does not have any substantial bearing on economic activity such as production, capital markets, financial markets, and retail markets. Consider an economy with a money supply of M dollars. Now imagine that every dollar is instantly converted into two dollars, increasing the money supply to 2M. The economy will continue as usual, but all prices will quickly double and all salaries will quickly double. Nothing has changed except the supply of money and corresponding price levels.

In perfect markets described by Adam Smith, an "invisible hand" quickly changes prices to reflect changes in the money supply. From this perspective, monetary policy is an ineffective way for a government to impact the economy. In reality, prices tend to be "sticky" and will take time to reflect changes in the money supply. This is largely due to fixed contracts such as leases. This sort of dynamic macroeconomic effect is not generally a factor in the determination of discount rates.

Investors expect to earn higher returns when exposing themselves to higher levels of risk. This means that higher risk requires the use of higher interest rates and/or higher discount rates. In finance, risk is related to the uncertainty of cash flows. Consider two possible $100 investments with different payment profiles:

Investment A: Pays back $110   (100% chance)

Investment B: Pays back $220   (50% chance)
                 Pays back $0   (50% chance)

The expected return of each investment is 10%. Investment A guarantees this expected return, while Investment B is much more of a gamble. Mathematically, the standard deviation of returns is higher for Investment B, which means that it is riskier from a financial perspective. Since both investments have the same expected return but Investment B is more risky, rational investors prefer Investment A.

Investments with a long duration financial commitment are riskier than investments with shorter durations. To understand this, recall that real returns are equal to nominal returns minus inflation. A long duration financial commitment has less predictable real returns since it is exposed to inflation uncertainty. Consider a US treasury note with a one-month maturity period. It is not likely that inflation levels will substantially change within one month. Compare this to a US treasury bill with a thirty-year maturity period. Clearly the real return of this bill has significantly more risk of being affected by changes in inflation. Therefore, investors require that treasury bills have higher nominal interest rates than treasury notes.

The relationship of investment maturity period to interest rate is called a *yield curve*. Examples of US treasury yield curves are shown in Figure 4.3. These yield curves can change dramatically from year to year, reflecting changing investor expectations about inflation. Occasionally, the yield curve will have a negative slope, referred to as an *inverted yield curve*. More often, part of the yield curve will have a negative slope, referred to as a *partially inverted yield curve*. These negative slopes are widely thought to be leading indicators of an upcoming recession. When investors expect a recession, they demand more long-term investments which drive prices up and yields down.

In most utilities, NPV calculations use the same discount factor for cash flows in all future years. From a yield curve perspective, it is appropriate to use smaller discount rates for cash flows in early years and higher discount rates for cash flows in later years. Think of NPV as the amount of money required to be put in the bank today to finance the entire project over its life. Since inflation is uncertain, more money is required today to make sure that future costs can be met if inflation turns out to be high.

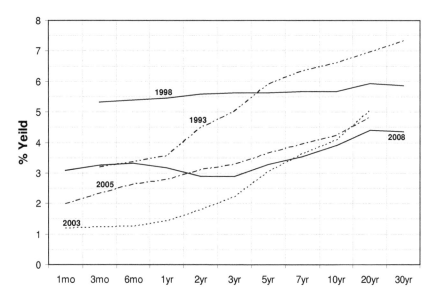

Figure 4.3. This figure shows yield curves for US treasuries ranging in maturity period from one month to thirty years. Since longer maturity periods have higher uncertainty with respect to inflation effects, yields curves are generally upward sloping. These yield curves for different years vary dramatically, reflecting changing investor expectations about inflation.

When a utility is borrowing money to finance a project, the interest rate is determined by the lender's perception of risk that the utility will not repay the loan. Within a utility, some potential projects will have highly predictable cash flows and some potential projects will have highly uncertain (i.e., risky) cash flows. These issues relate to the weighted average cost of capital and risk-adjusted discount rates. Both of these topics are addressed in the next section.

4.1.5 Weighted Average Cost of Capital

Utilities need to raise cash to fund their business. This cash is referred to as *financial capital,* or simply *capital.* Most of this capital is raised without a specific purpose in mind. Raised capital effectively goes into a treasury and is parceled out for specific activities. Separate capital is not raised for new customer connections. Separate capital is not raised for a new call center. Separate capital is typically not raised for specific construction projects.

As discussed in Chapter 2, utilities typically raise capital by selling common stock, by selling preferred stock, and by selling bonds. Utilities must pay a return to these investors, and this return is called the *cost of capital.* In a business, money is not free. Money needs to be borrowed from investors and these investors expect to earn money in return for the use of their financial capital.

In terms of priority, utilities are obligated to pay returns to bond holders first, preferred stockholders second, and common stockholders last. Therefore, bonds are the least risky investment, preferred stock is the next least risky investment, and common stock is the most risky investment. Because the levels of risk are different, these types of investments will have different financial return requirements from investors.

Consider a utility with 20 million shares of issued common stock that is currently trading at $50 per share. The market value of common stock is equal to 20 million x $50 = $1 billion. This utility has also issued 1 million shares of preferred stock at $100 per share with dividend payments of $10 per share per year. Last, the utility has issued $900 million worth of bonds at an interest rate of 8% per year. Taken together, these sources of cash define the *capital structure* of the utility, which is discussed further in Section 4.4. Assuming that investors expect a 12% return on common stock, the capital structure is summarized as follows:

Source of Cash	Amount ($ millions)	Required Return
Common Stock	1,000	12%
Preferred Stock	100	10%
Bonds	900	8%

This utility has raised $2 billion in capital, which can be used to meet all cash obligations. In return, the utility is expected to give each investor class the required return on their investment. The cost of capital for common stock is 12%. The cost of capital for preferred stock is 10%. The cost of capital for bonds is 8%. On average, the utility must pay investors a return of 10.1%. This is computed by taking the weighted average of the required returns, and is called the *weighted average cost of capital* (*WACC*, pronounced "wack").

Stated differently, WACC is the average profit that a company must achieve in order to meet all investor expectations. Mathematically, WACC is equal to the following

$$\text{WACC} = \frac{r_{cs} \cdot C_{cs} + r_{ps} \cdot C_{ps} + r_b \cdot C_b}{C_{cs} + C_{ps} + C_b} \tag{4.12}$$

where r_{cs} is the required return on common stock, r_{ps} is the required return on preferred stock, r_b is the required return on bonds, C_{cs} is capital from common stock, C_{ps} is capital from preferred stock, and C_b is capital from bonds.

The above equation for WACC assumes three sources of capital. It can easily be extended to include any number of capital sources. The numerator is the sum of the product of all amounts of capital multiplied by their required return.

The denominator is equal to the total amount of capital. Each utility will have a finance group capable of determining the company WACC, and all utility engineers should be familiar with this value.

WACC represents the overall financial risk of a utility as perceived by investors. Many activities within a utility have risks similar to the overall risk of the company in terms of cash flow uncertainty. It is appropriate to use WACC as a discount rate when examining these activities.

Some proposed projects will be riskier than the overall risk of the company. For example, a project may be subject to uncertain regulatory treatment. Similar projects may never have been done in the past. The cash flows of the projects may be highly sensitive to commodity prices, foreign exchange rates, or other factors. In these cases, it is appropriate to use *risk-adjusted discount rates* that are based on WACC plus a *risk premium*.

Consider the following example. A utility decides to use a new technology for equipment inspections. The utility expects that the use of this new technology will save money due to reduced inspection costs and fewer equipment failures. However, it is not certain if these reduced costs will be realized, since there is no historical evidence. Using WACC for the NPV analysis will overestimate the value of expected cost reductions since the risk of the project is higher than the overall risk of the company. Working with the finance department, it is decided to add a 10% risk premium to WACC when performing the NPV analysis. The risk-adjusted analysis allows this risky project to be fairly compared with traditional approaches with more predictable cash flows.

4.2   COMPANY VALUATION

Whether you own a company, work for a company, or are interested in purchasing a company, a natural question to ask is, "How much is the company worth?" One way to answer this question is to determine how much it would cost to purchase all owner's equity. For a publicly traded company, this amount is equal to the total number of outstanding shares of common stock multiplied by the trading price of the stock. For example, a company with 100 million outstanding shares currently trading at $50 per share has a value of $5 billion. This seems easy enough. But how did the stock price level get at its present trading value in the first place? Clearly, somebody must be examining the intrinsic value of a company.

The number of outstanding shares times share price is called *market capitalization*, or *market cap*. Market capitalization reflects the investor consensus on the value of a company. If investors feel that market cap is too low, they can purchase stock, which will drive up the stock price. If investors feel that market cap is too high, they can sell stock, which will drive down the stock price. Typically, investors group companies into *large cap* (more than $5 to $10 billion), *small cap* (less than $1 or $2 billion), and *mid cap* (in between).

Figure 4.4. The only possible reason for investors to give money to a company is so they will receive more money back in the future. Common stockholders receive this money through dividends. An increase in the expected NPV of dividends will raise stock price. A decrease in the expected NPV of dividends will lower stock price.

To determine whether market cap is too low or too high, investors must estimate the inherent value of common stock. This is typically done through a dividend analysis. In the long run, common stock is only valuable if dividends are paid. Investors initially purchase common stock with the expectation that company profits will be distributed to shareholders in the form of dividends. If a company cannot ever pay any dividends, its stock is worthless. This concept is graphically shown in Figure 4.4.

A straightforward way to determine the value of common stock is to perform an NPV analysis on expected future dividend payments. Analysts will typically perform a detailed analysis of a utility's financial prospects for the next few years. All dividend payments that occur beyond the time frame of the detailed analysis are included as a *residual value*, typically modeled as a growing annuity.

Consider a utility with expected dividend payments of D_0 this year, D_1 next year, D_2 in year 2, and D_3 in year 3. Dividends are expected to be D_4 in year 4, and then grow annually at a rate of g. The NPV of this dividend stream is equal to the following:

$$NPV = D_0 + \frac{D_1}{(1+r)} + \frac{D_2}{(1+r)^2} + \frac{D_3}{(1+r)^3} + \frac{D_{4+}}{(r-g)(1+r)^4} \tag{4.13}$$

Although most utilities pay regular dividends, many companies do not. For example, Microsoft became a publicly traded company in 1986, and was extremely profitable. However, Microsoft did not pay any dividends at all until 2003. Microsoft's market cap grew exponentially, raising the question of how investors were determining the company's inherent value.

If Microsoft intended *never* to distribute dividends, its common stock would have no value. It is clear that investors expected dividends to eventually be paid. Why then did owners of Microsoft stock not demand that, prior to 2003, dividends be paid? There are two answers to this question. First, most investors have to pay taxes on dividends. As long as Microsoft was taking good care of its cash, delaying dividend payments served as a tax shelter. Second, Microsoft consistently reinvested its profits into profitable growth opportunities. It chose to reinvest, or *plowback*, earnings so that even more dividends could be distributed in the future.

Consider a utility that generates $115 million in free cash flow over the course of a year. This utility identifies a profitable growth opportunity with a cost of $100 million to pursue. In Scenario A, the utility distributes all $115 to shareholders through dividend payments. The shareholders are taxed 15% on these dividends and are left with about $100 million. The utility then raises $100 million to fund the growth opportunity by issuing common stock. In Scenario B, this utility funds the growth opportunity with $100 million of its free cash flow and distributes the remaining $15 million to shareholders through dividend payments. The shareholders are taxed 15% on these dividends and are left with $12.75 million. In Scenario A, the growth opportunity is funded and investors are left without any additional cash. In Scenario B, the growth opportunity is funded and investors are left with $12.75 million.

Many investors prefer high dividends, even though they are taxed. These investors are often skeptical about the cost effectiveness of plowback and fear that executives with too much cash on hand will make wasteful spending decisions. In any case, it is typically easier to include both plowback and dividends when performing a company valuation rather than estimating the impact of plowback on future dividends.

The amount of cash available for plowback and/or dividends is equal to free cash flow. Recall from Chapter 2 that, in the long run, free cash flow is equal to net earnings. Therefore, a reasonable approximation of company value is the NPV of expected future net earnings.

Consider a company with stable net earnings of $100 million per year, with investor expectations that this value will stay constant for many years. If common stockholders expect a return of 10%, the NPV of the net earnings perpetuity is equal to $100 million ÷ 10% = $1 billion. If investors believe that net earnings will grow 5% per year, the NPV of the growing perpetuity is equal to $100 million ÷ (10% − 5%) = $2 billion. Clearly, growth expectations are critical to company valuation. For this reason, one of the most important issues for utility executives is earnings growth and the credible communication of earnings growth to investors and analysts.

Higher levels of debt result in lower market capitalization. Consider a utility that has a capital structure consisting of $2 billion in common stock and no debt. The market cap for this utility is $2 billion. Now assume that the utility issues $1 billion in debt and uses this money to buy back stock. Even though nothing has changed with regard to customers, infrastructure, or operations, the market cap of the utility has been reduced to $1 billion.

In order to fairly compare companies with different levels of debt, investors often use *total value*, which consists of the value of equity plus the value of debt. The total value of a company as seen by the market is equal to the value of common stock plus the value of debt. Total value is closely related to the amount of money available for distribution all investors, including lenders and owners. This amount is equal to EBIT minus income taxes, which is equal to NOPAT. From this perspective, the total value of a company is equal to the NPV of expected future NOPAT. The tax effects of debt on total value are discussed further in Section 4.5.

In summary, the inherent value of common stock (i.e., the value of owner's equity) is equal to the net present value of future net income using a discount rate equal to the expected return of common stockholders. Total value is equal to the NPV of future NOPAT, using a discount rate equal to WACC. Equations for these calculations are:

Value of Owner's Equity = NPV of future net income   (4.14)
(discount rate equal to required return of shareholders)

Total Value = NPV of future NOPAT   (4.15)
(discount rate equal to WACC)

What happens when an analyst computes a value of owner's equity that is much higher than the current market cap? A confident analyst will purchase stock at the "bargain price," keeping in mind that many other analysts have come to the conclusion that the stock is fairly priced. Does the analyst know something that others do not? Does the analyst have insights that others do not? Does the analyst have a better understanding of the company than others?

It is clear that analysts will never agree fully on the inherent value of a company. However, it is important to ask whether stock prices represent a fair approximation of inherent value in general. This question is the same as asking whether the market is efficient or not.

4.3   MARKET EFFICIENCY

Investors buy and sell stock all of the time. Stock for a company is sold if the investor thinks that it is overvalued or will decline in value. Stock for a company is bought if the investor thinks that it is undervalued or will increase in value. These buy and sell transactions are driven by new information from sources such as press releases, earnings reports, stockholder meetings, and rumors. Can these investors consistently outperform the market? Stated differently, is it possible for investors to consistently act on information so that returns exceed the average return of broad market indices? The *efficient market theory* claims that consistently beating the market is not possible.

Efficient market theory states that all information about a company is already incorporated into the price of the company. Therefore, it is impossible to use information about a company to see whether its stock price is too low or too high. The price is at its current price as a consequence of this information.

Recall that the stock price of a company is based on expected future earnings. Expected future earnings can be potentially impacted by a variety of factors that can be specific to the company, related to the industry, or affecting the overall economy. When new information becomes available, investors will disagree on its impact on expected future earnings. If an acquisition is announced, one investor may think that expected future earnings will increase and another may think that expected future earnings will decrease. The investor with the correct prediction will, in this case, "beat" the market while the investor with the incorrect prediction will "get beat" by the market. On average, the market is neither beat nor beaten.

In the above paragraph, it is clear that an investor that consistently interprets new information better than other investors can beat the market. However, investors that consistently misinterpret information will not be investors for long. Therefore, most professional investors, on average, cannot be expected to outperform one another by significant amounts.

To be sure, professional investors often disagree about the expected future earnings of companies. Therefore, these professional investors will also disagree on the proper value of the corresponding stock price. There is also a keen awareness that nobody is certain about what future earnings will eventually be in reality. The level of confidence in earnings predictions is a key factor in determining whether a stock will be purchased or sold. This uncertainty is also a strong motivation for portfolio diversification, which is discussed in the next chapter. Smart investors typically prefer to take many small bets rather than a few large bets. These bets determine stock prices, which can be viewed as an average consensus of typical investors.

When thinking about stock markets and stock price dynamics, it is helpful to understand that large investors set trading prices. Examples of these large investors are mutual funds and pension funds. When large investors trade large blocks of stock, prices adjust according to the laws of supply and demand. Consider a pension fund that wants to purchase 5 million shares of a stock that has a current trading price of $10.00 per share. To incentivize current shareholders to sell, this pension fund may have to offer $10.25 per share, raising the trading price. Similarly, the pension fund may wish to sell 2 million shares of another stock that has a current trading price of $20.00 per share. To encourage investors to buy this amount, the pension fund may initially offer a selling price of $19.75 per share, lowering the trading price. If there are not enough takers, the offer will have to withdrawn or further lowered. When small investors trade small blocks of stock, stock prices are not likely to change.

Determining the impact of new information on future earnings is quite different than looking at a stock price and deciding whether it is cheap or expensive. At this point, professional investors have already analyzed all relevant information, completed a future earnings analysis, bought and sold shares as necessary, and set the market price at a fair level. Amateur investors are free to buy and sell stock based on tips and hunches. The good news is that all of the transactions will be at a fair price. The bad news is that each transaction costs money (this is good news for brokers and online trading services).

The next sections examine various theories of market efficiency. The weak form of efficiency states that any useful information in historical prices is fully reflected in current price. The semi-strong form of efficiency states that all publicly-available information is reflected in current price. The strong form of efficiency states that all information is reflected in current price.

4.3.1   Weak Efficiency

The *weak form* of market efficiency states that all historical stock price information is reflected in the current stock price. If the weak form is true, it is impossible to use historical stock price movements to anticipate future stock price movements.

Analyzing historical stock price movements to anticipate future stock price movements is called *technical analysis*. Technical analysis is common practice, even though its value is questionable. Perhaps the reason that technical analysis remains popular is psychological. Sophisticated mathematics and regression analyses are able to create a seemingly credible story that can be easily visualized.

Consider Figure 4.5, which shows ten years of daily closing stock prices for a large US utility. Although a general upward trend in stock price is evident, there are frequent and seemingly random price movements superimposed on the upward trend, similar to high frequency noise on a waveform. The author has created a mathematical function that attempts to capture major trends in historical stock price movement. This simple function consists of a linear component plus a sinusoidal component.

Imagine a financial advisor making recommendations to investors based on Figure 4.5. According to the advisor, the stock price of the utility is approaching a valley, and the mathematical model shows that peaks and valleys occur approximately every five years. By investing now, the advisor promises a high probability of buying low and selling high, resulting in a return greater than the average long term return of the utility.

Will investors following the advice in the preceding paragraph achieve higher than normal returns? Extensive research has consistently and definitively shown that the answer is "no." There are a few idiosyncrasies that will be discussed later, but fluctuations in historical stock price movement cannot be used to predict future stock price movement. The basic reason is that any useful information about past prices is already reflected in the price today.

It is true that mathematical functions can often be used to accurately fit historical data, often much better than the simple example of Figure 4.5. However, past experience is not necessarily a good prediction of future outcomes. Making this assumption would violate the logical fallacy that "correlation does not imply causation." Causality can only be assumed if a believable theory explains why the past can be used to predict the future.

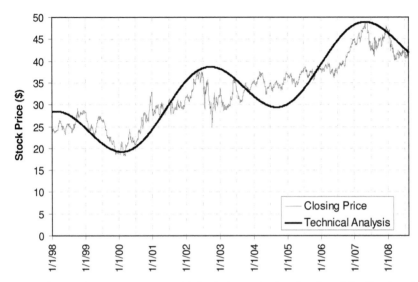

Figure 4.5. This figure shows ten years of daily closing stock prices for a large US utility. Superimposed is a mathematical function that attempts to capture major trends in historical stock price movement. This type of historical price modeling is called a technical analysis. Extensive research has shown that technical analysis cannot generally be used to predict future prices.

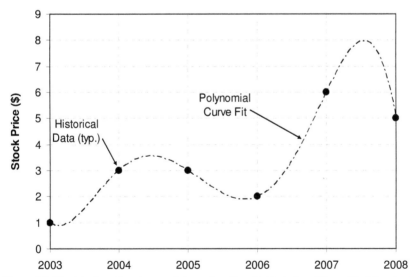

Figure 4.6. The dots show the outcome of six random die rolls. Each die roll is used to determine the price of a stock in a particular year. A sixth-order polynomial has been fitted to the randomly-generated stock prices. The mathematical function fits the historical data exactly but has zero predictive capability. This example illustrates the danger of assuming that historical patterns have predictive capability.

Consider the life of a turkey. Each day for three years, the turkey wakes up, is fed nice meals of grain, and is allowed to roam about freely. Using the same logic as technical analysis, this turkey assumes that his experience tomorrow will be the same as the past. Imagine the turkey's surprise when the next day arrives and its head is chopped off in preparation for a Thanksgiving meal.

Mathematical analysis can also provide a false sense of confidence. Consider a fair six-sided die. The dots in Figure 4.6 show the outcome of six random die rolls. Each die roll is used to determine the price of a stock in a particular year. In 2003 the die roll was 1 and the stock price is assumed to be $1. In 2004 the die roll was 3 and the stock price is assumed to be $3. A sixth-order polynomial has been fitted to the randomly-generated stock prices. *The mathematical function fits the historical data exactly but has zero predictive capability.* The 2009 stock price is completely independent of the stock prices of 2003 through 2008.

In summary, past prices cannot be used to predict future prices. Any useful information in past prices is already reflected in the current price. It may be possible to fit historical data with mathematical functions, but these functions cannot accurately predict future prices. At a minimum, the market exhibits the weak form of efficiency.

4.3.2 Semi-Strong Efficiency

The *semi-strong form* of market efficiency states that all publicly-available information is reflected in the current stock price. If the semi-strong form is true, it is impossible to use any type of research to determine whether a stock is overvalued or undervalued. According to the semi-strong form, any information that someone might find, or any analysis that someone might undertake, is already reflected in the current stock price.

Is it not possible for a smart person to examine company financial statements and gain new insights about stock price valuation? Of course this can happen. However, stock prices are average values on which investors disagree. Research and analysis is ongoing, and positive findings by some investors will tend to be balanced by negative findings by others.

Stock prices do change, and new public information has to be incorporated into stock price at some time. Is it possible to sit in front of a computer terminal, wait for news about a company, and then quickly buy or sell stock before anyone else gets a chance? This question has been extensively researched. The technique is to use an *event study* where the abnormal returns of a stock are examined. The abnormal return for a stock is equal to the percent change in stock price minus the percent change in the overall market. For example, if a stock increases in value by 0.8% in one day and the overall market increases in value by 0.6%, then the abnormal return of the stock is 0.8% – 0.6% = 0.2%. An event study averages abnormal returns for many similar events and plots the average abnormal returns on a timeline.

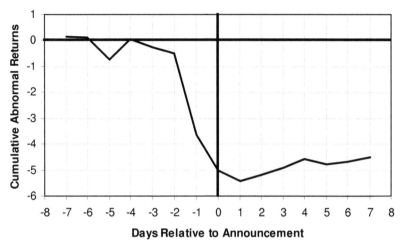

Figure 4.7. This figure shows the results of an event study for the announcement of dividend omission. Even though the announcement of dividend omissions occurs in Day 0, the stock price begins to decline several days earlier. The preemptive change occurs because professional investors are anticipating these announcements before they occur. When the actual announcement is made, there is no significant change in stock price.

An event study for the announcement of dividend omissions is shown in Figure 4.7. The announcement of dividend omissions occurs in Day 0. From the graph, it is clear that the announcement of dividend omissions is not good for the stock price of a company. Investors view this as sign of reduced future earnings and, therefore, a lower stock price. However, the full impact on the stock price has already occurred by the time of the announcement. Professional investors are skilled at anticipating announcements and start acting on them about three days ahead of time. When the actual announcement is made, there is no significant change in stock price.

Event studies for other types of announcements have similar results. By the time of the announcement, nearly all of the stock price change has already occurred. A summary of event study results is shown in Table 4.2. Dividend omissions are the worst announcement for a company, resulting in an average abnormal return of minus 5.2%. Being acquired is, by far, the best announcement for a company, resulting in an average abnormal return of plus 31%. An explanation of the events shown in Table 4.2 and their impact on stock price is now provided.

Dividend Omission. Dividends are the basis for company valuation. Lower dividends imply lower company value and therefore lower stock price. Since dividends are so important to investors, announcing a dividend omission is communicating that there is absolutely no possibility of freeing up enough cash to fund dividend payments. The company is cash strapped, typically because revenues are lower than expected and/or costs are higher than expected.

Table 4.2. Abnormal returns based on event studies (abnormal returns are changes in stock price minus the change in the overall market over the same time period). These events have a definite impact on abnormal returns, with acquisition announcements being the most dramatic. However, these abnormal returns are already reflected in the stock price by the time of the announcement. This occurs because professional investors are skilled at anticipating announcements before they occur.

Event	2-Day Cumulative Abnormal Return
Dividend Omission	− 5.2%
Equity Issues	− 3.2%
Lower than Expected Earnings	− 2.1%
Reduced Capital Expenditures	− 1.5%
Increased R&D Expenditures	+ 0.8%
Dividend Increase	+ 0.9%
Increased Capital Expenditures	+ 1.2%
Obtaining a Bank Loan	+ 1.7%
Higher than Expected Earnings	+ 1.7%
Stock Splits	+ 3.5%
Being Acquired	+ 31.0%

Equity Issues. Issuing new stock is generally not good news for investors. First, existing stock owners do not appreciate being diluted. Second, why does the company need more cash? If this cash is going to be used for positive NPV investments, why can't the money be borrowed? Investors typically interpret equity issues as a last resort for generating needed cash.

Lower than Expected Earnings. Corporate valuation is based in part on earnings forecasts. If earnings targets are not met, the NPV of expected future profits are less and company valuation is therefore lower. Also, not meeting earnings expectations demonstrates to investors that the company is not able to accurately determine achievable earnings targets.

Reduced Capital Expenditures. Businesses make capital expenditures to replace facilities and equipment that are wearing out, to make investments in technology, to invest in growth opportunities, and other items intended to increase future earnings. When capital expenditures are reduced, investors conclude that profit is not sufficient to replace aging equipment, growth opportunities are slowing down, or both.

Increased R&D Expenditures. Investors see increased research and development expenditures as positive. As long as earnings targets are being consistently met, increased spending on research and development shows a commitment to investing in the future so that profits can continue when current products, services, and technologies become obsolete.

Dividend Increase. Since dividends are the basis for company valuation, higher dividends result in higher stock prices. Increasing dividends communicates to investors that a company is confident that profits can support this higher level of dividend payment into the future.

Increased Capital Expenditures. Just as investors frown on a reduction in capital expenditures, they value increased capital expenditures. An increase in capital expenditures shows that the company is keeping up with the replacement of aging and obsolete equipment and is growing the business.

Obtaining a Bank Loan. When a company raises cash by obtaining a bank loan, its stock price typically rises. This is in contrast to raising cash by issuing stock, which typically causes stock price to fall. Investors know that it is not easy to obtain a bank loan. Banks perform careful due diligence on the company to ensure that interest payments can be made and the loan eventually repaid. In addition, investors know that the bank will continually monitor the financial health of the company while the loan is outstanding. In essence, obtaining a bank loan is the equivalent of getting a clean bill of health from the bank.

Higher Than Expected Earnings. Exceeding earnings targets increases stock price for the same reasons that not meeting earnings targets reduces stock price. If earnings targets are exceeded, the NPV of expected future profits are more and company valuation is therefore higher. Also, exceeding earnings expectations demonstrates to investors that the company is conservative when setting earnings targets.

Stock Splits. When a stock splits, existing shares are replaced with more shares valued at a proportionally lower price. For example, a two-to-one split replaces current shares with two new shares at half the price. Generally, stock splits are viewed as a sign of company confidence about its future. A study by the New York Stock Exchange showed that seven years after a split, stocks appreciated in value more than twice that of stocks that had not split. In addition, total dividend payout is almost always increased after a stock split. This said, there are several companies who have refused to split their stock, feeling that it is a waste of shareholder money. The most notable is Warren Buffet's Berkshire Hathaway (NYSE: BRK.A), which has traded for over $140,000 per share.

Being Acquired. In a non-hostile acquisition, the acquiring company must offer a price-per-share higher that current market value. The difference between the current market value and the offer price is called the *premium*. For example, assume that Utility A is currently trading for $50 per share. Utility B wishes to acquire Utility A, and negotiates a price of $65 per share, a 30% premium. Although the deal has not yet been approved by a shareholder vote, the stock of Utility A will quickly rise in price based on the premium and the probability of the acquisition being successful.

Event studies themselves are interesting, but also demonstrate the semi-strong form of efficiency. Savvy investors are typically able to anticipate public announcements before they occur, preventing the ability of public information to be used to predict future stock price movement. Large investment companies have dedicated analysts who "walk the halls" of companies and are skilled at gleaning information from company executives before public announcements are made. In general, markets exhibit the semi-strong form of efficiency.

4.3.3   Strong Efficiency

The *strong form* of market efficiency states that all relevant information is reflected in the current stock price. If the strong form is true, it is impossible to use any relevant information at all to determine whether a stock is overvalued or undervalued.

The strong form of market efficiency does not hold, as evidenced by the benefits of insider trading. Often, executives within a company will have information that has a material impact on future stock prices. These executives are prohibited by law from acting on this information, preventing the information from being reflected in stock price until the information is released to the public or anticipated by the public.

For example, an executive may be involved in an acquisition deal that is not public. If the executive purchases stock in the acquisition target, a profit of about 30% will typically be made once the merger is announced. Similarly, the executive may be aware of a product defect that will likely lead to a large class action lawsuit. The executive can avoid losses by selling stock before the product defect is announced publicly. Again, trading based on insider knowledge is illegal and the above actions are not recommended. These examples are simply provided to demonstrate that the strong form of market efficiency does not hold.

4.3.4   Summary of Market Efficiency

The hierarchy of market efficiency is shown in Figure 4.8. The strong form implies that the semi-strong form and the weak form also hold. The semi-strong form implies that the weak form also holds.

Most evidence shows that markets exhibit the semi-strong form of efficiency. This means that all publicly-available information is reflected in the current stock price. Historical prices cannot be used to predict future stock prices. Undiscovered insights cannot be gained by performing company or market research. Profits cannot be increased by quickly acting upon press releases or public statements.

Profits can be increased by acting upon insider information, but doing so is illegal. For example, the senior executives of a company about to be acquired for premium could act on the information and purchase shares prior to the announcement, knowing that the price will rise. The SEC is virtually certain to detect and prosecute this type of criminal behavior.

Can smart investors beat the market? Clearly the answer is yes. Some people will be able to better and more consistently determine the inherent value of companies than others. They will be able to find good deals and avoid bad deals better over time. Some people will be able to better and more consistently anticipate public announcements. They will be able to buy and sell preemptively more effectively than others.

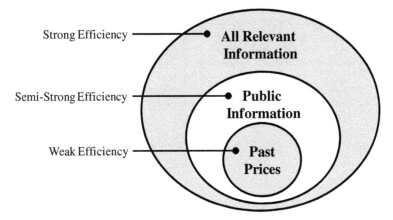

Figure 4.8. The three forms of market efficiency. The weak form states that any useful information in historical prices is fully reflected in current price. The semi-strong form states that all publicly-available information is reflected in current price. The strong form of efficiency states that all information is reflected in the current price. The semi-strong form implies that the weak form holds, and the strong form implies that the semi-strong form holds.

Presumably, many people would like these smart investors to manage their investments, with the smartest people managing the largest portfolios. Some of the largest investment portfolios are mutual funds. This begs the question, if extremely smart people are managing mutual funds, do they consistently beat the market? The short answer is "no." Figure 4.9 shows the returns of various classes of mutual funds when compared to the overall market (data is from 1963 to 1998).

On average, mutual funds slightly underperform the overall market. This underperformance is about equal to the maintenance fees of mutual funds. This means that purchasers of mutual funds effectively pay a fund manager to keep a stock portfolio that, for the most part, performs like the overall market. They do not pay fund managers that are able to buy underpriced stocks and sell over-priced stocks.

In accordance with the previous paragraph, most mutual funds attempt to track broad market indices like the S&P 500, which represents approximately 70% of US market capitalization, the Russell 3000, which represents approximately 98% of US market capitalization, and the Wilshire 5000, which attempts to measure the entire US market. These mutual funds constantly buy and sell stock so that their portfolios represent the tracking index, not because they feel that stocks are overpriced or underpriced.

Mutual funds that do not attempt to track broad market indices perform worse than the overall market. Income funds are particularly poor, with average returns of 40% to 50% less than the overall market. Income funds typically consist of less risky stocks, but it is unlikely that the reduction in risk is worth the drastically lower returns.

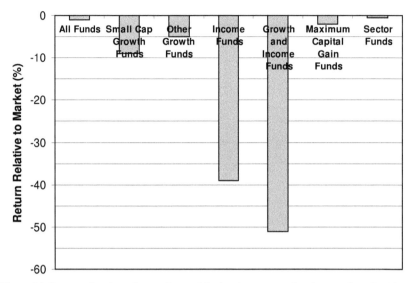

Figure 4.9. Returns of various classes of mutual funds when compared to the overall market (data is from 1963 to 1998). On average, mutual funds slightly underperform the market. This underperformance is about equal to the maintenance fees of mutual funds. Mutual funds that do not attempt to track broad market indices perform worse than the overall market. Income funds are particularly poor, with average returns of 40% to 50% less than the overall market.

This analysis does not imply that mutual fund managers are not smart. Most are probably very good at their jobs. The same is true for large pension fund managers. Because these people are the most active traders and deal in large dollar values, the trades that they make tend to set the price of stocks. Small investor trades do not impact price. Large investor trades can dramatically impact price. Therefore, the most informed and qualified people are the price setters and others are the price takers. A large pool of informed and sophisticated investors are quickly acting upon all public information, resulting in, for the most part, semi-strong market efficiency.

There are documented market abnormalities that slightly deviate from pure semi-strong market efficiency. These include small cap stocks, low price-to-earnings stocks, price momentum, market crashes, speculative bubbles, and wizards.

Small Cap Stocks. Historically, small cap stocks (those valued less than $1 billion) have had average annual returns of about 12%, compared to 11% for the overall market. Although this may seem to violate market efficiency, many economists argue that (1) this return reflects higher price volatility, and (2) broad market indices do not often include a large dollar volume of small cap stocks.

Low Price-To-Earnings Stocks. A common indicator of price level is the ratio of market capitalization to net earnings (price-to-earnings ratio, or P/E). When comparing similar companies, studies typically show that the lowest P/E companies have returns almost double those of the highest P/E companies. Many economists argue that this does not violate market efficiency because low P/E companies are viewed unfavorably by investors and are therefore riskier.

Price Momentum. Stock prices that are going up tend to continue going up, and stock prices that are going down tend to continue going down. This effect is called *price momentum*. Although the effect is small, it does appear to violate market efficiency. Some economists argue that investing based on price momentum is risky, and higher returns are compensation for this risk. Others see price momentum as evidence of irrational investor behavior such as herding (i.e., acting as a group).

Market Crashes. A market crash is a precipitous fall in price across a broad range of stocks. Towards the bottom of a crash, many rational investors will continue to sell even though the fundamental value of stock seems to indicate bargain prices. These rational investors seem to be aware of an "irrational market." At extremes, people tend to buy and sell stocks based on whether the price will rise or fall rather than on inherent value. The notable economist John Maynard Keynes famously commented, "Markets can remain irrational longer than you can remain solvent."

Speculative Bubbles. Speculative bubbles occur when the prices of stocks are higher than can be justified by expected future dividends, and continue to keep going up. Like a pyramid scheme, people buy high-priced stocks with the hope that more people will continue to buy and drive prices up further (the *greater fool theory*). Like price momentum, many see bubbles as evidence of investor herding behavior.

Wizards. A wizard is a person who can consistently identify overvalued or undervalued stocks. Wizards purchase undervalued stocks, patiently wait for the market to correct the price to a fair value, and sell at the correct time. Wizards also avoid purchasing overvalued stocks and will patiently hoard cash until bargains become available. The success of Warren Buffet is evidence that at least one wizard exists. Many economists argue that the existence of wizards does not violate market efficiency since the excess profits represent fair compensation for the wizards' analysis and judgment.

In summary, financial markets are mostly efficient in the semi-strong sense. Experts often disagree on the fair value of a specific company, but stock prices on average can be trusted. Certain anomalies exist, but prices reflect most public information and there is little point in trying to time the market.

It is important for utility engineers to have a good understanding of financial markets since engineering decisions should be made with the hope of maximizing stock price. Investors use public information to determine the net present value of future cash flows, and are generally not influenced by irrational thinking. If investors are convinced that profits will be higher than previously thought, the stock price will go up. If investors are convinced that profits will be lower than previously thought, the stock price will go down.

The intent of this section is not to advise the personal investor. However, many will not appreciate the implication that picking individual stocks for a personal portfolio is futile. If the reader has a better understanding than professional analysts of the expected future cash flows of a company, consider yourself a wizard and go for it. At worst, the stock was fairly priced. From this perspective, there are no bad stock purchases. But there are bad stock portfolios, which will be addressed in the next chapter. Personal investors should understand that proper portfolio balancing is much more important than individual stock selection, and bad combinations of stocks will almost always perform worse than good combinations of stocks.

4.4   CAPITAL STRUCTURE

As described in Chapter 2, financial capital is money available for funding a business. The sources of financial capital are owners and lenders. Financial capital provided by owners is called owner's equity, often shortened to equity. Financial capital provided by lenders is called debt. All sources of equity and debt are called the *capital structure* of a company.

Both the equity (e.g., common stock) and debt (i.e., bonds) of a company are typically valued by markets. The value of the total capital structure of a company is equal to the value of equity plus the value of debt:

$$\text{Value of Company} = \text{Value of Equity} + \text{Value of Debt} \qquad (4.16)$$

For example, a utility might have 100 million shares of common stock outstanding that are trading at $60 per share, for a total market capitalization of $6 billion. This same company may have a variety of outstanding bonds that have a total market value of $4 billion. The total value of this company is $10 billion, with a capital structure of 60% equity and 40% debt. Typical utilities have highly complex capital structures with a variety of equity, debt, and hybrid sources of financial capital.

If a company goes bankrupt, different sources of financial capital have different claim priorities on company assets. Consider a company with $1 billion in issued debt that goes out of business and liquidates all of its assets. The money raised by liquidation must first be used to pay off the debt. If a bankrupt company has no assets of value, all of its debt is worthless. If the company has assets than can be sold for $100 million, the issued debt is only worth $100 million (i.e., 10¢ on the dollar). If the company has assets that can be sold for $500 million, the issued debt is worth $500 million (i.e., 50¢ on the dollar). If the company can sell its assets for $1 billion or more, the issued debt is worth its full face value. The money remaining after full debt payment is available for owners.

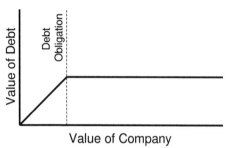

Figure 4.10. This figure shows the value of debt as a function of company value. The value of debt is equal to company value until the company value reaches the debt obligation. At this point, the value of debt does not rise as company value continues to rise.

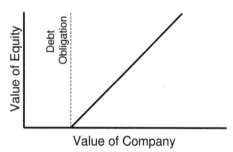

Figure 4.11. This figure shows the value of equity as a function of company value. The value of equity is zero until the company value reaches the debt obligation. At this point, the value of equity rises with company value.

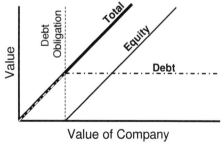

Figure 4.12. This figure shows the relationship of debt value, equity value, and total value. Total value is equal to debt value plus equity value. As expected, the total value line rises with unity slope, which it is required to do by definition.

The value of debt as a function of company value is shown in Figure 4.10. The value of debt is equal to company value until the firm value reaches the debt obligation. At this point, the value of debt does not rise as company value continues to rise. The value of debt is similar to a financial derivative called a *put option*, which is discussed in detail in the next chapter.

Equity owners do not get anything until all debts are paid off. Consider again a company with $1 billion in issued debt that goes out of business and liquidates all of its assets. If the firm is worth $1 billion or less, the equity in the firm is worthless since all of the company's value is owed to lenders. If the company is worth more than $1 billion, the value of equity is equal to the excess value. For example, if the company value is $1.5 billion, the value of equity is equal to $1.5 billion – $1 billion = 500 million.

The value of equity as a function of company value is shown in Figure 4.11. The value of equity is zero until the company value reaches the debt obligation. At this point, the value of equity rises with company value. The value of equity is similar to a financial derivative called a *call option*, which is discussed in detail in the next chapter.

The relationship of debt value, equity value, and total value is shown in Figure 4.12. Total value is equal to debt value plus equity value. As expected, the total value line rises with unity slope, which it is required to do by definition.

The preceding example is simple, but emphasizes the concept of *seniority*. Seniority is the order of repayment after a bankruptcy, and determines the relative risk of capital structure components. A debt that must repaid before another is said to be *senior*. A debt that must repaid after another is said to be *subordinate* to the senior debt. Debts that have the same priority are said to be *pari passu*. A common classification of seniority is the following:

Typical Seniority Classification
- Senior secured debt (highest seniority)
- Senior unsecured debt
- Subordinated debt
- Convertible debt
- Preferred stock (mezzanine)
- Common stock (lowest seniority)

Senior secured debt gets paid off first. The term *secured* means that the debt is guaranteed by the assets of the company. This is sometimes called *mortgage-backed debt*. If the secured debt is not repaid, the lender assumes ownership of the assets. Unsecured debts, also called *debentures*, are not secured by assets and only by the general credit worthiness of the issuer. Senior unsecured debt gets paid off after secured debt, followed by various levels of subordinate debt. Next to be paid is debt that can be converted into preferred or common stock. Mezzanine capital is senior only to common stock; in this case the mezzanine capital is preferred stock. Anything left after all senior obligations are fulfilled goes to common stockholders.

Table 4.3. Debt to value ratios for various industries as classified by the Standard Industrial Classification (SIC) code. On average, companies have a capital structure that has about 30% debt, but this varies widely. Telephone, gas, and electric utilities have among the highest levels of debt for all industries.

SIC Code	Industry	Debt to Value (%)	
		Mean	σ
2830-2840	Drugs & Cosmetics	9.07	9.50
3800	Instruments	11.90	8.60
1000	Metal Mining	13.47	9.90
2700	Publishing	15.52	16.90
3600	Electronics	15.79	12.10
3500	Machining	19.57	11.40
2000	Food	20.56	12.80
1300	Oil Exploration	22.58	15.10
1500-1700	Construction	23.84	15.10
2900	Oil Refining	24.36	12.10
3400	Metal Working	25.02	13.90
2800	Chemicals	25.44	13.50
2300	Apparel	26.03	12.30
2400	Lumber	26.05	18.20
3700	Motor Vehicle Parks	21.74	13.80
2600	Paper	28.95	11.40
2200	Textile Mill Products	32.57	13.30
3000	Rubber	32.62	16.70
5300	Retail Dept. Stores	34.33	15.00
5400	Retail Grocery	34.60	18.70
4200	Trucking	37.30	20.90
3300	Steel	38.19	19.50
4800	**Telephone**	**51.50**	**9.70**
4900	**Elec. & Gas Utilities**	**53.09**	**8.10**
4500	Airlines	58.25	17.10
Total		29.13	18.80

On average, companies have a capital structure that has about 30% debt, but this varies widely. Different industries tend to have different capital structures, as shown in Table 4.3. The reasons for these differences are discussed in Chapter 5. At this point, it is sufficient to point out that utilities have high levels of debt in their capital structure. In fact, telephone, gas, and electric utilities have the highest levels of debt for all industries except airlines.

Table 4.4. This table shows median corporate debt levels for seven major countries. Canada and the US have the highest use of debt, indicating business-friendly bankruptcy laws and favorable tax treatment for debt. Germany and the United Kingdom have the lowest use of debt, indicating lender-friendly bankruptcy laws and relatively unfavorable tax treatment for debt.

Country	Median Debt to Value (%)
Canada	32
United States	25
Japan	21
Italy	21
France	18
Germany	11
United Kingdom	10

Debt levels also vary by country. These differences typically relate to a combination of tax laws (to be discussed in the next section) and bankruptcy laws (to be discussed in the subsequent section). Median corporate debt levels for seven major countries are shown in Table 4.4. Canada and the US have the highest use of debt, indicating business-friendly bankruptcy laws and favorable tax treatment for debt. Germany and the United Kingdom have the lowest use of debt, indicating lender-friendly bankruptcy laws and relatively unfavorable tax treatment for debt.

4.5   TAX SHIELDS

Imagine a world without corporate taxes. In this world, companies would obtain financial capital, operate their business, and realize positive EBIT (earnings before interest and taxes). Since taxes are zero, the entire amount of EBIT could be distributed to sources of financial capital, including both equity and debt. In such a world, total investor return would be the same regardless of capital structure. In a world without corporate taxes, it would not matter how much debt a company issued.

Two famous economists, Franco Modigliani and Merton Miller, proved that capital structure does not matter if there in are no taxes, no bankruptcy costs, and an efficient market. This is typically called the *Modigliani–Miller theorem*, or the *capital structure irrelevance principle*.

Corporate taxes exist in the real world. This impacts capital structure because interest paid on debt is tax deductible. More interest payments lead to lower taxes, which increases the amount of EBIT that is available for investors overall.

Consider a utility with equity of E, debt of D, an interest rate on debt of r_d, and a corporate tax rate of r_t. The total interest that the utility will pay is $r_d D$. Because the utility is able to deduct this amount from taxable income, a tax benefit of $r_t r_d D$ is realized. This tax benefit is called an *interest tax shield*. If the tax benefit continues every year, the associated perpetuity is equal to $r_t r_d D \div r_d = r_t D$:

$$r_t D \qquad \text{; present value of interest tax shield} \qquad (4.17)$$

Modigliani and Miller go on to prove that the value of a levered company (i.e., has debt) is equal to the value of an unlevered company (i.e., has no debt) plus the present value of the interest tax shield. An unlevered company will pay EBIT x r_t in taxes, and will therefore have EBIT x $(1 - r_t)$ left for shareholders. If the required return on equity is r_e, the value of an unlevered and levered company with constant future EBIT is:

$$\frac{\text{EBIT}(1 - r_t)}{r_e} \qquad \text{; value of unlevered firm} \qquad (4.18)$$

$$\frac{\text{EBIT}(1 - r_t)}{r_e} + r_t D \qquad \text{; value of levered firm} \qquad (4.19)$$

And so, the value of a firm increases as debt increases. Typical utilities have between 50% and 60% debt. Will increasing this level increase the total value of the utility? The short answer is "no," due to an increased probability of bankruptcy and associated bankruptcy costs.

4.6   BANKRUPTCY

Bankruptcy occurs when a company cannot meet its debt obligations. Even if a company is profitable from an earnings perspective, high debt obligations and poor cash flow may lead to bankruptcy. In the US, distressed companies typically file for *Chapter 11 bankruptcy*, referring to the eleventh chapter of Title 11 of the US Bankruptcy Code. Under Chapter 11, the company continues to operate under the supervision of a bankruptcy court. The company is obligated to propose a restructuring plan that must be voted on and approved by creditors. The percentage of companies that successfully confirm a restructuring plan under Chapter 11 is less than 30%.

Consider a utility with $3 billion in issued bonds. If the average interest rate on the bonds is 8%, the utility is obligated to pay $240 million in interest payments every year. On average this is $20 million per month. This company risks bankruptcy unless it can generate at least $20 million in cash per month, every month.

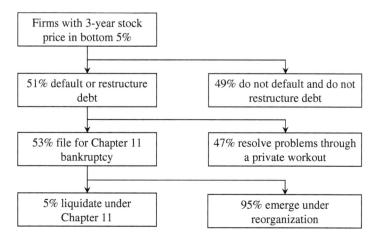

Figure 4.13. This figure shows statistical outcomes of financially distressed companies. About half of companies having very poor stock price performance over a three year period (as measured by price-to-earnings) become financially distressed and have to restructure or default on their debt. Of these, half file for bankruptcy and some are forced to go out of business.

Chapter 11 outcomes are shown in Figure 4.13. About half of companies having very poor stock price performance over a three year period become financially distressed and have to restructure or default on their debt. About half of the distressed companies file for Chapter 11 bankruptcy. Of those filing for Chapter 11, most emerge in some form or another, but about 5% go out of business.

A good example of the realities of bankruptcy is the case of Pacific Gas & Electric (PG&E), one of the largest utilities in the US. PG&E has billions of dollars of debt, and massive debt payment obligations each month. Typically this is not a problem since the large customer base of PG&E provides strong free cash flow each month. In 2001, the price of wholesale electricity spiked and PG&E was forced to purchase a large amount of power at prices higher than it could charge customers. PG&E was no longer able to generate enough free cash flow to meet its debt obligations and filed for Chapter 11 bankruptcy on April 6th 2001. The State of California (whom many blame as the root cause of the bankruptcy) bailed out PG&E for a taxpayer cost of $450 million, one of the most expensive bankruptcies in state history.

Another example of utility bankruptcy is Entergy New Orleans, a subsidiary of Entergy Corporation. On August 29th 2005, Hurricane Katrina struck the Louisiana coast causing major wind damage, storm surge damage, and flooding. The storm surge breached the New Orleans levee system and flooded 80% of the city. Entergy New Orleans had a lot of damaged infrastructure to repair. However, it no longer had customers using significant amounts of electricity. Revenue quickly reduced to near zero, and bankruptcy was filed on September 23rd. In this case, Entergy New Orleans became a *debtor-in-possession* of its parent company, allowing Entergy Corporation to make loans and ensure that restoration efforts were not stalled.

Although most companies survive after bankruptcy, the costs are high. Direct costs include lawyer fees, court fees, and management time spent dealing with all of the bankruptcy issues. Indirect costs include lost customers (Do I want to do business with a company that may not be around very long?), hesitant suppliers (Do I want to sell stuff to a customer who may never pay?), and the possibility of sub-optimal investment decisions.

What should management do after filing for bankruptcy? The first option is to use cash reserves to pay off debts. Lenders will certainly appreciate this, but does it benefit shareholders? Maybe management should use the cash to take large risks. If the risks do not pay off, shareholders are no worse off. And who knows – maybe the risks will pay off and fix the financial problem. Perhaps management should distribute cash reserves as dividends or big bonuses. Why not just use the cash reserves to throw a big party?

Lenders would never lend money if they did not have first rights to cash and other assets. These rights are enumerated in *restrictive covenants* that state what a company can and cannot do during both normal situations and during financial distress. If covenants are broken, lenders can typically take aggressive action to ensure repayment.

Financial distress can also result in lost opportunities. Even if management finds good investment opportunities, it is unlikely that stockholders will provide additional funds so that the opportunities can be pursued. It is almost certain that existing restrictive covenants prevent the issue of new debt.

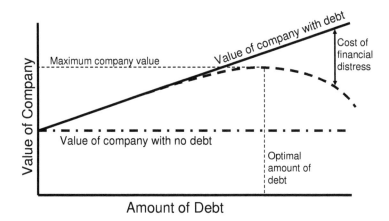

Figure 4.14. The value of a company increases with increasing debt. At a certain point, the expected cost of financial distress begins to reduce the benefits of further increased debt. The optimal amount of debt occurs when the incremental benefit of additional debt is equal to the incremental cost of financial distress.

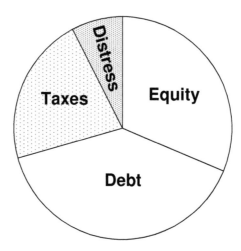

Figure 4.15. This pie chart visually represents the value of a company. The volume of the pie represents present value of all future EBIT. Taxes must be paid, which takes away a slice of pie. Financial distress may occur, which takes away another piece. The remaining pie (the unshaded part) is equal to the value of the company. Debt payment takes away another piece, and the remainder is left for equity investors.

A company with no debt obligations will never go bankrupt. This is definitely a positive characteristic from a risk perspective. However, a debt-free company forgoes the benefits of an interest tax shield, which is a negative characteristic from a profitability perspective. As debt increases, fixed cash obligations increase, increasing the likelihood that these obligations cannot be met. If they cannot be met, bankruptcy occurs and bankruptcy is very expensive. Therefore, additional debt should only be issued if the tax benefits are greater than the expected cost of financial distress.

The benefits and detriments of debt are graphically illustrated in Figure 4.14. The value of the company increases with increasing debt. At a certain point, the expected cost of financial distress begins to reduce the benefits of further increased debt. The optimal amount of debt occurs when the incremental benefit of additional debt is equal to the incremental cost of financial distress. This point is where the total value of the company is maximized. As stated previously, the optimal debt level for US utilities is typically between 50% and 60%.

Figure 4.15 shows the value of a company as a pie chart. Each year, a company achieves a certain level of EBIT. The present value of all future EBIT represents the entire pie. Taxes must be paid, which takes away a slice of pie. Financial distress may occur, which takes away another piece. The remaining pie (the unshaded part) is equal to the value of the company. Payments to lenders take away another piece, and the remainder is left for owners.

4.7   SUMMARY

Corporate finance is the mechanism through which companies raise financial capital to operate and grow their business. Without financing, businesses could not exist within a capitalistic economy. Equity financing is achieved by selling common stock to investors, who then become owners. Debt financing is achieved by selling bonds or obtaining loans. The mix of equity financing and debt financing is called the capital structure of a company.

Because financing involves investments and returns, corporate finance must always consider the time value of money. Cash today is worth more than cash promised in the future. To account for this, the present value of future cash is discounted based on the risk of the future cash being higher or lower than expected. If future cash payments are risky, they are worth less today.

Most utilities raise financial capital from a variety of sources. Together, these sources constitute the capital structure of the utility. Each source of capital has its own required rate of return, and the weighted average of these sources determines the overall cost of money, referred to as WACC. Utilities, more than any other industry, use high levels of debt financing. Since the required return on debt is lower than the required return on equity, the cost of money for utilities is lower than most other industries.

Companies use debt financing because interest payments are tax deductable. Higher debt results in higher interest payments which results in lower tax bills. Since less is paid in taxes, more money is left for investors. The downside of high debt is the resulting high interest payment obligations which, if not paid, results in bankruptcy.

The market capitalization of a utility is equal to its stock price times its number of shares. The total value of a utility is equal to its market capitalization plus the value of its outstanding long-term debt. Analysts determine fair market capitalization and total value by analyzing expected future earnings. This analysis relies heavily on public data and uses WACC when performing present value calculations. The intensive efforts of many analysts lead to a market that can be best characterized as efficient in the semi-strong sense. All public information is reflected in current share prices, and there is little to be gained by trying to time the market or by trying to find undervalued stocks.

4.8   STUDY QUESTIONS

1. Why is money today worth more than money in the future? What factors determine what discount rate to use when performing net present value calculations?
2. What does the term WACC stand for and what is its importance? What factors will impact the value of WACC for a utility?

3. What is the relationship between dividends and stock price? How can expected future dividends be used to determine the fair price of a stock?

4. What is the difference between market capitalization and total company value? How could two utilities have similar total values but different market capitalizations?

5. What are the three forms of market efficiency? Which form of efficiency has been shown to best characterize the actual market?

6. What is an event study? What types of events result in lower stock prices and what types of events result in higher stock prices?

7. What is meant by the term "capital structure?" What constitutes the typical capital structure of a utility?

8. What is meant by debt seniority? What is the relationship between debt seniority and interest rates?

9. What is the relationship between capital structure and WACC?

10. What are some of the advantages and disadvantages of debt financing? How is the capital structure of utilities different from other industries?

5
Risk

Risk is a loaded term in the utility world. Most utility engineers have the intuition that less risk is better and that more risk is worse. Undesirable events happening are undesirable, and are ideally avoided. Much discussion revolves around unlikely but bad events and targeted spending to either reduce the probability of these events happening or reducing their impact should they occur. When things go wrong, fingers start pointing to indicate blame.

Risk means something different in the business world. Financial risk relates to the predictability of financial outcomes. Less risk is preferable to high risk in the sense that investors will pay more for more predictable outcomes. A certain payout of $100 in one year might cost $90 today. A payout that could range between $90 and $110 might cost $85 today. A payout that could range between $50 and $150 might cost $70 today. The expected payout for each of these options is the same, but risk considerations result in different valuations. The tools of financial risk analysis allow these types of risk-based valuations to be treated quantitatively.

Financial risk is also concerned about indemnification and options, both of which can be used to lessen risk exposure. Should I purchase an insurance policy today to financially mitigate something that may or may not happen in the future? Should I pay more for a long-term fuel contract that provides future fuel price certainty, or should I purchase fuel on day-to-day markets, which, on average has lower prices but sometimes spikes to very high prices? How much should I pay for the rights to purchase a parcel of land that I may or may not need in the future? The theoretical framework behind financial risk management answers these and other questions in a rigorous and theoretical manner.

Mathematical analyses based on expected values and average outcomes are insufficient for financial risk analysis. Financial risk analysis is not interested in averages. Rather, it is interested in knowing the predictability of outcomes, which requires knowledge of all possible outcomes and their likelihood of occurrence. This type of analysis can either be backward-looking or forward-looking. A backward-looking financial risk analysis examines historical data and uses the tools and techniques of statistics. A forward-looking financial risk analysis creates predictive models and uses the tools and techniques of probability theory.

5.1   PROBABILITY AND STATISTICS

Practicing professional engineers will have had, at a minimum, a basic college course on probability and statistics. Even so, many at this point in their careers will be quite rusty with much of the subject matter. Basic statistical calculations using spreadsheets may be commonplace, but many readers may struggle with the distinction between a covariance and a correlation coefficient.

This section provides a brief refresher on probability and statistics for engineers. It also serves as a primer and reverence for readers without any background in these areas. The intent of this section is not to be comprehensive. Rather, this section covers the specific topics of probability and statistics that are required for basic financial risk analysis as presented in the remainder of the chapter.

5.1.1   Statistics

Consider a set of numbers. The value of the first member of the set is denoted x_1. The value of the second member of the set is denoted x_2. The value of the ith member of the set is denoted x_i. If the size of the data set is n, the data set is denoted as follows:

Statistics from Data Set
Size of Data Set $= n$
Data Set $= [x_1, x_2, ..., x_n]$

The *mean value* of the data set, also called the *expected value*, is defined as the sum of all member values divided by the size of the population. This value is, by far, the most commonly used statistical measure. Unfortunately, it is not very useful when performing risk analyses. The mean value of a population is mathematically defined as:

$$\text{Mean Value} = \bar{x} = \frac{1}{n} \sum_{i=1}^{n} x_i \qquad (5.1)$$

The mean value is not necessarily the most likely outcome, nor is it necessarily a typical outcome. It is simply the mathematical average of all possible outcomes. Consider the following data set with ten members:

Data Set = [1, 1, 1, 1, 1, 1, 11, 11, 11, 11]

This data set has six entries with value one and four entries with value eleven. The sum of all entries is fifty, making the mean value equal to five. No member has a value equal to the mean. In fact, no value is even close to the mean. The mean value is not necessarily representative of specific members of the data set.

The most common member value of a data set is called the *mode*. In the above example, the mode is one, since this value appears more than any other. A data set with two very common values is called *bimodal*. More generally, a data set with multiple common values is called *multimodal*.

The *median* is the value where half of members are larger and half of members are smaller. For certain purposes, median is preferred to mean since median is less impacted by extreme outliers. Consider the following data sets:

Data Set A = [1, 2, 3, 4, 5, 6, 7, 8, 9]
Mean = 5
Median = 5

Data Set B = [1, 2, 3, 4, 5, 6, 7, 8, 99]
Mean = 15
Median = 5

Data Set A has all values within a narrow range and, consequently, the mean is close in value to the median (in this case identical). Data Set B has one value that is much larger than all other values, causing a dramatic increase in the mean but not affecting the median.

The most basic measure of statistical predictability is called *variance*. Variance is based on the distance of each member from the mean value. Variance is equal to the average of the square of these values. Many values far away from the mean result in a high variance. Many members close to the mean result in a low variance. The mathematical definition for the variance of a population is:

$$\text{Variance} = \frac{1}{n} \sum_{i=1}^{n} (x_i - \bar{x})^2 \qquad (5.2)$$

The units of variance are the square of the units of the population. It is therefore convenient to use the square root of variance as a measure of data predictability. The square root of variance is called *standard deviation*, which is often represented by the Greek letter sigma, σ. Standard deviation is mathematically equal to the following:

$$\text{Standard Deviation} = \sigma = \sqrt{\frac{1}{n}\sum_{i=1}^{n}(x_i - \bar{x})^2} \qquad (5.3)$$

Standard deviation is commonly expressed as a percentage of the mean value, referred to as *volatility*. Consider Data Set A above with a mean of 5 and a standard deviation of 2.58. Often this type of data will be presented as a mean of 5 and a volatility of 52%.

The above definitions for standard deviation and variance are for entire populations. For mathematical reasons, definitions are slightly different for a sample of a population. Instead of dividing the summation by n, the formulae for samples uses $n - 1$. This difference does not matter for reasonably large populations and samples, but can cause confusion when using statistical packages. For example, the formula STDEV in Microsoft Excel uses the standard deviation formula for samples, and STDEVP uses the standard deviation formula for populations.

Up to this point, the statistical discussion has addressed single sets of numbers. Financial risk assessment often requires the examination of multiple sets that are related. Consider the daily stock price for a utility and the daily value of a broad market index. If the broad market index goes up, the utility stock price will also tend to go up. The extent to which values move together is most commonly measured by *covariance* and *correlation coefficient*.

Recall that the variance of a data set is based upon the square of the difference between the value of a member and the mean of the data set. The covariance between two data sets is similar except it is based on the product of the differences between the value of a member and the mean of each data set. The equation for covariance between a data set with members $[x_1, x_2, \ldots, x_n]$ a data set with members $[y_1, y_2, \ldots, y_n]$ and is the following:

$$\text{Covariance} = \frac{1}{n}\sum_{i=1}^{n}(x_i - \bar{x})(y_i - \bar{y}) \qquad (5.4)$$

Like variance, covariance has units equal to the square of the units of the data set. It is common to normalize covariance by dividing it by the product of the standard deviation of each data set. The result is the correlation coefficient, typically represented by the Greek letter rho, ρ. If the standard deviations for the two data sets are σ_x and σ_y, the equation for correlation coefficient is the following:

$$\text{Correlation Coefficient} = \rho = \frac{1}{n \cdot \sigma_x \cdot \sigma_y} \sum_{i=1}^{n} (x_i - \bar{x})(y_i - \bar{y}) \tag{5.5}$$

A positive correlation coefficient means that two data sets tend to move together – when one goes up the other tends to go up and when one goes down the other tends to go down. A negative correlation coefficient means the opposite; when one goes up the other tends to go down. A correlation coefficient of one shows a perfect positive linear relationship and a correlation coefficient of minus one shows a perfect negative linear relationship. A correlation coefficient of zero means that the movements of the data set are independent. Example data set plots and their associated correlation coefficient are shown in Figure 5.1.

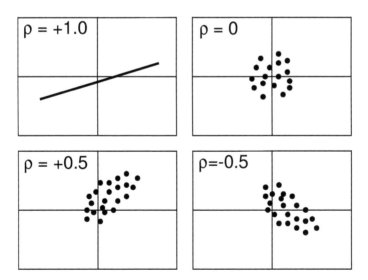

Figure 5.1. Each plot shows two data sets plotted against each other to demonstrate correlation coefficient values (ρ). A correlation coefficient of $\rho = 1$ of one shows a perfect positive linear relationship (upper left). A correlation coefficient of $\rho = 0$ means that the data sets are independent (upper right). A positive ρ means that two data sets tend to move together (lower left), and a negative ρ means the two data sets tend to move in opposite directions.

5.1.2 Probability

Many values cannot be known with certainty. Examples include future financial quantities such as stock prices and bond prices. Guesses about future values can be made, but these guesses are likely to be higher or lower than the values that actually occur in the future. A numerical quantity with an uncertain value is called a *random variable*. Random variables are represented by *probability distribution functions*.

Probability distribution functions are mathematical equations allowing a large amount of information, characteristics, and behavior to be described by a small number of parameters. A probability distribution function has an associated *probability density function*, f(x), which represents the likelihood that a random variable x will be a particular value. For example, the probability density function for a random coin toss is:

f(heads) = 0.5
f(tails) = 0.5

There are two important characteristics associated with each probability density function. First, each function always returns a value between (and including) zero and one. A value of zero indicates no probability of occurrence and a value of one indicates certain occurrence. The next characteristic is that the integral of the function over all possible outcomes must equal unity because each random event will have precisely one actual outcome. These characteristics are mathematically represented as follows:

$$f(x) \in [0,1] \quad ; f = \text{probability density function} \tag{5.6}$$

$$\int_{-\infty}^{\infty} f(x)\,dx = 1 \tag{5.7}$$

A close cousin to the probability density function is the cumulative distribution function, F(x). The cumulative distribution function is the integral of the probability density function, and reflects the probability that f(x) will be equal to or less than x. The equation for cumulative distribution function is the following:

$$F(x) = \int_{-\infty}^{x} f(y)\,dy \quad ; F = \text{cumulative distribution function} \tag{5.8}$$

To illustrate the concept of probability density functions and cumulative distribution functions, consider a random toss of two six-sided dice. There are 36 possible combinations resulting in 12 possible scores. Rolling double ones will only happen once every 36 times, making f(2) = 1/36 = 0.028. Since two is the lowest possible score, F(2) = f(2) = 0.028. Rolling the number 3 will happen twice as often: f(3) = 0.056. The cumulative distribution function at this point will be the sum of all previous density function values: F(3) = f(2) + f(3) = 0.083. This computation continues until the cumulative distribution function is equal to unity. The probability density function and the cumulative distribution function for this example are shown in Figure 5.2.

Notice that the probability density function shown in Figure 5.2 is a triangle with a peak at the value seven. As is the case for statistics, this high point is called the *mode* of the probability density function.

Probability distribution functions can be, and often are, characterized by statistical measures such as the expected value, variance, and standard deviation. As in statistics, the expected value is the mean of the function. Since the integral of all density functions is equal to unity, the expected value is equal to the integral of first-order moments. The mathematical equation for expected value, typically represented by the Greek letter mu, μ, is the following:

$$\text{Expected Value} = \mu = \int_{-\infty}^{\infty} x \cdot f(x) dx \qquad (5.9)$$

Variance is a measure of how a function varies about the mean. A small variance indicates that a value close to the mean is likely, and a large variance indicates that a value close to the mean is unlikely. The equation for the variance of a probability density function is:

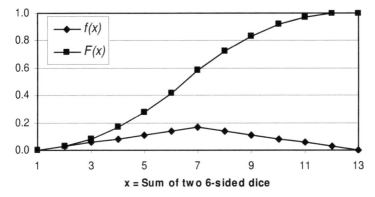

Figure 5.2 Probability density function, f(x), and the cumulative distribution function, F(x) for a random toss of two six-sided dice. The probability density function can assume any value between zero and one. The cumulative distribution function is equal to the integral of the density function, begins with a value of zero, ends with a value of 1, and is monotonically increasing.

$$\text{Variance} = \int_{-\infty}^{\infty} (x - \mu)^2 f(x) dx \tag{5.10}$$

As is the case for statistics, the standard deviation is the square root of the variance, and is often represented by the Greek letter sigma, σ. Standard deviation is often convenient because it has the same units as the density function and the expected value. The equation for standard deviation is the following:

$$\text{Standard Deviation} = \sigma = \sqrt{\int_{-\infty}^{\infty} (x - \mu)^2 f(x) dx} \tag{5.11}$$

As described in the section on statistics, standard deviation is commonly expressed as a percentage of the mean value, referred to as volatility.

$$\text{Volatility} = \frac{\sigma}{\mu} \times 100\% \tag{5.12}$$

Now that the basic theory and vocabulary for probability distribution functions have been discussed, two common and useful distributions will be presented. These distributions, normal and lognormal, are the most commonly used distributions in financial risk analysis.

5.1.3   Normal Distribution

The most well known and commonly used probability distribution function is the normal distribution. The normal distribution is commonly referred to as a "bell curve" because its probability density function resembles the cross section of a bell. The normal distribution is mathematically characterized by two parameters: an expected value, μ, and a standard deviation, σ. Because of the popularity of the normal distribution, μ is commonly used to represent any mean value, and σ is often used to represent any standard deviation.

The equation for the probability density function for the normal distribution function is the following:

Normal Distribution

$$f(x) = \frac{1}{\sigma\sqrt{2\pi}} \exp\left[-\frac{(x-\mu)^2}{2\sigma^2} \right]; \quad -\infty \le x \le \infty \tag{5.13}$$

Expected value = μ
Standard deviation = σ

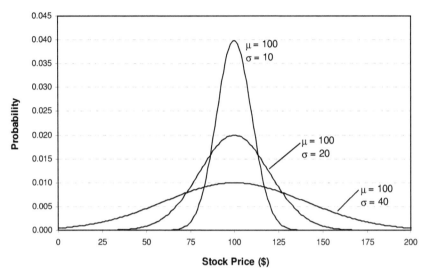

Figure 5.3. Three example normal distributions. Each distribution has a mean (μ) of 100, but different standard deviations (σ). A normal distribution will always have a symmetrical bell curve shape. The bell curve is flattened as the standard deviation increases, indicating a higher probability of values far from the mean (i.e., higher risk).

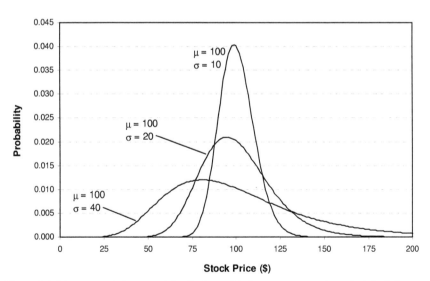

Figure 5.4. Three example lognormal distributions with the same mean and standard deviations as the normal distributions in Figure 5.3. With a standard deviation of 10, the normal and lognormal curves look similar. With a standard deviation of 20, skew becomes noticeable for the lognormal distribution. With a standard deviation of 40, the lognormal distribution no longer resembles the normal distribution.

Normal distributions are symmetrical about their mean; it is just as likely to be a certain percentage above the mean as the same percentage below the mean. In addition, the mean, median, and mode of a normal distribution are the same value. Outcomes cluster around the mean, with 68% falling within one standard deviation, 95% falling within two standard deviations, and 99.7% falling within three standard deviations.

Three example normal distributions are shown in Figure 5.3. Each of these distributions has a mean of 100, but different standard deviations.

The normal distribution is convenient because its parameters can be easily computed from experimental data. For any data set, computing the mean and standard deviation produces the μ and σ of the normal distribution. However, comfort with the normal distribution does not mean that it is always appropriate for financial risk analysis. In fact, a more common probability function for use in financial risk analysis is the lognormal distribution.

5.1.4 Lognormal Distribution

The lognormal distribution, as one may guess, is a close relative to the normal distribution. The primary differences are that the lognormal distribution uses the natural logarithm of the independent variable, and the independent variable is constrained to be nonnegative. Formulae for the lognormal distribution are:

Lognormal Distribution

$$f(x) = \frac{1}{x \cdot b\sqrt{2\pi}} \exp\left[-\frac{(\ln x - a)^2}{2b^2}\right]; \quad x \geq 0 \tag{5.14}$$

$$\text{Expected Value} = \mu = \exp\left[a + \frac{b^2}{2}\right] \tag{5.15}$$

$$\text{Standard Deviation} = \sigma = \sqrt{\exp(2a + 2b^2) - \exp(2a + b^2)} \tag{5.16}$$

The lognormal distribution has a skewed shape; its peak will always be to the left of its mean. Depending upon its parameters, this skew can be slight or severe. Lognormal distributions with small standard deviations look similar to normal distributions. Lognormal distributions with high standard deviations are noticeably skewed. Three example lognormal distributions are shown in Figure 5.4. These have the same mean and standard deviations as the three normal distributions in Figure 5.3, so that similarities and differences can be easily examined. With a mean of 100 and a standard deviation of 10, the normal and lognormal curves look similar. With a mean of 10 and a standard deviation of 20, skew becomes noticeable for the lognormal distribution. With a mean of 10 and a standard deviation of 40, the lognormal distribution no longer resembles the normal distribution.

Computing lognormal parameters from a data set is straightforward. First, take the natural logarithm for each member of the data set (all data is required to be greater than zero). The mean and standard deviation of the transformed dat set correspond to the lognormal parameters as follows:

Size of Data Set $= n$

Data Set $= [x_1, x_2, \ldots, x_n]$

$$a = \frac{1}{n} \sum_{i=1}^{n} \ln(x_i) \qquad ; \text{mean of transformed data} \qquad (5.17)$$

$$b = \sqrt{\frac{1}{n} \sum_{i=1}^{n} (\ln(x_i) - a)^2} \qquad ; \text{standard deviation of transformed data} \qquad (5.18)$$

Sometimes it is easier to compute the lognormal parameters directly from the mean (μ) and standard deviation (σ) of the untransformed data set. Using this method does not require knowledge of individual data values, and is somewhat easier from a computational perspective. The lognormal parameters can be computed directly from statistical parameters as follows:

$$b = \sqrt{\ln\left(\sigma^2 + e^{2\ln\mu}\right) - 2\ln\mu} \qquad (5.19)$$

$$a = \ln\mu - \frac{1}{2} b^2 \qquad (5.20)$$

There is potential parameter confusion when using the lognormal distribution. Often the symbols μ and σ correspond to the mean and standard deviation of the probability density function, as is done in this section. Other times, μ and σ will correspond to the mean and standard deviation of the transformed data set, represented by the parameters "a" and "b" in this section. In any case, care should be taken so that the parameters used in a lognormal distribution correspond to the above equations.

There is much more to probability and statistics than is presented in this short section. However, these basics are sufficient for a full understanding of the risk topics that are addressed in the remainder of the chapter.

5.2   STOCK PRICE MOVEMENT

Financial risk assessment requires knowledge about the movement of prices. Take stock prices as an example. A share of stock can be purchased today at the market price. What is the likelihood that the stock will increase in value? Decrease in value? Increase in value a lot? Decrease in value a lot? To address these questions quantitatively, a mathematical model of stock price movement is required.

The most common model of stock price movement assumes that stock price returns are equally likely to increase or decrease in value. After a short time step, the initial stock price of P_0 has a 50% chance of increasing in value to $P_0 \times (1 + r)$ and a 50% chance of decreasing in value to $P_0 \div (1 + r)$. After the next time step, P_1 has a 50% chance of increasing in value to $P_1 \times (1 + r)$ and a 50% chance of decreasing in value to $P_1 \div (1 + r)$. The process continues, resulting in a *recombinant binary tree*. It is a binary tree because each incremental stock price movement follows one of two paths. It is recombinant because upward movement followed by a downward movement results in the original price. The tree computes the probability of any stock price at any future time.

An example binary tree is shown in Figure 5.5. The initial stock price at Step 0 is $100 and r is equal to 2%. At Step 1, the stock price has a 50% chance of being multiplied by 1.02 to $102, and a 50% chance of being divided by 1.02 to $98.04. Each stage adds a possible higher price and a possible lower price, but most likely the price will remain near its starting point.

Completely random movement is called *Brownian motion*. Stock price returns as described above follow Brownian motion. As time steps grow smaller, Brownian motion results in a normal distribution of returns at each point in time, with the variance of outcomes being proportional to elapsed time.

If stock price returns are normally distributed, it can be shown that the underlying stock price is lognormally distributed. Consider the binary tree shown in Figure 5.5, but expanded out fifty, one hundred, and two hundred steps. The resulting probability distribution curves of stock prices are shown in Figure 5.6. The curves are almost perfectly lognormal, and will converge to perfect lognormal distributions as the time step becomes infinitesimally small.

Recall that the variance of stock price returns increases in proportion to elapsed time. This results in the lognormal distribution of underlying stock price value also spreading out as time elapses. In Figure 5.6, this is demonstrated by the relatively tight distribution after fifty time steps, a wider spread after one hundred time steps, and an even wider spread after two hundred time steps.

The intent of stock price models is not to predict future stock prices. Rather, it is to compute the range and likelihood of future prices so that financial risk can be better understood and managed. Stock price models are valuable in their own right, but are also the foundation of models for *financial derivatives,* whose values are tied to the value of an underlying asset (in this case a share of stock). Financial options are discussed further in Section 5.6.

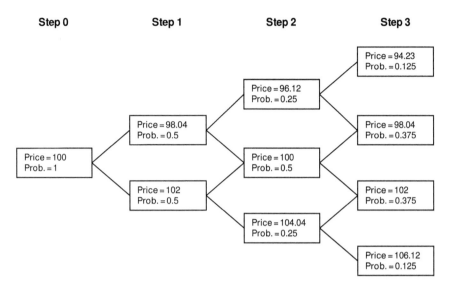

Figure 5.5. A recombinant binary tree showing stock price movement. The initial stock price at Step 0 is $100. At Step 1, the stock price has a 50% chance of being multiplied by 1.02 to $102, and a 50% chance of being divided by 1.02 to $98.04. Each stage adds a possible higher price (highest so far multiplied by 1.02) and a possible lower price (lowest so far divided by 1.02). Most likely the price will remain near its starting point.

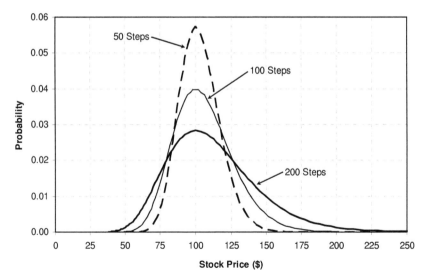

Figure 5.6. A lognormal distribution of stock prices corresponding to a normal distribution of stock price returns. This curve results from the binary tree in Figure 5.5 expanded out fifty, one hundred, and two hundred steps. The curves are almost perfectly lognormal, and will converge to perfect lognormal distributions as the time step becomes infinitesimally small.

In summary, the probability distribution of future stock prices is, for the most part, lognormal. Due to its simplicity and ease of application to quantititative risk management, the lognormal assumption has been and remains popular. It remains a model, and does not perfectly describe stock price movements. In the late 1980s, a stock market crash showed the limitations of the lognormal assumption in reflecting sudden changes in stock price variance. This led to a variety of expanded models where future stock price variance depends upon historical stock price variance. These models are beyond the scope of this section and the lognormal assumption is used for the remainder of this chapter.

5.3   DIVERSIFICATION

The essence of diversification can be summed up by the old adage, "don't put all of your eggs in one basket." Consider an investor that purchases one thousand shares of a company. After one year, should this investor be upset if the stock price goes down? Has the investor somehow failed if his purchase does not result in positive returns?

Recall from the previous chapter that the market is efficient and stocks are, for the most part, priced fairly. The discussion on stock price movement states that after a short period of time, a stock has essentially the same chance of increasing or decreasing in value. Therefore, the investor should expect that many stock picks will decrease in value, and not be upset.

Good investors do not care whether the value of a particular stock goes up or down, but are extremely interested in whether their overall portfolio value goes up or down. Consider a stock market where the returns of each individual stock after one year are characterized by a normal distribution with a mean return of 12% and a standard deviation of 12%. If an investor only purchases a single stock, the standard deviation of returns is 12%. For normally distributed independent variables, standard deviation decreases with the square of the number of variables. Therefore, if the investor purchases equal amounts of four separate stocks (with independent price movement), the mean return is still 12%, but the standard deviation of the portfolio is only 6%. If the investor purchases equal amounts of one hundred separate stocks, the mean return is still 12%, but the standard deviation of the portfolio is only 1.2%. Diversifying a portfolio reduces the standard deviation of future returns, resulting in lower financial risk.

Consider the possibility of a portfolio losing money. For the situation described in the preceding paragraph, an investment in a single stock will result in a one-year loss when the price is one standard deviation below the expected value, amounting to 15.9% of the time. Equal investments in four stocks will result in a one-year loss about 2.3% of the time. After investing in ten stocks, the risk of an annual loss falls below 0.1%. Clearly, there are dramatic benefits in not putting all of your financial eggs in one basket.

Why do some stocks do well and other stocks do poorly? Based on the principles of a semi-strong efficient market, the reasons must be either (a) random chance, or (b) factors not knowable through public information. Consider a small startup company developing a new product for utilities. To be commercially successful, this product must be successfully designed, survive patent challenges, be accepted for use by the industry, and be produced for an acceptable cost. These uncertainties are all specific to this company, and are called *idiosyncratic risks.*

To a large extent, the outcomes related to idiosyncratic risk are based on random chance. The basic research and product development of many new technology companies will not result in commercially viable products. Some of the companies will get lucky and develop a successful product or technology. Even though most of these firms have excellent engineers and scientists, many fail because they are unlucky and many succeed because they are lucky. There are other, more controllable factors that influence business success, but luck is extremely important in many cases.

Idiosyncratic risks can be diversified away. Investing in one small startup company is risky, since it may succeed and it may not succeed. Investing in many small startup companies is much less risky since a certain percentage will fail, a certain percentage will succeed, and the actual return of the portfolio is likely to fall near the expected return of the portfolio (i.e., the distribution of return has a small standard deviation). In a developed market, idiosyncratic risk can be effectively eliminated by purchasing thirty to forty assets with similar idiosyncratic risk characteristics.

Not all risk is idiosyncratic. There are some risk factors that will impact all investments at the same time. This is similar to the engineering concept of a *common mode* root cause. Something external is impacting many things at once. Whereas idiosyncratic risks are diversifiable, common mode risks are *non-diversifiable.* Another common term for non-diversifiable risk is *systematic risk.* Purchasing a large number of disparate investments does not impact non-diversifiable risk because all of the investments will be impacted by the common mode events in a similar way. For example, higher interest rates will reduce the value of all stock prices in a similar way. Lower consumer confidence will have a similar effect on the overall market. Some systematic risk will be specific to certain industries (such as the development of a substitute product or service), while others will be broader in scope.

An example of diversification is shown in Figure 5.7. Diversification refers to the number of dissimilar investments in the portfolio. The far left represents a portfolio consisting of only a single investment, where exposure to idiosyncratic risk is the highest possible. Total risk is equal to idiosyncratic risk plus non-diversifiable risk. As more investment types are added to the portfolio, idiosyncratic risk is reduced, causing total risk to be reduced as well. As diversification increases, idiosyncratic risk eventually becomes negligible and the investor is only exposed to non-diversifiable risk.

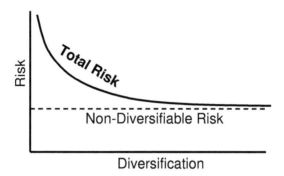

Figure 5.7. Risk exposure versus diversification. The far left represents a single investment, where exposure to diversifiable risk is the highest possible. Total risk is equal to diversifiable risk plus non-diversifiable risk. As more investment types are added, diversifiable risk is reduced, causing total risk to be reduced as well. As diversification increases, diversifiable risk eventually becomes neglig-ible and the investor is only exposed to non-diversifiable risk.

The diversification of idiosyncratic risk is analogous to the average value of six-sided dice rolls. The expected value of each dice roll is 3.5. However, the average value of a small number of rolls can easily be much higher or much lower than the expected value. The average value of a large number of rolls is almost certain to be close to the expected value.

Highly specialized financial quantititative risk analysts (commonly referred to as *quants*) have always addressed risk in a structured and theoretically sound manner. Unfortunately, the methods and mathematics used by quants are very difficult for most investors to understand and utilize when making investment decisions. In 1997, the US Securities and Exchange Commission ruled that pub-lic corporations must disclose quantitative information about their risk exposure. Most major banks and investment brokers chose to implement the rule by utiliz-ing a straightforward measure, *value at risk* (VaR).

VaR represents the risk of losing more than a specified amount of money after a certain period of time. Each measure of VaR is associated with both a time period and percentage. Typical time periods are one day and two weeks. Typical percentages are 1% and 5%. For example, a VaR of 2 weeks at 5% represents an investment that, after two weeks, losses would exceed the VaR value only 5% of the time.

Consider Figure 5.8, where the bell-shaped curve represents the probability distribution of expected losses for an investment after two weeks. Notice that the expected value of losses is negative, indicating that the investment on average will have a positive return. The 5% VaR level occurs where the area under the curve to the left is equal to 95% and the area under the curve to the right is equal to 5%. For this example, the 5% VaR is $1.2 million.

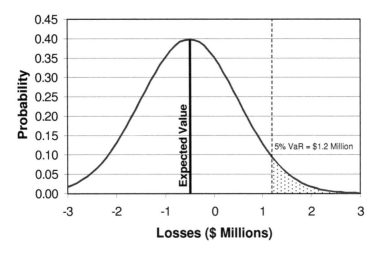

Figure 5.8. The bell-shaped curve represents the probability distribution of expected losses for an investment after two weeks. The expected value of losses is negative, indicating that the investment on average will have a positive return. The 5% VaR level occurs where the area under the curve to the right is equal to 5%. In this figure, the 5% VaR is $1.2 million.

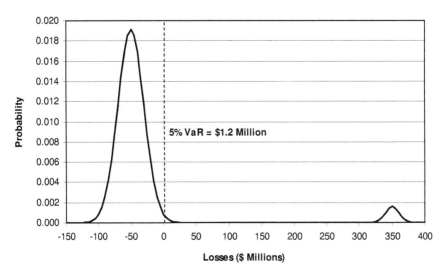

Figure 5.9. VaR does not reflect rare but catastrophic events. This figure has the same 5% VaR as the investment returns represented by Figure 5.8 ($1.2 Million), but much higher risk exposure. In Figure 5.8, losses are grouped close to the VaR level. In this figure, losses are likely to be much higher than the VaR level.

VaR was designed to be simple, and has been the target of criticism due to its simplicity. Opponents argue that VaR gives investors a false sense of confidence because it does not reflect rare but catastrophic events. A 5% VaR identifies the 5% threshold, but does not indicate how high potential losses might be beyond this threshold.

Consider Figure 5.9. The investment returns represented by this figure have the same 5% VaR as the investment returns represented by Figure 5.8 ($1.2 Million). In each case, losses above $1.2 million will occur 5% of the time. In Figure 5.8, these losses are grouped closely to the VaR level. In Figure 5.9, losses are likely to be much higher than the VaR level; when things go bad, they tend to go extremely bad. In other words, VaR is insensitive to the magnitude of losses beyond the VaR level. It does not reflect the risk of unlikely but severe negative outcomes.

Author and former trader Nassim Taleb predicted that VaR would give investors a false sense of confidence and would be exploited by traders. Two years before the financial institution Fannie Mae was bailed out by the federal government due to poor risk management, Taleb wrote in his book *The Black Swan*, "Fannie Mae, when I look at its risks, seems to be sitting on a barrel of dynamite, vulnerable to the slightest hiccup. But not to worry: their large staff of scientists deems these events 'unlikely.'" David Einhorn, the well-known hedge fund manager, compares VaR to "an airbag that works all the time, except when you have a car accident." He famously sold short (betting that the price would go down) the stock of the financial institution Lehman Brothers before it filed for bankruptcy in September 2008 due to poor risk management.

The criticism of Taleb, Einhorn, and others goes beyond VaR. These criticisms generally argue that (1) mathematically rigorous risk assessment results in a false sense of confidence since these methods do not fully reflect reality, and (2) these methods rely on historical data, which cannot accurately predict future risk. In other words, the models do not reflect the risk of the models themselves, and cannot reflect what they do not know.

Despite the criticism, VaR and other approaches to risk quantification are common and potentially useful if taken in context. A good approach is to view a quantitative risk assessment as a lower bound on risk that may not be based on perfect models and may not reflect the risk of rare events. Risk exposure is at least as much as the models show.

In any case, diversification is effective at reducing idiosyncratic risk. Once all idiosyncratic risk is eliminated, a group of investments is said to be *fully diversified*. Full diversification is good, but not the only characteristic of a smart set of investments. This becomes evident through the study of portfolio theory.

5.4 PORTFOLIO THEORY

Portfolio theory studies how rational investors can select an optimal mix of investments. While diversification encourages investors to purchase a wide variety of investments, portfolio theory concludes that some combinations of investments are superior to others.

Portfolio theory assumes that rational investors are interested in maximizing investment returns for a given level of risk. Investment returns typically assume lognormally distributed future prices, as discussed in Section 5.2. Investment return is defined as the expected value of this distribution divided by current price. Investment risk is defined as the standard deviation of this distribution. Definitions of investment risk and return are:

Investment Return. The expected value of future investment price divided by current investment price.

Investment Risk. The standard deviation of investment return, typically measured as a percentage of investment return (i.e., volatility).

Portfolio theory builds on diversification theory; all of the mathematics and derivations assume a fully diversified portfolio with no idiosyncratic risk. All investments portfolios in the realm of portfolio theory are only exposed to systematic risk. Undiversified portfolios will not necessarily behave in the manner discussed in the remainder of this section.

For clarity, the discussion on portfolio theory will assume that investments consist of companies that have publicly traded stock. Portfolios are created by buying and selling stock through the stock market. Portfolio theory can be easily extended to include any type of asset such as bonds, real estate, and commodities. Terms that are used synonymously in this section are investment, company, firm, security, and asset. Definitions of these investment types are provided to avoid confusion.

Firm. A legally recognized business, corporation, partnership, or proprietorship.

Company. A firm. The terms "company" and "firm" can typically be used interchangeably.

Security. A negotiable contract representing financial value. Examples of securities include common stock, bonds, bank notes, and financial options.

Asset. Something of value. A portfolio is comprised of individual assets such as securities.

As has been previously discussed, investors prefer lower risk. This means that investments with more predictable returns have higher value than investments with less predictable returns. Consider two companies with publicly traded stock. The price of each stock is characterized by an expected return, μ. The risk of each company is defined as the standard deviation of returns, σ. Company A is a low risk and low return investment. Company B is a high risk and high return investment.

Assume that the stock price returns of Company A and Company B are perfectly correlated. When Company A increases in value, Company B also increases in value. When Company A decreases in value, Company B decreases in value. Since it is riskier, the stock price of Company B will always move more (as a percentage) than Company A, but the ratio of returns is always the same. For example, assume that when Company A increases by 1%, Company B increases by 2%. If they are perfectly correlated, Company B will always increase (or decrease) by twice the amount of Company A.

Now consider three portfolios. The first portfolio consists just of Company A, and has a risk and return equal to the stock price of Company A. The second portfolio consists just of Company B, and has a risk and return equal to the stock price of Company B. The third portfolio has a half of its value invested in Company A and has a half of its value invested in Company B. These three portfolios are graphically represented in Figure 5.10.

What is the risk and return associated with the third portfolio? Remember that portfolio theory assumes a fully diversified portfolio; there will be no benefit due to the reduction of idiosyncratic risk. Calculating expected return is easy. If 50% of the portfolio has the expected return of Company A and 50% of the portfolio has the expected return of Company B, the expected return of the portfolio must be the average.

Risk can be treated in a similar manner. If Company A increases by a certain amount, Company B will also increase by a certain amount due to the assumption of perfect correlation. Therefore, the risk of the portfolio (as measured by standard deviation) is equal to the average risk of the two companies.

The two-asset portfolio with perfect correlation has a return equal to the weighted sum of returns and a standard deviation equal to the weighted average of standard deviations. This relationship between risk and reward is shown as the line connecting the one-asset portfolios in Figure 5.10. No portfolio is superior to any other, but rational investors can choose the mix of portfolio investments that maximizes returns while satisfying risk requirements.

Now consider two companies with perfect negative correlation. When Company A increases in value, Company B decreases in value. When Company A decreases in value, Company B increases in value. The stock price of Company B will move more than Company A, but the ratio of returns is always the same negative. For example, assume that when Company A increases by 1%, Company B decreases by 3%. If they are perfectly correlated, Company B will always decrease by three times the amount that Company A increases.

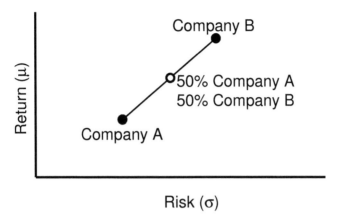

Figure 5.10. A two-asset portfolio with perfect correlation. The risk versus return relationship for different portfolio mixes is shown as the line connecting the one-asset portfolios (either 100% Company A or 100% Company B). With perfect correlation, portfolio return is equal to the weighted sum of company returns and a portfolio standard deviation equal to the weighted average of company standard deviations.

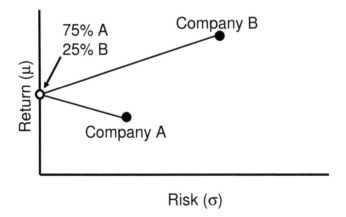

Figure 5.11. Portfolio risk and return characteristics for two perfectly negatively correlated stocks. A rational investor will never invest just in Company A. By investing part of the portfolio in Company B, returns are increased and risk is decreased. As the percentage of Company B increases, risk continues to decrease until all risk is eliminated (shown by the white dot).

What are the characteristics of a portfolio with 75% of its value in Company A and 25% of its value in Company B? Expected returns are still equal to the weighted average of the two companies. The risk exposure is much different. If Company A increases in value by 4%, the total portfolio value increases by three fourths of this amount, or 3%. Simultaneously, Company B will decrease in value by 12%, causing the total portfolio value to decrease by one fourth of this amount, or 3%. The portfolio gain due to Company A is exactly offset by the portfolio loss due to Company B. This offset will happen for all random price movements.

Portfolio risk return characteristics for two perfectly negatively correlated stocks are shown in Figure 5.11. In this scenario, a rational investor will never invest just in Company A. By investing part of the portfolio in Company B, returns are increased and risk is decreased. This portfolio benefit is beyond diversification since the analysis already assumes full diversification. As the percentage of Company B increases, risk continues to decrease until all risk is eliminated. With two perfectly negatively correlated stocks, the upward movement of one stock will be exactly offset by the downward movement of the other stock, assuming that the portfolio is properly balanced.

The ability of negatively correlated stocks to cancel out risk is the reason investors spend a lot of time trying to identify securities with this relationship. Although perfect negative correlation has never been found, there has been some evidence of a negative correlation between stocks and real estate, between stocks and commodities, and between stocks and long-term treasury bills. From this perspective, professional investors do not include treasury bills in their portfolio because they are less risky than stocks, rather, they include treasury bills because of positive portfolio effects. Similarly, they do not include real estate in their portfolios because it is a good stand-alone investment, but due to its positive portfolio effect.

Finding assets with negative correlation is difficult. Even bonds have shown a slight positive correlation with stocks. However, positive portfolio effects occur even if assets are uncorrelated or weakly correlated. Consider Figure 5.12. The straight dotted line between Company A and Company B represents portfolios where the companies are perfectly positively correlated. The two remaining dotted lines represent portfolios where the companies are perfectly negatively correlated. The uncorrelated case lies within this dotted line triangle, and is shown by the solid curved line.

In the uncorrelated case, a rational investor will never invest just in Company A. By investing part of the portfolio in Company B, returns are increased and risk is decreased, just as in the negatively correlated case. As the percentage of Company B increases, risk continues to decrease until a minimum value is reached. The combination of stock that corresponds to this point is called the *minimum variance portfolio*. Beyond this point, investing higher percentages in Company B will increase both portfolio risk and return.

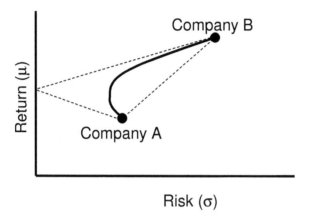

Figure 5.12. Portfolio risk and return characteristics for two uncorrelated stocks The straight dotted line between Company A and Company B represents portfolios where the companies are perfectly positively correlated. The two remaining dotted lines represent portfolios where the companies are perfectly negatively correlated. The uncorrelated case is shown by the solid curved line.

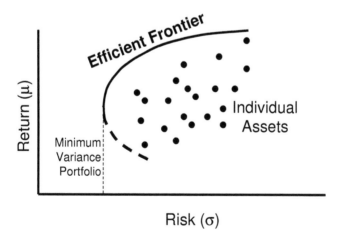

Figure 5.13. The efficient frontier. For each asset (a dot on the graph), there is a point on the efficient frontier that has the same return but lower risk. The dashed line represents levels of return that are irrational since higher returns could be achieved for lower levels of risk. The point of the efficient frontier with the lowest level of risk is called the minimum variance portfolio.

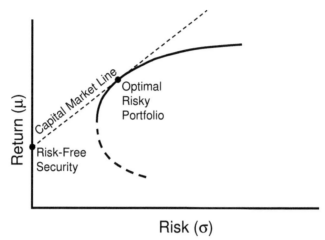

Risk (σ)

Figure 5.14. The capital market line. There is a single point on the efficient frontier where a tangent line can be drawn through the risk-free security. This point is called the optimal risky portfolio, and the line is called the capital market line. According to portfolio theory, the only portfolio a rational investor should consider is the optimal risky portfolio. Risk is reduced by purchasing risk-free securities, and risk is increased by borrowing against risk-free securities and using this money to purchase more of the optimal risky portfolio.

Portfolio effects do not end with two assets. If the correlations of a large number of assets are known, the risk and return of any mix can be easily computed. Consider a large pool of assets from which a portfolio can be built. A computer can randomly build a large number of portfolios from these assets and determine their risk and return. For a given level of return, one of the randomly generated portfolios will have the lowest level of risk, and is therefore called *efficient*. The set of all efficient portfolios creates a curved line on the risk versus return graph called the *efficient frontier*.

An example of an efficient frontier generated from a set of individual assets is shown in Figure 5.13. For each asset (a dot on the graph), there is a point on the efficient frontier that has the same risk but a higher return. For each asset, there is also a point on the efficient frontier that has the same return but lower risk. The dashed line represents levels of return that are irrational since higher returns could be achieved for lower levels of risk. The point of the efficient frontier with the lowest level of risk is called the *minimum variance portfolio*, which is equivalent to the lowest possible standard deviation.

When a portfolio is on the efficient frontier, the tradeoff between risk and return is binding. Higher portfolio returns cannot be achieved without increasing risk. Similarly, portfolio risk cannot be reduced without accepting lower levels of return.

The efficient portfolio represents the best risk versus return possibilities for combinations of risky assets – assets with uncertain returns. Portfolio theory is

extended by consideration of a risk-free security – a security with a highly predictable return. An example of a nearly risk-free security is a US treasury bill,
which has nominal returns only exposed to the risk of a collapse of the US government.

Since a risk-free security has no random price movements, it is perfectly
correlated with risky assets that do have random price movements. It is also perfectly correlated with any portfolio of risky assets. When a risk-free security is
combined with a risky portfolio, the return of this combination is the weighted
average return, and the risk (i.e., standard deviation) is the weighted average
risk.

A risk-free security is graphically represented in Figure 5.14. There is a single point on the efficient frontier where a tangent line can be drawn through the
risk-free security. This point is called the *optimal risky portfolio*, and the line is
called the *capital market line*. The capital market line represents the highest
possible risk versus return potential for combinations of risky assets and risk-
free securities.

Rational investors will start by purchasing the optimal risky portfolio. If the
investors prefer less risk, they will not purchase a higher percentage of lower
risk assets. Rather, they will reduce risk by adding zero-risk securities to their
portfolio. If the investors prefer more risk, they will not purchase a higher percentage of higher risk assets. Rather, they will borrow money against the risk-
free rate and use this leverage to purchase more of the optimal risky portfolio.
According to portfolio theory, the only portfolio a rational investor should consider is the optimal risky portfolio. The portfolio decision is separate and distinct
from the risk decision.

What does the optimal risky portfolio contain? The answer is surprisingly
simple. It can be mathematically proven that the optimal risky portfolio must
represent the weighted sum of every asset in the market, with weights being
proportional to market value. A portfolio with these characteristics is called the
market portfolio. In theory, the market portfolio must consist of everything of
value in the global market including stocks, bonds, real estate, and commodities.
Practically, the makeup and weights of various components can be estimated by
various indices that are used by the financial community. A reasonable estimate
of the market portfolio can be achieved through a combination of the components shown in Table 5.1. A representative market index is provided with each
component. These indices will each contain a large number of assets with well-
specified weights.

For those not familiar with the term, a market index is a weighted average
value that attempts to track the price level of a market sector. For example, the
S&P 500 consists of five hundred stocks, selected by the company Standard &
Poor's, that closely resemble the overall US stock market. Each included company contributes to the value of the S&P 500 based on its stock price and its
number of publicly traded shares. Weights are adjusted to account for events
such as share issuance, share repurchase, mergers, and so forth.

Table 5.1. The market portfolio consists of everything of value in the global market including stocks, bonds, real estate, and commodities. A reasonable estimate of the market portfolio can be achieved through a combination of the following components. The types of assets and weights associated with each component can be estimated through representative indices. The indices in this table are provided as examples, not recommendations.

Component	Representative Index
US stocks	Russell 3000
US bonds	Lehman Aggregate Bond Index
Foreign stocks in developed markets	Morgan Stanley Capital International (MSCI) EAFE (Europe, Australasia, and Far East)
Foreign stocks in emerging market	Morgan Stanley Capital International (MSCI) Emerging Markets
Foreign bonds in developed markets	Citigroup WGBI (World Group Bond Index)
Foreign bonds in emerging markets	Citigroup ESBI (Emerging Sovereign Bond Index)
US Real Estate	US Real Estate; DJ Wilshire REIT (Real Estate Investment Trust)
US Treasuries	Lehman Brothers US TIPS (Treasury Inflation Protected Securities)
Commodities	Dow Jones AIG Commodities (American International Group)

There are a large number of broad market indices such as the Dow Jones Industrial Average (30 stocks), the aforementioned S&P 500 (500 companies), and the less popular but broader Russell 3000 (3000 companies). There are also a large number of indices that track specific sectors of the market, typically listed as follows:

Stock Market Sectors
- Basic Materials
- Capital Goods
- Communications
- Consumer Cyclical
- Energy
- Financial
- Health Care
- Technology
- Transportation

Portfolio theory is critical for financial investing, but the lessons of portfolio theory generalize. For any set of correlated risky choices, portfolio theory shows how to create an efficient frontier where the tradeoff between risk and performance is binding.

5.5 CAPITAL ASSET PRICING MODEL

Chapter 4 introduced the concept of required returns for equity investors. If an investor purchases $1000 of common stock in a company, the expected return will be more for higher risk companies and less for lower risk companies. The Capital Asset Pricing Model (CAPM) provides a methodology for determining what these expected returns should be.

Like portfolio theory, CAPM assumes that investors are fully diversified and are only exposed to systematic risk. It also recognizes that rational investors will want to both invest in the market portfolio with the expected market return (r_m) and invest in risk-free securities with the risk-free rate of return (r_f).

Why would an investor want to invest in the market portfolio rather than risk-free securities? The reason, of course, is for a higher expected return. The difference between the expected market return and the risk-free rate is called the *market risk premium*. In the US, the stock market risk premium has averaged about 8.4%.

Historical market returns can be easily determined by examining returns of broad market indices. Historical market risk can be similarly determined by examining the standard deviation of returns for these same indices. These represent weighted averages of stock price returns within the market portfolio. Therefore, the average return of stocks must be equal to the overall market, and the average systematic risk of stocks must also be equal to the overall market. Riskier stocks will require higher returns than the market, and less risky stocks will require lower returns than the market.

Stock price risk is most commonly measured by comparing it to the overall risk of the market using covariance. The covariance of market movement and stock price movement, when divided by market variance, is called *beta* (β). Beta is mathematically defined as:

$$\beta = \frac{\text{Covariance}(S, M)}{\sigma_m^2} \qquad (5.21)$$

S   =   Individual stock prices

M   =   Overall market prices

σ_m^2 =   Variance of market price

The beta of a company describes how its price moves relative to the overall market. If $\beta = 1$, a stock market increase (or decrease) of one percent results in the company stock tending to increase (or decrease) by one percent. If beta is greater than one, stock price movement tends to be greater than the overall market. For example, if $\beta = 2$, a stock market increase by one percent results in the company stock tending to increase by two percent. If beta is less than one, stock price movement tends to be less than the overall market. For example, if $\beta = 0.5$, a stock market increase of one percent results the company stock tending to increase by one half of a percent.

Another way to compute beta is to through linear regression. First, create a scatter plot with daily market returns (r_m) on the x-axis and stock returns (r_s) on the y-axis. For example, a particular day may result in a market return of 1% and a stock return of 2%. This results in a data point of (1%, 2%) on the scatter plot. After all data points are plotted, linear regression is used to find the best fit line. The y-intercept of the line is called alpha (α), and represents the expected return due to idiosyncratic risk. The slope of the line is beta, and is a relative measure of stock risk compared to market risk. An example of computing alpha and beta through linear regression is shown in Figure 5.15. Most computer spreadsheet applications will automatically generate a scatter plot and perform the linear regression.

Alpha is elusive, and represents market inefficiencies that can lead to high profit opportunities. Investors relying on profits through exposure to systematic risk can rely on broad portfolios and are called *passive investors*, or *beta investors*. Investors relying on profits through exposure to idiosyncratic risk must constantly be on the lookout for underpriced stocks (i.e., wizards) and market anomalies (i.e., hedge funds). These types of investors are called *active investors*, or *alpha investors*.

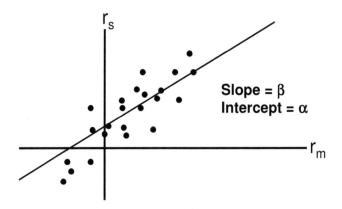

Figure 5.15. Beta can be calculated through the use of linear regression. First, create a scatter plot with daily market returns (r_m) on the x-axis and stock returns (r_s) on the y-axis. After all data points are plotted, linear regression is used to find the best fit line. The y-intercept of the line is called alpha (α), and represents the expected return due to idiosyncratic risk. The slope of the line is beta (β), and is a relative measure of stock risk compared to market risk.

For a fully diversified portfolio, it can be mathematically shown that the required return on a stock is a linear function of beta. This relationship, called the
Capital Asset Pricing Model (CAPM), was first published by William Sharpe in
the 1960s. For his work on CAPM, he was awarded the Nobel Prize in Economic Science in 1990. CAPM states that the required return of a stock, r_s, is the
following:

$$r_s = r_f + \beta(r_m - r_f) \quad ; \text{Capital Asset Pricing Model} \tag{5.22}$$

r_s	Required return on a stock
r_f	Risk-free rate
β	Stock price relative risk
r_m	Expected market return
$(r_m - r_f)$	Market risk premium

CAPM states that the required return on a stock is equal to the risk free rate
plus a risk premium. If the stock has the same risk as the overall market ($\beta = 1$),
the risk premium is equal to the market risk premium. If the stock is twice as
risky as the market ($\beta = 2$), the risk premium is twice the market risk premium.
If the stock is half as risky as the market ($\beta = 0.5$), the risk premium is half the
market risk premium.

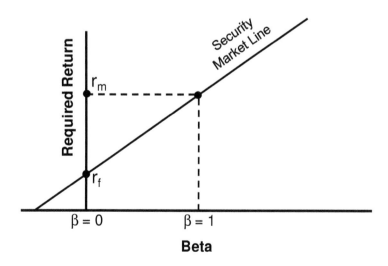

Figure 5.16. The security market line is based on the CAPM equation. One point corresponds to $\beta =$
0 and the risk-free rate. A second point corresponds to $\beta = 1$ and the expected market return. The
security market line is drawn through these two points, and represents the required expected return
for a security with any specified beta.

Plotting required return versus beta results in the *security market line*. The security market line is easily drawn. One point corresponds to a beta of zero and the risk-free rate. A second point corresponds to a beta of one and the expected market return. The security market line is drawn through these two points, and represents the required expected return for a security with any specified beta. The security market line is shown in Figure 5.16.

The security market line theoretically accounts for securities with negative beta. Since these types of securities are very effective at reducing overall portfolio risk, they require even lower returns than a risk-free security. Finding negative beta securities is highly unlikely.

Computing beta from historical stock price movements includes both industry risk and financial risk. Industry risk (also called *asset risk*) is based on factors such as product demand, prices of raw materials, availability of substitutes, and so forth. Financial risk is based on debt levels and the resulting exposure to financial distress and bankruptcy. Since it includes both types of risk, beta computed from historical stock price movements is called *levered beta* ($\beta_{levered}$). The beta for a company that has no debt is called *unlevered beta* ($\beta_{unlevered}$). Unlevered beta is useful for examining the inherent asset risk of different companies and industries that may have different levels of debt. Unlevered beta is computed by first computing levered beta and then using the following formula:

$$\beta_{unlevered} = \beta_{levered} \div \left(1 + \frac{(1-r_t) \cdot D}{E}\right) \tag{5.23}$$

r_t Tax rate
D   Value of debt
E   Value of equity

Beta in the CAPM equation refers to levered beta and, more generally, the generic use of beta almost always refers to levered beta (i.e., beta based on the covariance of stock price and overall market prices).

Historical levered and unlevered betas for a range of industries are shown in Table 5.2. Results are based on monthly stock price returns from 2002 to 2006. Notice that the unlevered betas for utilities are among the lowest for all industries. The only industries with unlevered betas consistently lower than utilities are banks and financial institutions, which are unique in that they do not provide consumable goods or services.

Unlevered beta shows that investors view the asset risk of utilities as very low. Utilities must be sensitive to this because beta drives stock price. Consider an electric utility with $1 billion in equity value and $600 million in debt value. The utility has 25 million shares of outstanding common stock, resulting in a share price of $40. Assume that the corporate tax rate is 29.3%, the risk-free rate is 4%, and the market risk premium is 8%. A levered beta of 0.88 results in a required return of 11%. This existing utility situation is shown in the before column in Table 5.3.

Table 5.2. Levered and unlevered betas for various industries. Results are based on monthly stock price returns from 2002 to 2006. Unlevered betas for utilities are among the lowest for all industries. The only industries with unlevered betas consistently lower than utilities are banks and financial institutions, which are unique in that they do not provide consumable goods or services.

Industry	# Firms	Unlevered Beta	Levered Beta	Market D/E Ratio	Tax Rate
Semiconductors	154	4.85	5.10	13.4%	37.3%
E-Commerce	56	2.02	2.08	3.4%	15.8%
Wireless Networking	74	1.96	2.20	14.1%	11.5%
Internet	266	1.94	1.97	1.8%	6.8%
Telecom Equipment	124	1.89	1.98	5.5%	12.8%
Computers & Peripherals	144	1.77	1.86	5.8%	8.4%
Integrated Steel	14	1.73	1.97	18.9%	23.9%
Merchant Power	58	1.69	1.87	12.0%	8.1%
Drugs	368	1.66	1.78	7.8%	6.0%
General Steel	26	1.59	1.71	11.0%	30.2%
Precision Instruments	103	1.53	1.66	10.4%	16.5%
Coal	18	1.52	1.71	14.1%	10.4%
Computer Software	376	1.51	1.56	3.4%	10.7%
Shoes	20	1.44	1.47	2.6%	32.2%
Biotechnology	103	1.38	1.51	9.1%	4.1%
Basic Chemicals	19	1.35	1.52	15.6%	17.0%
Medical Supplies	274	1.34	1.43	8.0%	11.5%
Educational Services	39	1.26	1.27	1.4%	20.7%
Recreation	73	1.25	1.54	28.2%	18.2%
Retail Automotive	16	1.24	1.58	42.1%	36.0%
Heavy Construction	12	1.21	1.25	3.6%	29.8%
Advertising	40	1.19	1.60	40.8%	14.9%
Electronics	179	1.17	1.32	14.2%	13.3%
Entertainment	93	1.17	1.53	37.2%	15.5%
Metal Fabricating	37	1.10	1.19	10.2%	18.3%
Retail Building Supply	9	1.10	1.23	19.4%	37.2%
Aerospace & Defense	69	1.06	1.19	16.2%	21.1%
Cable TV	23	1.06	1.56	59.8%	20.0%
Diversified Chemicals	37	1.05	1.16	13.8%	26.2%
Publishing	40	1.05	1.35	34.3%	15.6%
Auto Parts	56	1.05	1.45	46.7%	17.7%
Railroads	16	1.04	1.23	26.1%	32.2%
Precious Metals	84	1.04	1.11	7.2%	4.9%
Telecom Services	152	1.03	1.34	33.7%	12.4%
Industrial Services	196	1.02	1.22	23.8%	16.1%
Machinery	126	1.01	1.19	23.7%	22.7%
Air Transport	49	1.00	1.40	48.1%	17.4%
Metals & Mining	78	0.99	1.05	6.6%	7.7%

Table 5.2 (cont.). Levered and unlevered betas for various industries. Unlevered betas for utilities are among the lowest for all industries.

Industry	# Firms	Unlevered Beta	Levered Beta	Market D/E Ratio	Tax Rate
Pharmacy Services	19	0.99	1.07	9.8%	21.3%
Retail Store	42	0.98	1.11	17.3%	24.5%
Information Services	38	0.97	1.05	10.1%	19.9%
Petroleum (Integrated)	26	0.97	1.02	8.0%	32.2%
Reinsurance	11	0.96	1.01	6.4%	10.7%
Hotel/Gaming	75	0.96	1.25	35.5%	13.0%
Foreign Electronics	10	0.95	1.08	20.9%	34.0%
Medical Services	178	0.95	1.10	19.2%	16.2%
Chemical (Specialty)	90	0.93	1.06	17.9%	19.0%
Office Supplies	25	0.92	1.13	32.2%	27.7%
Home Furnishings	39	0.89	1.10	31.4%	24.0%
Automotive & Trucks	28	0.88	1.54	98.2%	22.9%
Newspaper	18	0.87	1.21	50.0%	22.3%
Building Materials	49	0.87	1.07	29.5%	23.8%
Petroleum (Producing)	186	0.86	1.00	19.1%	15.0%
Securities Brokerage	31	0.85	1.66	123.2%	22.9%
Restaurants	75	0.83	0.93	14.3%	20.7%
Homebuilding	36	0.83	1.64	128.8%	23.6%
Healthcare Information	38	0.82	0.91	12.7%	14.6%
Groceries	15	0.81	0.99	30.2%	28.2%
Beverages	44	0.81	0.89	11.5%	16.7%
Home Appliances	11	0.81	0.95	21.7%	16.0%
Electrical Equipment	86	0.80	1.35	80.9%	14.9%
Household Products	28	0.80	0.89	17.1%	29.5%
Packaging & Containers	35	0.79	1.12	52.5%	21.0%
Trucking	32	0.78	1.04	48.8%	31.5%
Apparel	57	0.76	0.87	19.9%	25.2%
Toiletries/Cosmetics	21	0.75	0.85	16.2%	18.3%
Natural Gas Production	31	0.75	0.93	30.0%	22.4%
Environmental	89	0.71	1.00	45.3%	11.4%
Paper Products	39	0.69	0.93	40.9%	14.3%
Food Processing	123	0.67	0.77	18.0%	19.3%
Tobacco	11	0.66	0.70	7.5%	27.6%
Electric Utilities	**69**	**0.62**	**0.88**	**59.8%**	**29.3%**
Food Wholesalers	19	0.60	0.79	48.3%	31.9%
Water Utilities	**16**	**0.58**	**0.78**	**51.0%**	**33.2%**
Oil & Gas Distribution	15	0.52	0.72	39.4%	5.2%
Natural Gas Utilities	**26**	**0.52**	**0.78**	**66.3%**	**25.9%**
Banking	504	0.48	0.63	42.2%	27.9%
Financial Services	294	0.44	1.14	193.5%	17.4%

Table 5.3. The impact of increased asset risk. In this case, a utility starts with an unlevered beta of 0.62. Investors conclude that future asset risk is 15% higher, corresponding to an unlevered beta of 0.72. Higher risk requires higher returns, lowering stock price. A lower stock price increases the debt-to-equity ratio, increasing levered beta and further reducing stock price. The cycle eventually converges; the 15% increase in asset risk results in a 33% reduction in stock price.

	Before	**After**
Debt	$600,000,000	$600,000,000
Equity	$1,000,000,000	$673,300,000
D/E	0.60	0.89
$\beta_{unlevered}$	**0.62**	**0.72**
$\beta_{levered}$	0.88	1.55
Required return	11.0%	16.4%
Shares	25 million	25 million
Stock Price	$40.00	$26.93

Consider a situation where investors feel that the utility is assuming more asset risk. This could be because equipment is getting older, systems are more heavily loaded, engineers are in short supply, regulatory relations are strained, and so forth. Assume that investors feel that this additional asset risk exposure increases unlevered beta by 15% from 0.62 to 0.72. This may seem modest, but the financial impact to the utility is dramatic.

The new levered beta can be easily calculated from the new unlevered beta. This new levered beta is used to compute a new required return. This increase in required return will result in a corresponding drop in stock price. An increase in asset risk directly results in a drop in stock price, but the process is not done yet.

The drop in stock price increases the debt-to-equity ratio. This increase in leverage further increases the levered beta and required return. The additional increase in required return further lowers stock price, continuing the cycle. Eventually the process will converge, but not until stock price is substantially reduced. Final values after convergence are shown in the after column of

Table 5.3. In this case, an increase in asset risk of 15% results in a stock price decrease of 33%.

Perceived asset risk by investors is an important driver of stock price. This is why risk management is so important to top utility executives. In addition to effectively managing risk throughout the company, it is equally important to communicate these activities in a credible manner to investors. If investors think that a utility has a good understanding of its asset risks and does a good job of managing these risks, beta will be lower and stock price will be higher.

One might ask, "If the market already sets stock prices, why should I care about CAPM calculations?" There are several common uses. First, CAPM can be used to compare alternative investments. Second, CAPM can be used to make an educated guess about the price of companies not yet traded on the market.

Third, CAPM can be used to evaluate whether presently traded stocks are priced correctly. Last and most important for this book, CAPM can be used to determine the required rate of return for utilities in a rate case.

The following CAPM application examples assume a risk-free rate of 4% and a market risk premium of 8%.

Consider Company A with a beta of 2.0 that is projected to have equity returns of 19%. Compare this to a Company B with a beta of 1.50 that is projected to have equity returns of 17%. According to CAPM, Company A should have equity returns of 4% + 2.0 x 8% = 20%, which is higher than the projected return of 19%. Company B should have equity returns of 4% + 1.5 x 8% = 16%, which is lower than the projected return of 17%. According to CAPM, Company A is overvalued and Company B is undervalued.

Now consider the acquisition of a privately held company with expected future net earnings of $100 million per year with no growth. Similar companies that are publicly traded have betas of 0.75. Therefore, a good estimate of the required expected return of the company is 4% + 0.75 x 8% = 10%. Recall that the present value of perpetuity is the annual amount divided by the discount rate. As such, the company is worth around $100 million ÷ 0.1 = $1 billion.

Last, consider a utility applying for a rate case. Historically, this utility has experienced a beta of 0.5. Its present market capitalization is $5 billion. According to CAPM, equity investors expect a return of 4% + 8% x 0.5 = 8%, which corresponds to net earnings of $400 million. If the regulators want the utility stock price to remain the same, it should set revenue requirements so that the utility can achieve this level of net earnings.

The mathematical formula of CAPM is able to compute the required return on equity investments. There are some practical shortcomings in this regard. First, CAPM assumes that stock prices are lognormally distributed. As discussed in Section 5.2, this is generally a good assumption but may not properly account for rare but significant risk factors. Second, CAPM requires a value for beta, and historical beta may not be a good indication of future beta. For these two reasons, CAPM will usually slightly underestimate the required return on equity investments.

5.6   FINANCIAL OPTIONS

The world of finance is replete with contracts and other legal documents that have monetary value. There are innumerable possibilities such as stocks, bonds, and bank loans. Legal documents like these that have monetary value are called *financial instruments.*

Financial instruments are broadly categorized as either *cash instruments* or *derivative instruments.* Cash instruments do not relate to hard currency *per se.* Rather, the monetary value of cash instruments is directly determined by the market. Cash instruments are further divided into *negotiable cash instruments* and *non-negotiable cash instruments.* Negotiable cash instruments, also called *marketable securities* and *transferable securities*, can be transferred from one owner to another owner without restrictions. Examples of negotiable cash instruments include stocks and bonds, which are freely traded in markets.

Non-negotiable cash instruments cannot be freely transferred between owner. They are also known as *non-marketable securities* and *non-transferable securities*. An example of a non-negotiable cash instrument is a government savings bond. These can only be redeemed by the owner of the bond and are not allowed to be sold to other parties. The most common non-negotiable cash instruments are bank loans and bank deposits. When a person deposits money in a bank account, rights to this deposit cannot be easily transferred to another person.

Unlike cash instruments, the monetary value of derivative instruments is derived from the underlying value of one or more cash instruments (hence the name). The most common derivative instruments are options, futures, and swaps. For example, the owner of an option on a utility stock does not have any ownership rights to the utility. Rather, the value of the option rises and falls as the value of the utility stock rises or falls. It is a pure financial mechanism. Today, new and complex derivative instruments are constantly being developed through *financial engineering*.

Derivative instruments are typically designed to address investor risk concerns. For example, a utility concerned about an increase in the price of fuel could purchase a futures contract that requires a fixed amount of fuel to be purchased at a future date at a fixed price. The futures contract is a derivative since its value is derived from the market value of the associated fuel. It addresses risk by providing future price certainty to the utility. A *forward contract* is closely related to a futures contract, but is a fixed deal between two parties and is not traded on an exchange.

A swap is an agreement between two parties to exchange a stream of cash flows. Consider a loan with a floating interest rate. The holder of the loan can purchase an *interest rate swap* that will exchange the floating interest rate payments with fixed payments. Purchasing the swap removes the risk of interest rate fluctuations. Other common types of swaps include foreign currency swaps, credit swaps, commodity swaps and equity swaps.

A financial option is the right, but not the obligation, to buy or sell an underlying security at some time in the future. A *call option* is the right to buy, and a *put option* is the right to sell. Financial options will be discussed in detail throughout the remainder of this section.

The hierarchical relationship of financial instruments is shown in Figure 5.17. As mentioned previously, the value of cash instruments is determined directly by markets. Therefore, the values of cash instruments are risky in the sense that they can go up and down in the marketplace (like stock and bond prices). Since the value of derivative instruments is based on underlying cash instruments, they can be used to reduce the risks inherent in portfolios only consisting of cash instruments.

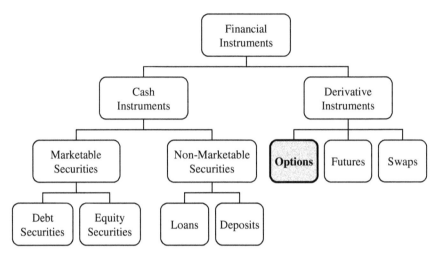

Figure 5.17. The hierarchical relationship of financial instruments. The value of cash instruments is determined directly by markets, and the value of derivative instruments is derived from underlying cash assets. Derivative instruments are typically used to reduce the risks inherent in portfolios only consisting of cash instruments.

The remainder of this chapter discusses the use of financial options for risk management. The logic, valuation methodologies, and applications for futures, swaps, and more complicated derivative instruments are similar, but beyond the scope of this book.

A call option is the right to buy a share of stock (or other cash instrument) at a specified price, called the *strike price*, at a future date. Consider a utility with stock trading at $50 per share. Now consider a call option with a strike price of $55 with an expiration of one year. Before the option expires, the holder has the right, but not the obligation, to *exercise* the option and purchase a share of stock for $55. If the utility stock is trading below the strike price, it does not make sense to exercise the call option since a share of stock can be obtained for less money through a direct purchase. If the utility stock is trading above the strike price, the call option is *in the money*; the holder can exercise the option and obtain the stock for a price below current market value. If the stock were trading at $60, the holder could obtain a share of stock for the strike price of $55, immediately sell the share back to the market at $60, and retain the difference of $5.

Investors can both buy and sell call options. A purchased call option is called a *long call*. A pure investment in a call option is a bet that the price will increase above the strike price in the future. A sold call option is called a *short call*. A pure sale of a call option is a bet that the price will not increase above the strike price in the future.

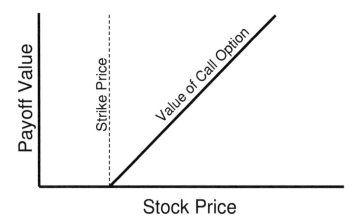

Figure 5.18. A call option is the right, but not the obligation, to buy a share of stock at a specified strike price before a future expiration date. The call option has no value if the stock price is below the strike price. If the stock price is above the strike price, the call option is in the money with a payoff value equal to stock price minus strike price.

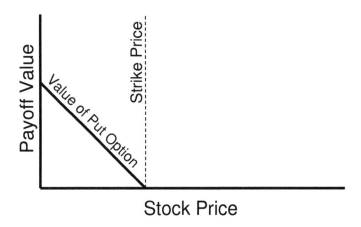

Figure 5.19. A put option is the right, but not the obligation, to sell a share of stock at a specified strike price before a future expiration date. The put option has no value if the stock price is above the strike price. If the stock price is below the strike price, the put option is in the money with a payoff value equal to strike price minus stock price.

The payoff value of a call option is demonstrated in Figure 5.18. The call option has no value if the stock price is below the strike price. If the stock price is above the strike price, the payoff value is equal to stock price minus strike price.

A put option is the right to sell a share of stock at a specified strike price before a future expiration date. Consider a utility with stock trading at $50 per share. Now, consider a put option with a strike price of $45 with an expiration of one year. Before the option expires, the holder can exercise the option and sell a share of stock for $45. If the utility stock is trading above the strike price, it does not make sense to exercise the put option since a share of stock can be sold for more money in the market. If the utility stock is trading below the strike price, the put option is in the money; the holder can exercise the option and sell the stock for a price above current market value. If the stock were trading at $40, the holder could purchase a share of stock on the market for $40, immediately exercise the option and sell the share for $45, and retain the difference of $5.

Like call options, investors can both buy and sell put options. A purchased put option is called a *long put*. A pure investment in a put option is a bet that the price will decrease below the strike price in the future. A sold call option is called a *short put*. A pure sale of a put option is a bet that the price will not decrease below the strike price in the future.

The payoff value of a put option is demonstrated in Figure 5.19. The put option has no value if the stock price is above the strike price. If the stock price is below the strike price, the payoff value is equal to strike price minus stock price.

In finance, the *going long* typically refers to strategies that bet on increasing prices and *going short* typically refers to strategies that bet on decreasing prices. Therefore, an investor could go long by purchasing more stock, by purchasing long calls, and by selling short puts. An investor could go short by *selling short*, by selling long calls, and by purchasing short puts. Selling short occurs when an investor sells stock at today's market price that is not yet owned, but must be delivered sometime in the future. The investor profits if the price declines before the delivery date, and loses if the price increases before the delivery date.

There are two common ways to treat the expiration date of an option. An *American option* allows the option to be exercised at any date up to and including the expiration date. A *European option* only allows the option to be exercised on the expiration date.

Although the use of financial options to go long or short on a stock can be considered speculative, financial options are typically used to reduce portfolio risk. Consider an investor with a share of utility stock valued at $50. This investor wishes to reduce the risk of the stock significantly declining in value. To do this, the investor buys a put option with a strike price of $45 and an expiration date in one year. This option will cost money. To offset the cost of the put option, the investor sells a call option with a strike price of $55 and an expiration date in one year. This raises cash to offset the put, but eliminates the upside benefit of the stock increasing above $55. The resulting portfolio consists of a share of stock, a long put, and a short call. By selling the call option and buying the put option, the investor has reduced portfolio risk. If the stock price remains between $45 and $55, return on the portfolio is the same as the return on the stock. However, losses and gains are both capped at $5. The portfolio characteristics of this situation are shown graphically in Figure 5.20.

How should an investor decide how much a financial option is worth? Before answering this question, it is important to consider the intimate relationship between put options and call options. Consider two portfolios. Portfolio A consists of a share of stock plus a put option with a strike price equal to the current stock price. If the stock price goes down, the payout value of the portfolio remains equal to the strike price. If the stock price goes up, the payout value of the portfolio is equal to the stock price.

Portfolio B consists of a call option with a strike price equal to the current stock price, and an amount of cash equal to the current stock value. If the stock price goes down, the call option is worthless and the payout value of the portfolio remains equal to the cash amount. If the stock price goes up, the payout value of the portfolio is equal to the cash amount plus the increase in stock price.

At the expiration date of the put option and call option, Portfolio A and Portfolio B have the exact same payout. Therefore, these portfolios must have the exact same value upon purchase. This equivalence is called *put-call parity*. Put-call parity says that a put option plus a share of stock must equal a call option plus the cash value of the stock (properly invested at a risk-free rate). This relationship is graphically shown in Figure 5.21.

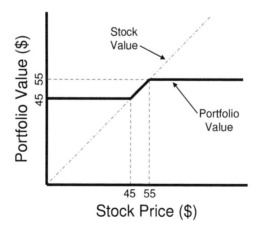

Figure 5.20. A portfolio consisting of a share of stock, a long put with a strike price of $45, and a short call with a strike price of $55. If the stock price remains between $45 and $55, return on the portfolio is the same as the return on the stock. However, the put option limits the portfolio value from dropping below $45 and the call option limits the portfolio value from increasing above $55.

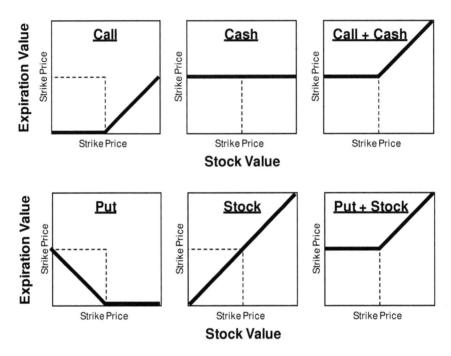

Figure 5.21. Put-call parity. The top portfolio consists of a call option plus an amount of cash equal to the value of the underlying stock. The bottom portfolio consists of a put option plus a share of the underlying stock. Upon expiration of the call option and put option, the value of both portfolios will be identical. This equivalence is called put-call parity; a put option plus a share of stock must equal a call option plus the cash value of the stock (properly invested at a risk-free rate).

Put-call parity is mathematically described by the following equation. The strike prices of the put and call option are both equal to the current stock price, and both have the same expiration date. The present value of cash, PV(Cash), is determined by investing an amount of cash equal to the current stock price at a risk-free rate for a period corresponding to the option expiration date.

$$P_{put} + P_{stock} = P_{call} + PV(Cash) \quad ; \text{put-call parity} \tag{5.24}$$

The above equation allows the value of a corresponding put or call option to be determined if the other is known. Interestingly, the equation can also be rearranged to show the required price of a stock in terms of puts and calls. This equation is:

$$P_{stock} = P_{call} + PV(Cash) - P_{put} \quad ; \text{synthetic stock} \tag{5.25}$$

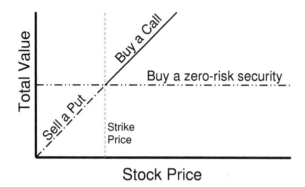

Figure 5.22. A synthetic stock. Buying a call option and selling a put option, both with strike prices equal to the underlying stock price, has total value movement equal to the underlying stock price movement. Investing cash in the amount of the underlying stock price in a zero-risk security results in a synthetic stock. The value of a synthetic stock is identical to the value of the underlying stock.

An investor should be indifferent about whether to purchase a share of stock or whether to hold the components of the right-hand side of the above equation. Because the price is mathematically identical to a stock, a long call plus a short put plus cash invested in a zero risk security is called a *synthetic stock*. A graphical representation of a synthetic stock is shown in Figure 5.22.

Imagine an investor that discovered a synthetic stock that was not identical in value to its underlying stock. This situation presents an *arbitrage* opportunity for the investor. Arbitrage is an opportunity for a guaranteed positive cash flow investment, and therefore risk-free profit.

Assume that the stock price was trading slightly higher than the value of the synthetic stock. An investor could sell a share of the stock, purchase a share of the synthetic stock, and keep the price difference. When the price difference eventually goes away, the investor can repurchase the share of stock, sell the synthetic stock, and return the portfolio to its original state with a risk-free profit included. Arbitrage opportunities tend to correct price imbalances quickly.

Since the payout value of an option is based on the price of the underlying asset, it can be probabilistically assessed using the binary tree approach discussed in Section 5.2. The underlying asset value is known at each point in the binary tree, allowing the option payout value to be computed at each point in the binary tree. This approach allows the expected payout, payout variance, and option value to be determined.

A more elegant way to value financial options was developed by applied mathematician Fischer Black and financial economist Myron Scholes. Their work resulted in the famous *Black–Scholes Option Pricing Model*, which provides a closed-form equation for the price of a call option. The Black–Scholes equation is easily computed, and is mathematically expressed as follows:

Black–Scholes Option Pricing Model

$$P_{call} = P_{stock} \cdot N(d_1) - P_{strike} \cdot e^{-r \cdot t} \cdot N(d_2)$$ (5.26)

$$d_1 = \frac{1}{\sigma\sqrt{t}}\left(\ln\left(\frac{P_{stock}}{P_{strike}}\right) + t \cdot \left(r + \frac{\sigma^2}{2}\right)\right)$$

$$d_2 = d_1 - \sigma\sqrt{t}$$

P Price
t Time until option expiration
r Risk-free interest rate
σ Volatility of underlying stock (standard deviation as a % of mean)
N Cumulative standard normal distribution

Black–Scholes states that the price of a call option is equal to the price of the underlying stock modified by a factor, minus a discounted strike price also modified by a factor. According to Black–Scholes, the value of a call option will increase with the price of the underlying stock, with the volatility of the underlying stock, with the maturity period, and with the risk-free interest rate. The value of a call option will decrease as strike price increases.

Black–Scholes requires some key assumptions for its equation to be valid. These include the following:

Key Black–Scholes Assumptions
- Stock price is lognormally distributed
- The stock pays no dividends
- European exercise terms
- Markets are efficient
- The risk-free interest rate is constant and known

Most of these assumptions are reasonable. Stock returns are generally lognormally distributed as described in Section 5.2. Dividends and exercise terms can be compensated for separately in the price. And markets are generally efficient. The big assumption is that risk-free interest rates are constant and known. In fact, future risk-free interest rates are not known and will not be constant. Because of this weak assumption, actual option prices tend to be slightly higher than the Black–Scholes model.

It is helpful to understand the Black–Scholes model through an example. Consider a call option for a stock with a current price of $50 and a price volatility of 10%. This option has a strike price of $55 and expires in one year. The risk-free interest rate is 6%. The price calculation for the call option is as follows:

Call Option Pricing – Example A

Input Parameters:
P_{strike} = $55
P_{stock} = $50
σ = 0.1
t = 1 year
r = 0.06

Intermediate Values:
d_1 = – 0.303102
d_2 = – 0.403102

Results:
P_{call} = $1.26

And so, Black–Scholes says that a call option with the above characteristics should be priced at $1.26. That is, a rational investor should be willing to pay $1.26 to have the option to purchase a share of stock at $55 in one year.

It is interesting to see how option prices change with input parameters. The best way to do this is to set up a spreadsheet and begin adjusting parameters. This process reveals that several parameters can have a significant impact. The below example prices a call option similar to Example A, but with a higher stock price volatility and a much longer expiration period:

Call Option Pricing – Example B

Input Parameters:
P_{strike} = $55
P_{stock} = $50
σ = 0.2
t = 5 year
r = 0.06

Intermediate Values:
d_1 = 0.681307
d_2 = 0.234094

Results:
P_{call} = $13.46

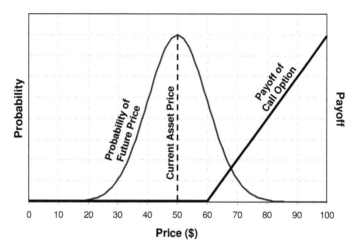

Figure 5.23. Payoff characteristics of a call option. The current price of the underlying stock is $50 and the bell curve shows the probability of the stock price at the option expiration date. If the bell curve were very taller and thinner, the probability of exceeding the strike price of $60 would be low. As the bell curve spreads out, the probability of exceeding the strike price increases, thereby increasing the value of the call option.

By increasing the expiration period from one year to five years and doubling the stock price volatility, the price of the call option increases by more than tenfold. Interestingly, an increase in stock price volatility will reduce the price of the stock, but will increase the price of options on the stock. The reason lies in the asymmetry of payoff. Consider Figure 5.23. The current price of the stock is $50 and the bell curve shows the probability of the stock price at the option expiration date. If the bell curve were very tall and thin (i.e., the stock price has low volatility), the probability of exceeding the strike price of $60 would be low. As the bell curve spreads out (i.e., the stock price has higher volatility), the probability of exceeding the strike price increases, thereby increasing the value of the option. It is true that a higher stock price volatility has a correspondingly higher chance of a large stock price drop. However, the payout of an option does not depend on how much the price of a stock drops; it only depends on whether it is below the strike price or not.

Black–Scholes allows the price of a call option to be computed directly. The price of a put option is calculated by first computing the price of the corresponding call option and then using the put-call parity relationship as follows:

$$P_{put} = P_{call} + PV(Cash) - P_{stock} \tag{5.27}$$

Computing the price of more complicated financial options is beyond the scope of this section. The general approach is to model the price movement of the underlying security (or securities) as a binary tree, determine the payout value for each outcome, and then determine the price based on the probability of each payout value.

In summary, a financial option is financial instrument whose payoff value is derived from the value of an underlying asset. It gives the holder the option, but not the obligation, to exercise the option at some point in the future. A call option allows an investor to purchase an asset at a specific strike price at a future date. A put option allows an investor to sell an asset at a specific strike price at a future date. As the saying goes, "Options are always valuable." The value of call options and put options can be directly computed through the Black–Scholes Options Pricing Model.

5.7   REAL OPTIONS

Just as there is value in having the option to make a financial transaction, there is value in having the option to make a non-financial transaction. Qualitatively, the value of having real options is well known and intuitively obvious. There is value in having multiple candidates for a job opening. There is value in having multiple suppliers of equipment from which to choose. The list of valuable business options is innumerable, but the term *real option* typically refers to a specific type of business option that is roughly analogous to a financial option and can be priced using similar techniques.

For example, a utility may be considering building a new facility in an urban area. There is a vacant parcel of land for sale that is necessary for the new facility. The utility may choose not to pursue the project for a variety of reasons, but would not be able to pursue the project if the land were sold or otherwise became unavailable. In this situation, there is value in the utility paying for the option to purchase the land through the date when the pursuit of project will be determined. The utility is paying money for the right, but not the obligation, to make a real transaction sometime in the future.

Real options are real in the sense that something tangible happens, such as ownership changing or activities being performed. Most real options are analogous to call options, where the real option allows a costly activity to occur sometime in the future.

The value of real options can be determined using the Black–Scholes Options Pricing Model. Care must be taken to make correct analogies for equation parameters. This is best understood through the classic real option example of oil drilling rights.

Imagine a parcel of land with known oil reserves. Extraction of the oil will cost $5 million. At current crude oil prices, this oil can be sold on the market for $3 million. Clearly, it does not make financial sense to extract the oil given current oil prices. However, oil prices are volatile and could rise to a level that makes drilling profitable. How much should an investor pay for drilling rights on the parcel of land for the next five years?

In this example, the underlying asset (the item of value) is the oil, which is valued based on the market price of oil. The market value of the oil therefore corresponds to stock price in financial options. The volatility of the underlying asset in this case is the historical volatility of oil prices. In this example, oil price volatility is assumed to be 20%. The strike price is the amount the investor would have to pay to exercise the option. In this case, the cost to extract the oil is $5 million. The risk-free rate is assumed to be 6% and the expiration time is five years. The real call option price calculation is as follows:

Real Option Pricing Example

Input Parameters:
P_{strike} = $5,000,000
P_{stock} = $3,000,000
σ = 0.2
t = 5 years
r = 0.06

Intermediate Values:
d_1 = – 0.247814
d_2 = – 0.695027

Results:
P_{call} = $304,401

Based on this analysis, a rational investor would be willing to pay $304,401 for the five-year drilling rights on the land.

Real option valuation has a high potential for misuse and abuse. The key to proper application is to follow several guidelines. First, money must be paid purely to give future flexibility that otherwise would not exist. Second, there must be clear underlying value that has demonstrated volatility. Last, there must be a specific cost that must be incurred to access the underlying value.

5.8   SUMMARY

Financial risk deals with uncertain future cash flows and the management of uncertain future cash flows. Rational investors prefer predictable cash flows, and will therefore demand higher rates of return for less predictable cash flows. This is true even if the predictable and unpredictable cash flows have the same expected value.

Since future asset values are uncertain, they cannot be represented by a single number. Instead, they are assigned a probability distribution function that describes the likelihood of each potential future value. In finance, asset returns are typically assumed to be normally distributed, which leads to lognormally distributed future asset values.

Asset price risk can broadly be categorized into idiosyncratic risk and systematic risk. Idiosyncratic risk is specific to an asset and can be diversified away by investing in many assets. Some assets will perform well, and some will not. But a collection of assets will tend to cancel out the positive upside and negative downside associated with idiosyncratic risk. A collection of assets that has nearly eliminated all idiosyncratic risk is said to be fully diversified. A fully diversified portfolio is still subject to systematic risk.

Portfolio theory, which assumes full diversification, describes the best combinations of risky assets so that expected returns are maximized for a given level of risk. The optimal combinations of risk versus return make up the efficient frontier; if a portfolio is not on the efficient frontier, either a higher return can be achieved for the same risk or the same return can be achieved for a lower risk. The capital market line is tangent to the efficient frontier and runs through a point corresponding to a risk-free asset. This tangent point is representative of the overall market, and implies that risk-versus-return is optimized by investing in a combination of the overall market plus risk-free securities.

The capital asset pricing model describes how to compute the required expected returns that investors have for an asset based on the riskiness of returns for that asset. Required expected return is equal to the risk-free rate plus the market risk-premium multiplied by the beta of the asset, with beta being a measure of how much the asset price moves in relation to the overall market. The unlevered beta of utilities has been historically one of the lowest of all industries, implying that investor asset return expectations for utilities are lower than for most other industries. Levered betas for utilities are higher, since utilities typically use a high amount of debt, increasing the financial risk associated with financial distress and possible bankruptcy.

A financial option is the right, but not the obligation, to buy or sell an underlying security at some time in the future. A call option is the right to buy, and a put option is the right to sell. The prices of call options and put options with the same strike price and expiration date are related through the put-call parity relationship. This relationship also shows how a synthetic stock can be built with a long call, a short put, and cash. The price of a call option can be calculated with the Black–Scholes Option Pricing Model. Non-financial transactions that are analogous to financial options are called real options, and can be priced using Black–Scholes and related techniques.

An in-depth understanding of financial risk requires advanced mathematics (but not too advanced). Fortunately, most utility engineers have a strong background in mathematics and will be reasonably comfortable with probability, statistics, and complicated equations with a lot of Greek symbols. Financial risk is truly one of the most important topics in business, and a good understanding of this chapter will help utility engineers view assets and portfolios in terms of risk versus return.

5.9 STUDY QUESTIONS

1. What is the difference between mean, median, and mode? How are these values related to the normal and lognormal probability density functions?
2. If stock price returns are assumed to be normally distributed, what will the distribution of the underlying stock price be?
3. What is the difference between systematic and idiosyncratic risk? About how many assets with similar idiosyncratic risk characteristics must be purchased to effectively eliminate idiosyncratic risk?
4. What is the efficient frontier? Are there situations where a rational investor would prefer a portfolio of assets that are not on the efficient frontier? Explain.
5. What is the capital market line? What is special about the portfolio of assets related to the tangent point of the capital market line and the efficient frontier?
6. Explain what is meant by the beta of a stock? What is the difference between levered beta and unlevered beta? What is interesting about the unlevered beta of most utilities?
7. What is the purpose of the capital asset pricing model? How does the capital asset pricing model relate to the security market line?
8. What is the definition of the market risk premium? How are the market risk premium and beta related in terms of the capital asset pricing model?
9. What is a financial instrument? What are the two main categories of financial instruments?
10. In terms of the Black–Scholes option pricing model, what are some factors that will increase the value of a call option?

6
Financial Ratios

Since companies come in many sizes, it is often difficult to compare financial performance using raw accounting numbers. For example, it is difficult to know whether a company with $100 million in earnings is doing better than a company with $500 million in earnings without knowing the size each company. It is equally difficult to compare the stock price of different companies without knowing the total number of issued shares. To help with the interpretation of accounting numbers, financial analysts often use financial ratios.

Assume in the above paragraph that the company with $100 million in earnings has revenues of $500 million and the company with $500 million in earnings has revenues of $5 billion. As a percent of revenues, the first company has earnings of 20% and the second company has earnings of 10%. Even though the absolute earnings of the first company are lower, it has a much higher profit margin. Ratios have the ability to normalize measures so that they can be more easily interpreted and more fairly compared.

There are a large number of financial ratios. Some are commonly used and others are rarely used. This chapter does not attempt to be comprehensive and address every obscure ratio that will occasionally be encountered. It is more important to focus on important ratios. Even with this approach, the number of important ratios is large and ratios are therefore categorized by function.

When going through the definitions of financial ratios, it is important to remember that usage in the industry is not always consistent. For example, some sources define the ratio "return on sales" as operating income divided by sales. Others define it as pretax income (IBT) divided by sales. Still others define it as net income divided by sales. To avoid confusion, the remainder of this chapter only provides the most frequently used definition for each financial ratio. When actually examining and using financial ratios, the reader is encouraged to always verify the precise definitions that are being used.

To help give the reader a better feel for financial ratios, beyond definitions, typical ratio values are provided. Most are based on 2007 financial statements for all US publicly traded electric utilities and combined electric and gas utilities. The exceptions are market ratios, which are based on traded market values as of January 2009. These data sets are used since they are large and publicly available.

6.1 PROFITABILITY RATIOS

Profitability ratios reflect how much money a company is making compared to some measure of company size. There are many different profitability ratios using different measures of profit for the numerator and different size measures for the denominator. Some of the more common profitability ratios are now presented.

$$\text{Gross Profit Margin} = \frac{\text{Operating Revenue - Cost of Goods Sold}}{\text{Operating Revenue}} \qquad (6.1)$$

Gross profit margin, often referred to as *gross margin*, is the average amount of profit made per sale only considering the direct cost of producing the goods and/or services. In the above equation, operating revenue is equivalent to sales. Sales (operating revenue) minus the cost of goods sold is called *gross profit*. If gross margin is negative, money is lost on each sale (on average). If gross margin is positive, the company still might not be profitable, but the positive gross margin can be used to pay for non-operational costs such as interest payments.

In 2007, the average gross profit margin for publicly traded US energy utilities was 55%, which was identical to the 2006 values. This represents the difference between energy sales and the direct cost of energy, including fuel for electricity generation, purchased gas, and purchased wholesale electricity. Gross profit margin is high, indicating a high amount of incremental profit for additional energy sales. Of course, additional energy sales will eventually lead to the need for additional infrastructure and associated costs.

$$\text{Operating Profit Margin} = \frac{\text{Operating Income}}{\text{Operating Revenue}} \qquad (6.2)$$

Operating profit margin, often referred to as *operating margin*, is the average amount of profit made per sale considering all operating expenses. If operating margin is positive, the company is fundamentally profitable not considering non-operational expenses such as interest payments, taxes, and other possible non-operational items.

In 2007, the average operating profit margin for publicly traded US energy utilities was 16%, which was identical to the 2006 value. This value is much lower than the 55% gross margin value. The difference represents added operational costs not associated with fuel and wholesale energy purchases. Examples include infrastructure depreciation expenses, field operation expenses, maintenance expenses, and overhead expenses. Roughly, direct energy expenses represent about 45% of operating revenue and indirect operating expenses represent an additional 39% of operating revenue.

$$\text{Pretax Profit Margin} = \frac{\text{Income Before Taxes (IBT)}}{\text{Total Revenue}} \qquad (6.3)$$

Pretax profit margin, often referred to as *pretax margin*, is the average amount of profit made per sale considering all revenue sources (including non-operational revenue) and all expenses except income taxes. If pretax margin is positive, the company is making money, and will owe taxes on these pretax profits.

In 2007, the average pretax profit margin for publicly traded US energy utilities was 12.6%, which was slightly higher than the 2006 value of 11.5%. This is lower than the operating profit margin of 16%, primarily due to interest expenses on long-term debt, although other non-operating revenue and expenses are also considered.

$$\text{Net Profit Margin} = \frac{\text{Net Income}}{\text{Total Revenue}} \qquad (6.4)$$

Net profit margin, often referred to as *net margin*, is the average amount of profit made per sale considering all revenue sources and all expenses. Net margin represents the amount of profit that is left for common shareholders, and can be either kept as retained earnings or distributed as dividends. Net profit margin is best used to compare companies within the same industry. Some healthy companies can have very low net margins but very high revenues (e.g., Wal-Mart). Other healthy companies can have very high net margins but relatively low revenues (e.g., Rolex). Even utilities can have varying net profit margins depending upon the mix of revenue sources (e.g., a pure electric distribution company compared to an integrated utility with generation, transmission, and distribution).

In 2007, the average net profit margin for publicly traded US energy utilities was 8.3%, which was slightly higher than the 2006 value of 8.2%. The difference between this value and pretax margin is income taxes. In 2007 these utilities paid an average of 4.2% of operating revenue in income taxes, which is somewhat higher than the 2006 amount of 3.3%. More importantly, utilities earned over 8% of operating revenue in net earnings, which increases owner's equity and is available for distribution to common shareholders as dividends.

$$\text{Return on Sales (ROS)} = \text{Net Profit Margin} = \frac{\text{Net Income}}{\text{Operating Revenue}} \qquad (6.5)$$

Return on sales (*ROS*) is a measure of how profitable a company is as a percentage of net sales (i.e., operating revenue). As mentioned previously, the industry is not consistent in its specific usage of ROS. The above definition is used since it is consistent with the definition used in the DuPont analysis, which is discussed in Section 6.6. Perhaps the most common usage of ROS uses operating income in the numerator, which is equivalent to operating margin. The terms operating income margin and net profit margin should be used instead of ROS to avoid confusion.

$$\text{Return on Assets (ROA)} = \frac{\text{Net Income}}{\text{Assets}} \qquad (6.6)$$

Return on assets (*ROA*) is a measure of how effectively a company is utilizing its assets to generate profits for its shareholders. Since assets are a good measure of the overall amount that has been invested in a utility, ROA provides a good measure of profitability normalized by size. The problem with ROA is that it normalizes profits available for shareholders by investments made by both shareholders and lenders (recall that assets are equal to liabilities plus owner's equity). Therefore, a highly leveraged company will have a relatively low ROA (since it has high interest payments), even though shareholder returns will be higher precisely due to this leverage. Because of this problem, ROA should be used with caution when comparing companies with differing capital structures.

In 2007, the average ROA for publicly traded US energy utilities was 3.4%, which was slightly higher than the 2006 value of 3.2%. This may seem low, but reflects the dual issues of utility assets and utility debt. Utilities are very asset intensive, which increases the denominator in ROA. Utilities are also highly levered, resulting in high interest payments and a lowering of the numerator in ROA.

$$\text{Return on Capital Employed (ROCE)} = \frac{\text{EBIT}}{\text{Assets - Current Liabilities}} \qquad (6.7)$$

Return on capital employed (*ROCE*, pronounced "Rocky") is a measure of how efficiently a company is utilizing net invested capital to generate profits for all stakeholders (i.e., interest, taxes, and dividends). Of the ratios examined so far, ROCE is probably the best for comparing the profitability of different companies. It is somewhat insensitive to capital structure and tax situation, since it includes both interest payments and taxes in the numerator. It is also a good measure of capital efficiency, since the denominator subtracts current liabilities (which reduce capital requirements) from assets.

In 2007, the average ROCE for publicly traded US energy utilities was 8.6%, which was slightly higher than the 2006 value of 8.1%. This means that the capital employed by utilities (assets minus current liabilities) returned 8.6% for distribution as interest payments, income taxes, retained earnings, and dividends. Like ROA, ROCE for utilities may seem low due to the high value of utility assets.

$$\text{Return on Equity (ROE)} = \frac{\text{Net Income}}{\text{Owner's Equity}} \qquad (6.8)$$

Return on equity (*ROE*) is a measure of how efficiently a company is utilizing equity investments to generate profits for shareholders. It is the best measure of "bottom line" profitability for the book value of shareholder's equity, which represents retained earnings plus the original paid-in capital from the issuance of common stock. It is important to remember that the market value of common stock (i.e., market capitalization) will be different from the book value of common stock.

In 2007, the average ROE for publicly traded US energy utilities was 12.8%, which was slightly higher than the 2006 value of 12.1%. This means that for each dollar originally invested by shareholders (including retained earnings), 12.8% was generated for additional retained earnings and/or dividend distribution.

$$\text{Payout Ratio} = \frac{\text{Dividend Payments}}{\text{Net Income}} \qquad (6.9)$$

Payout ratio is a measure of how much net income is distributed as dividends (the rest being kept as retained earnings). Companies that do not distribute any dividends will have a payout ratio of zero. Companies with a negative net income that still distribute dividends will have, confusingly, a negative payout ratio. Investors expect mature companies to have stable net incomes and stable payout ratios. Since the stock price of a company is based on expected future dividends, payout ratio is an important ratio that investors consider when valuing companies.

In 2007, the average payout ratio for publicly traded US energy utilities was 49.4%, which was slightly less than the 2006 value of 53.3%. This is a high payout ratio compared to most other industries, reflecting a mature industry, predictable earnings, and an ability to raise growth capital when needed through new debt and equity offerings (i.e., utilities typically do not have to use retained earnings to fund growth opportunities).

6.2   ACTIVITY RATIOS

Activity ratios provide information about the rate of various financial activities within a company. They do not directly address issues related to business health, but help to explain how a company is achieving its profits and managing its working capital through its daily activities.

$$\text{Asset Turns} = \frac{\text{Operating Revenue}}{\text{Assets}} \tag{6.10}$$

Asset turns is a measure of how much a company sells compared to its assets. Assuming typical profitability, a company can be classified based on asset turns versus asset intensity. It is difficult for asset-intensive companies, like most utilities, to have high asset turns due to the large denominator in the asset turns equation. The only way to do this is to have very high operating revenue. Most utilities have high amounts of assets and high depreciation expenses, resulting in low asset turns. Companies with low amounts of assets will have low depreciation expenses and can achieve high asset turns with moderate levels of operating revenue (assuming typical profitability). The relationship of asset turns to asset intensity is shown in Figure 6.1.

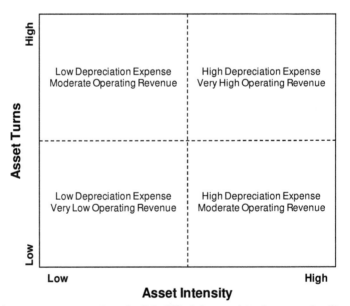

Figure 6.1. Asset turns versus asset intensity. It is difficult for asset-intensive companies, like most utilities, to have high asset turns. Most utilities have high amounts of assets and high depreciation expenses, resulting in low asset turns. Companies with low amounts of assets will have low depreciation expenses and can achieve high asset turns with moderate levels of operating revenue.

In 2007, the average asset turns for publicly traded US energy utilities was 41%, which was slightly higher than the 2006 value of 39%. This means that operating revenue is less than half of asset book value. Even though utilities are very asset intensive, depreciation and amortization expenses amounted to only about 8% of operating revenue.

$$\text{Days Receivable} = \frac{\text{Accounts Receivable}}{\text{Operating Revenue}} \times 365 \qquad (6.11)$$

Days receivable is a measure of how long it takes a company to collect payment from customers. Fewer days receivable is preferable since it results in less working capital and correspondingly less financial capital. Days receivable is sometimes referred to as *collection period*.

In 2007, the average days receivable for publicly traded US energy utilities was 42, which was slightly higher than the 2006 value of 41. Although this number may seem high, it is important to consider the atypical billing cycle of utilities. Most utilities bill customers monthly. The bill is for utility services provided from approximately 30 days to 1 day prior to the date of the bill. On average, this means that utility services are not billed until fifteen days after they are provided, adding fifteen to the days receivable ratio.

$$\text{Days Payable} = \frac{\text{Accounts Payable}}{\text{Operating Revenue}} \times 365 \qquad (6.12)$$

Days payable is a measure of how long it takes a company to pay its bills owed to suppliers. It specifically measures the number of days of revenue that amount to unpaid bills. More days payable is preferable since it results in less working capital and correspondingly less financial capital. Days payable is similar to the ratio *payable-to-sales*, which normalizes payables to operating revenue rather than days:

$$\text{Payable to Sales} = \frac{\text{Accounts Payable}}{\text{Operating Revenue}} \qquad (6.13)$$

In 2007, the average payable-to-sales ratio for publicly traded US energy utilities was 15%, which was slightly less than the 2006 value of 16%. This corresponds to days payable of 55 and 58 days, respectively. Even with a high value for days receivable, utilities managed to pay their bills to suppliers more slowly than they receive payment from customers.

$$\text{Payment Period} = \frac{\text{Accounts Payable}}{\text{Operating Expense}} \times 365 \qquad (6.14)$$

Payment period is similar to days payable, but normalizes accounts payable to operating expense rather than operating revenue. It specifically measures the weighted average number of days it takes a utility to pay its bills. Like days payable, a higher payment period is preferable since it results in less working capital and correspondingly less financial capital. For a profitable company, operating expense will be less than operating revenue, making payment period longer than days receivable.

In 2007, the average payment period for publicly traded US energy utilities was 67 days, which was less than the 2006 value of 71 days. These numbers are longer than the respective values of 55 and 58 days for days payable, which is expected, since utilities were profitable in these years.

$$\text{Days Inventory} = \frac{\text{Inventory}}{\text{Operating Expense}} \times 365 \tag{6.15}$$

Days inventory is a measure of how many days of inventory supply are in stock. In operations management, inventory is often said to be the root of all evil. From this perspective, fewer days inventory is better than more. However, there must sufficient inventory to ensure operational efficiency and the ability of utilities to respond to problems in a timely manner.

In general, days inventory is problematic as a financial ratio since the inventory account can vary widely due to accounting practices. For example, inventory accounts will be different for utilities using a first-in-first-out approach when compared to utilities using a last-in-first-out approach. For utilities, days inventory is further complicated since many assets traditionally considered inventory are classified as construction work in progress, making days inventory comparatively short.

In 2007, the average days inventory for publicly traded US energy utilities was 25, which was slightly less than the 2006 value of 27. This means that utilities have about a month's worth of supplies on hand to support daily operations. Things appear quite different if construction work in progress is included as part of inventory. Including construction work in progress, the average days inventory was 72 in 2007 and 65 in 2006.

$$\text{Inventory Turns} = \frac{\text{Operating Expense}}{\text{Inventory}} \tag{6.16}$$

Inventory turns is a measure of how many times per year inventory is used up and restocked, which is essentially the reciprocal of days inventory. As a financial ratio, inventory turns has all of the same complications as days inventory, including the impact of construction work in progress.

In 2007, the average inventory turns for publicly traded US energy utilities was 15, which was slightly higher than the 2006 value of 14. This means that utilities are using up their stock of inventory about fifteen times per year. Including construction work in progress, average inventory turns was 5.1 in 2007 and 5.6 in 2006.

6.3   LEVERAGE RATIOS

Leverage ratios provide information about the capital structure of companies. Specifically, leverage ratios indicate the extent to which companies use debt to fund business operations. When examining leverage ratios, it is important to remember that debt values and equity values are based on book value, not market value.

$$\text{Debt-to-Equity} = \frac{\text{Long Term Debt}}{\text{Owner' s Equity}} \qquad (6.17)$$

Debt-to-equity means precisely what its name implies. It is the amount of long term debt of a company divided by the owner's equity of a company. A higher debt-to-equity ratio corresponds to higher financial leverage, higher financial risk, higher levered beta, and higher required returns for investors.

In 2007, the average debt-to-equity ratio for publicly traded US energy utilities was 115%, which was slightly less than the 2006 value of 118%. This means that these utilities used slightly more long term debt in their capital structure than owner's equity, which is high compared to most other industries.

$$\text{Debt Ratio} = \frac{\text{Liabilities}}{\text{Assets}} \qquad (6.18)$$

Debt ratio is a measure of total liabilities to total assets. The components of debt ratio correspond to the basic accounting equation of assets being equal to liabilities plus owner's equity. Debt ratio is more general than debt-to-equity since it considers all liabilities, not just long-term debt.

In 2007, the average debt ratio for publicly traded US energy utilities was 73%, which was about the same as the 2006 value. Like debt-to-equity values, these are high compared to most other industries.

$$\text{Financial Leverage} = \frac{\text{Assets}}{\text{Owner' s Equity}} \qquad (6.19)$$

Financial leverage provides the same information as debt ratio, but does it by comparing total assets to owner's equity. Like debt ratio, the components of financial leverage correspond to the basic accounting equation. Without any financial leverage (i.e., financial leverage = 1), a company will have no liabilities and assets must be equal to owner's equity. With financial leverage, a company is able to have more assets than owner's equity, resulting in the potential for more revenue and higher shareholder returns (and correspondingly higher financial risk).

In 2007, the average financial leverage ratio for publicly traded US energy utilities was 376%, which was slightly less than the 2006 value of 383%. This means that for every dollar invested by shareholders, a typical utility has almost four dollars in assets. Like debt-to-equity values and debt ratio values, these amounts are high compared to most other industries.

6.4 LIQUIDITY RATIOS

Liquidity ratios reflect the ability of a company to meet its upcoming debt obligations like interest payments and unpaid invoices. Conversely, liquidity ratios indicate the likelihood that a company will not be able to meet these upcoming debt obligations. Poor liquidity ratios are a signal to investors that the risks of financial distress and bankruptcy are high. This higher risk corresponds to a higher required return and a corresponding lower stock price.

$$\text{Current Ratio} = \frac{\text{Current Assets}}{\text{Current Liabilities}} \qquad (6.20)$$

The *current ratio* is a measure of the value of current assets a company has compared to current liabilities. The usefulness of this ratio requires the assumption that all current assets can be quickly and easily converted to cash (i.e., they are highly liquid). With this assumption, a current ratio greater than one shows that a company, if required, could liquidate its current assets and have enough cash to pay off all of its current liabilities. A current ratio less than one is problematic in that short term debt obligations are more than a company's liquid assets.

In 2007, the average current ratio for publicly traded US energy utilities was 88%, which was slightly less than the 2006 value of 90%. These values are typical, and show that utilities (unlike most other industries) are comfortable having current asset values less than current liability values. This is due to strong monthly free cash flow, which can be used to pay for current liabilities.

$$\text{Quick Ratio} = \frac{\text{Cash} + \text{Cash Equivalents}}{\text{Current Liabilities}} \qquad (6.21)$$

The *quick ratio* is a similar to the current ratio, but more conservative, since it only looks at the amount of cash and cash equivalents that can be used to pay off current liabilities. A quick ratio greater than one shows that a company, could, at any time, pay off all of its current liability quickly using only cash and cash equivalents. A quick ratio greater than one indicates a very low possibility of a company becoming insolvent in the short run. On the other hand, a higher quick ratio implies higher working capital and the resulting lower return for equity investors. The quick ratio also called the *acid test ratio*.

In 2007, the average quick ratio for publicly traded US energy utilities was 9.1%, which was slightly less than the 2006 value of 9.4%. This value is far lower than most other industries, many of whom try to keep the value of cash and cash equivalents higher than the value of current liabilities. Like the current ratio, these low quick ratio values reflect the strong cash flow characteristics of utilities, limiting the need for liquid assets to back liabilities.

Many analysts question the value of the current ratio and quick ratio since, for a healthy company, bills and interest expenses are paid through free cash flow and not by cannibalizing assets on the balance sheet. Therefore, many prefer to use liquidity ratios based on earnings rather than assets.

$$\text{Times Interest Earned} = \frac{\text{EBIT}}{\text{Interest Expense}} \qquad (6.22)$$

Times interest earned is a measure of how much income is available to pay interest expenses. As such, times interest earned is an indication of how likely it is that a company will not be able to meet its interest expense obligations. From an accrual accounting perspective, EBIT is available to cover interest payments. That is why times interest earned typically uses EBIT in the numerator. From a cash flow perspective, a company (in the short term) has approximately EBITDA available to pay interest expenses, since depreciation and amortization are non-cash expenses. Therefore, sometimes EBITDA is used instead of EBIT in the definition of times interest earned.

In 2007, the average times interest earned for publicly traded US energy utilities was 341%, which was slightly higher than the 2006 value of 321%. Clearly, utilities are earning enough profit to cover their interest even though their current ratios and quick ratios are low. Using EBITDA instead of EBIT, times interest earned increases further to 499% in 2007 and 477% in 2006.

6.5 MARKET RATIOS

A common criticism of the ratios discussed so far is that they are based on accounting values and not market values. Analysts and investors in agreement with this criticism tend to use market ratios, which are ratios based wholly or partly on values determined by the market. This section describes the most commonly used market ratios.

$$\text{Price-to-Earnings (P/E)} = \frac{\text{Stock Price} \times \text{Shares}}{\text{Net Income}} \qquad (6.23)$$

Price-to-earnings (*P/E*, pronounced "P to E", or "Pee Eee Ratio") is a measure of how the share price of a stock compared to per-share earnings. Equivalently, it compares market capitalization (share price times shares outstanding) to net income. Price-to-earnings is one of the most important financial indicators of a company, especially when compared to industry peers. A high P/E indicates a high stock price, most likely due to strong financial performance and high earnings growth expectations. A low P/E indicates a low stock price, most likely due to weak financial performance and low earnings growth expectations. Low P/E ratios may indicate a bargain, since very low P/E stocks have historically outperformed very high P/E stocks (one of the few documented violations of market efficiency – see Section 4.3.4 of Chapter 4).

In January 2009, the average price-to-earnings ratio of energy utilities was 14.2, with pure electric utilities having slightly higher values than pure gas utilities. A typical P/E range for electric and combined utilities is between ten and fifteen. The P/E range for gas utilities is similar, but with a higher variance (i.e., it is more common to see gas utilities with very low P/E, below five, and a very high P/E, above twenty).

$$\text{Price-to-Book (P/B)} = \frac{\text{Stock Price} \times \text{Shares}}{\text{Owner's Equity}} \qquad (6.24)$$

Price-to-book (*P/B*) is a measure of how market capitalization compares to owner's equity as shown on the balance sheet. Market capitalization is equal to the trading price of common stock multiplied by shares outstanding. Owner's equity is equal to paid-in capital plus retained earnings. A price-to-book ratio greater than one shows that the market values the company more than what investors have provided in equity.

The book value of a company is equal to the book value of assets minus the book value of liabilities. From this perspective, book value is a rough approximation of the value of a company if it were liquidated (i.e., all assets were sold and all liability were paid off). A price-to-book ratio less than one indicates that the liquidation value is potentially higher than market value. If true, a corporate raider could buy the company at market value, sell off the assets, use the proceeds to pay off debts, and have money left over to keep. Since utilities have a large amount of infrastructure assets that are difficult or impossible to liquidate, they are almost never the target of corporate raiders.

In January 2009, the average price-to-book ratio of energy utilities was 2.1, with gas utilities having higher values than electric and diversified utilities. The price-to-book ratio of energy utilities is low compared to most other industries. This is due to the high amount of money, and corresponding equity investment, required to build utility infrastructure. Industries requiring minimal fixed assets will tend to have much higher price-to-book ratios.

$$\text{Dividend Yield} = \frac{\text{Dividends}}{\text{Stock Price} \times \text{Shares}} \times 100\% \tag{6.25}$$

Dividend yield is a measure of dividends per share compared to stock price. For example, if a company's stock is trading at $50 per share and it distributes $5 per share in dividends, the dividend yield is 10%. Dividend yield is an important measure for mature companies with stable dividend payments. This is certainly true for utilities, where many utility stock owners count on dividend payments to supplement other sources of personal income.

In January 2009, the dividend yield of energy utilities was 4.4%, which is higher than most other industries. This means that a 2008 investment in utilities resulted in dividend payments equaling 4.4% of the initial investment. Over time, dividend payments are likely to rise in absolute terms. However, as the stock price rises, the dividend yield may not change significantly.

$$\text{Earnings per Share (EPS)} = \frac{\text{Net Income}}{\text{Number of Common Shares}} \tag{6.26}$$

Earnings per share (*EPS*) is, as its name implies, the amount of net earnings divided by the number of outstanding common shares. Earnings per share is difficult to compare across companies since the number of common shares outstanding may not be proportional to company size. Regardless, it is common for earnings targets and earnings reports to be reported on a per-share basis. It is also common for earnings results to be stated relative to the target, such as "the utility beat its quarterly earnings target by two cents a share."

Since earnings per share is not directly comparable between companies, typical earnings per share values are now presented based on the 2008 results for ten large energy utilities. These include American Electric Power, Consolidated Edison, Edison International (Southern California Edison), Entergy, Exelon (Commonwealth Edison and PECO Energy), FirstEnergy, FPL Group (Florida Power & Light), Pacific Gas & Electric, Progress Energy, and Sempra (San Diego Gas & Electric). The average earnings per share of these companies was $4.0, with a low of $2.8 and a high of $6.3. Companies with relatively high earnings per share are not necessarily better financial performers since the number of outstanding shares can vary widely. For example, earnings per share for a utility will be cut in half after a stock split, even though financial performance remains unchanged.

Table 6.1. Market ratios for energy utilities and selected other industries. There are other industries with a few market ratios that are similar to utilities. However, there is no industry that has all market values similar to the utility values, demonstrating the unique investor characteristics of utilities.

Industry	Market Cap ($B)	Price-to-Earning	Price-to-Book	ROE	Dividend Yield
Diversified Utilities	135	14.9	2.0	12.4%	4.6%
Electric Utilities	275	14.2	1.9	11.3%	4.5%
Gas Utilities	87	12.9	3.1	17.9%	4.0%
Utility Weighted Avg.	497	14.2	2.1	12.8%	4.4%
Aerospace/Defense	54	11.7	19.5	26.3%	3.1%
Appliances	8	11.4	11.8	15.0%	2.7%
Biotechnology	308	4.7	8.1	5.0%	0.0%
Business Software	79	25.3	16.8	13.4%	2.0%
Clothing	22	24.7	4.6	8.6%	2.4%
Credit Services	75	73.0	2.1	2.4%	3.6%
Department Stores	35	16.7	2.5	9.5%	3.1%
Drug Manufacture (major)	965	14.4	8.6	20.9%	4.1%
Grocery Stores	48	13.8	4.3	12.5%	2.2%
Heavy Construction	21	14.1	3.4	19.4%	0.6%
Long Distance Carriers	19	9.7	7.0	67.3%	3.4%
Oil & Gas Production/Refining	1,337	6.8	2.0	27.4%	4.0%
Property Management	18	21.9	2.2	9.5%	5.2%
Railroads	241	15.1	2.5	16.1%	1.0%
Restaurants	108	16.4	8.4	32.0%	2.9%
Semiconductors (ICs)	82	29.1	2.7	10.4%	4.5%
Staffing & Outsourcing	21	19.9	5.1	13.8%	3.6%
Steel & Iron	163	5.7	2.9	31.5%	4.2%

A summary of market ratios is show in Table 6.1. Market ratios for energy utilities and selected other industries. There are other industries with a few market ratios that are similar to utilities. However, there is no industry that has all market values similar to the utility values, demonstrating the unique investor characteristics of utilities.. In addition to utility market ratios, data is provided for a variety of other industries for comparison. In terms of individual market ratios, there are other industries with values similar to utilities. However, there is no industry that has all market values similar to the utility values, demonstrating the unique investor characteristics of utilities. For example, the oil and gas industry (production and refining) has a price-to-book ratio and a dividend yield similar to utilities. However, it has a much higher return on equity value and a much lower price-to-earnings ratio. In contrast, grocery stores have similar return on equity value and price-to-earnings ratio, but have a higher price-to-book ratio and a lower dividend yield.

6.6   DUPONT ANALYSIS

It is difficult to gain useful information from looking at a single financial ratio. It is better to look at a number of key ratios together. A *DuPont analysis* goes a step further by looking at the relationship of financial ratios as they contribute to overall profitability as measured by return on equity.

As the reader may have guessed, the DuPont model of financial analysis originated within the chemical company with the same name. The model was developed within its treasury division in the early 1900s to assess and improve the finances of General Motors, of which DuPont had a large ownership stake. As the finances of General Motors improved, the DuPont model gained in popularity and became used at virtually all major US corporations.

In its most simple form, the DuPont model states that return on equity is equal to the product of three primary factors. The first factor is return on sales, which is equal to net income divided by sales. The second factor is asset turns, which is equal to sales divided by assets. The last factor is financial leverage, which is equal to assets divided by owner's equity. The simple version of the DuPont model corresponds to the following mathematical equation (sales are equivalent to operating revenue):

Simple DuPont Model

$$\frac{\text{Net Income}}{\text{Owner's Equity}} = \frac{\text{Net Income}}{\text{Sales}} \times \frac{\text{Sales}}{\text{Assets}} \times \frac{\text{Assets}}{\text{Owner's Equity}} \qquad (6.27)$$

The DuPont model shows that profit for equity investors is determined by three factors. The first, return on sales, is how much profit is made on each sale. The second, asset turns, is how many sales are made. The third, financial leverage is how much debt is used to fund asset purchases. From this perspective, profits can be increased by making more money on each sale, by selling more, and by borrowing more.

Statistics for energy utilities (based on 2007 financial statements) as they relate to the DuPont model are the following:

$$\text{ROE} = \text{Return on Sales} \times \text{Asset Turns} \times \text{Financial Leverage} \qquad (6.28)$$

$$12.8\% = 8.36\% \times 40.9\% \times 375\%$$

Return on equity for energy utilities is 12.8%. This is based on a net return on sales of 8.36%; for every $100 in sales, a utility makes a net profit of $8.36. Return on sales is then multiplied by sales volume. The volume of sales for a utility is equal to 40.9% of total asset value (recall that total asset value for a typical utility is very high); for every $100 million in assets, a utility sells $40.9

million in services. Last, profit is multiplied by financial leverage; for each dollar of owner's equity, a utility owns $3.75 of asset value.

The simple DuPont model can be expanded to include five terms instead of three terms. This is done by converting return on sales to include factors related to tax burden, interest burden, and EBIT margin. The expanded version of the DuPont model corresponds to the following mathematical equation (typical energy utility values are shown below their corresponding component):

Expanded DuPont Model

$$\text{ROE} = \frac{\text{Net Income}}{\text{NOPAT}} \times \frac{\text{NOPAT}}{\text{EBIT}} \times \frac{\text{EBIT}}{\text{Sales}} \times \frac{\text{Sales}}{\text{Assets}} \times \frac{\text{Assets}}{\text{Equity}} \quad (6.29)$$

$$12.8\% = 72.8\% \times 64.6\% \times 17.8\% \times 40.9\% \times 375\%$$

This equation presents similar information to the simple DuPont model, but shows the factors that determine return on sales. In this case, EBIT margin is equal to 17.8%; before interest and taxes, utilities are making $17.80 for each $100 in sales. Not all of this profit is available to shareholders. EBIT is reduced by 35.4% (to 64.6%) to pay for interest expenses. This value is further reduced by 27.2% (to 72.8%) to pay for income taxes.

Based on the expanded DuPont model, shareholder profit can be improved in the following five ways.

Reduce Tax Burden. Paying less in taxes is always desirable from a corporate perspective (if done legally). Without adversely affecting other profitability factors (like reducing EBIT margin), reducing tax burden is best done by taking advantage of tax credit and accelerated depreciation opportunities in the tax code (assuming these tax strategies have benefits in financial accounting.)

Reduce Interest Burden. Like taxes, paying less in interest expenses is always desirable from a corporate perspective. Of course, interest burden can be reduced by using less debt, but this defeats the goal of increasing profits through financial leverage. To increase profits, utilities should make sure that all debt is at the lowest possible interest rates. This typically involves retiring old debt with high interest rates and issuing new debt with lower interest rates.

Improve EBIT Profit Margin. EBIT profit margin can be improved by either increasing price or decreasing operating expenses. For utilities, prices can typically only by increased with regulatory approval, often justified to offset increasing expenses. Price changes can also affect sales volume, although utility demand tends to be somewhat inelastic. A more common approach for utilities is to increase profits by reducing operating expenses. Profits gained through this approach will continue at least until the next rate case.

Increase Sales Volume. With high fixed costs, utilities could significantly increase profits by increasing sales volume, as long as these increases do not require infrastructure additions or enhancements. However, it is generally unseemly for utilities to encourage increased usage from existing customers. In fact, utilities are commonly under regulatory pressure to conserve, become more efficient, and generally use less. This presents a very real conflict that requires legal and regulatory thoughtfulness; utility management has an obligation to maximize profits for shareholders, but is being pressured by regulators to take actions that will decrease profits. Different states have different approaches to this conundrum. In any case, a good way to increase sales volume is to encourage new customers, especially large customers, to relocate or build new facilities in the utility's service territory.

Increase Financial Leverage. As discussed in Chapter 4, more debt leads to higher equity returns, but these returns are required due to an increase in financial risk. The true ability of financial leverage to increase profits lies in the ability to deduct interest expenses from taxable income. Therefore, increasing financial leverage is beneficial from a DuPont perspective since it reduces tax burden.

A DuPont analysis is a straightforward and simple way to gain insight into the profitability of a company. Although it is not as commonly used as in the past, it remains a good framework for profit analysis and profit management. Two of the biggest weaknesses of a DuPont analysis are related to accounting numbers and the cost of capital. Financial accounting numbers do not necessarily reflect fair market value, which is a drawback of a DuPont analysis and every financial ratio using accounting values. The cost of capital involves investor expectations and financial performance compared to these expectations; it is discussed further in the next section.

6.7 RESIDUAL INCOME AND EVA

As discussed in Section 4.1.5, the weighted expected return of equity investors and debt investors is called the weighted average cost of capital (WACC). The net capital used by a company, capital employed, therefore requires a return of WACC to satisfy investor expectations. The capital employed by a company (equal to assets minus current liabilities) multiplied by WACC is defined as the *capital charge* of the business.

$$\text{Capital Charge} = (\text{Assets} - \text{Current Liabilities}) \times \text{WACC} \qquad (6.30)$$

Profit levels above the capital charge indicate that the business is exceeding investor expectations and profit levels below the capital charge indicate that the business is not meeting investor expectations. The correct measure of profit to use is NOPAT (net operating profit after taxes), which represents all profits available for distribution to equity investors and debt investors.

The difference between NOPAT and capital charge is commonly called *economic value added* (EVA). This term recognizes that capital employed has an opportunity cost. If the company were not using this capital, it could be deployed in another profit-making opportunity. Therefore, nominal profits that equal the capital charge are said by economists to have zero *economic profit*. EVA borrows from this terminology by recognizing that economic value is created when nominal profits exceed the capital charge.

Economic Value Added is a registered trademark of the management consulting company Stern Stewart & Company. However, the underlying concept of EVA is not new. In 1890, the English economist Alfred Marshall described a new accounting performance measure based on operating profit in excess of a capital charge. Marshall referred to this measure as *residual income*. Residual income and EVA are equivalent, but EVA is by far the more commonly used term today. As such, EVA will be used for the remainder of this section with the understanding the nomenclature, but not the concept, is protected intellectual property.

The mathematical equation for EVA is simply the difference between NOPAT and Capital Charge:

$$EVA = NOPAT - \text{Capital Charge} \tag{6.31}$$

An EVA of zero is an indication that investors are achieving their expected rate of return. A business with positive EVA is said to be "creating value" since returns are higher than investor expectations. Conversely, a business with negative EVA is said to be "destroying value" since returns are lower than investor expectations. In the jargon of EVA, it is good to be a value creator and bad to be a value destroyer. A profitable company is still classified as a value destroyer if its profits are not enough to cover its capital charge.

Readers may have noticed at this point that EVA is not a ratio. Instead of normalizing an accounting value to some measure of size, EVA subtracts the opportunity cost of capital employed. It would be simple to define a ratio equal to NOPAT divided by capital charge, but this is not desirable. Part of the compelling nature of EVA is its ability to measure economic profit in dollars. Consider a business with a NOPAT of $80 million and a capital charge of $100 million. A ratio approach would state that the business is meeting 80% of investor expectations. EVA states that the business is destroying $20 million of investor value. From a managerial perspective, the EVA approach is more compelling because it describes the situation in absolute terms while still compensating for investor expectations.

Improving EVA requires an increase in economic profit. This does not necessarily require an increase in accounting profits. The fundamental ways to increase EVA are to increase NOPAT, reduce capital employed, and reduce WACC.

Increase NOPAT. Increasing NOPAT involves standard business operational improvement and involves pricing, efficiency improvement, expense reduction, tax management, and similar issues related to accounting profit. Most businesses already have a strong focus on these issues without the use of EVA.

Reduce Capital Employed. One of the values of EVA is its ability to focus management attention on capital employed. Even if operating profits remain the same, a reduction in capital employed means that less investment is required and returns on investments are higher. A typical approach to reduce capital employed is to focus on working capital. For example, collecting unpaid customer bills and otherwise reducing accounts payable will increase EVA. In the same way, delaying invoice payments until the last possible day will increase current liabilities, reduce capital employed, and improve EVA.

Reduce WACC. Most companies are always trying to optimize their cost of capital by having an appropriate level of debt and by insuring that issued debt is at the lowest possible interest rate. It is true that reducing WACC will improve EVA, but this is not typically a viable option. If it is, actions are typically assigned to the finance and treasury functions.

In summary, EVA is a useful way for a company or a profit center to look at profitability from the perspective of investors. Unlike financial ratios, EVA is not typically used to analyze and compare the financial performance of industries or companies. Rather, EVA is used as a managerial accounting tool to identify whether investor expectations are being met and to help guide and control efforts to increase economic profit.

6.8   SUMMARY

Financial ratios are common jargon in the business world. They are used to describe aspects of financial performance without requiring information about the size of the company. It is not necessarily helpful to say that a company had net earnings of $10 million. Is this good or bad? The answer depends upon the size of the company. It is good for a company with revenues of $20 million. It is bad for a company with revenues of $1 billion. It is much clearer to use financial ratios and state that a net profit margin of 50% is good and a net profit margin of 1% is bad. Other financial ratios have similar advantages.

There are a large number of financial ratios. As such, it is helpful to group them into categories such as profitability, activity, leverage, liquidity, and market ratios. The first four categories are computed primarily from financial accounting data, which is fine as long as those using these values remember that accounting numbers do not necessarily reflect market value, and are not necessarily representative of what will happen in the future. In contrast, market ratios rely primarily on market data, closely reflecting investor opinions of expected future financial performance.

Financial ratios are most useful when many are examined together. This advantage is heightened when the mathematical relationship of ratios are considered. The best example of this approach is the DuPont analysis, which recognizes that return on equity is equal to the product of return on sale, asset turns, and financial leverage.

Financial ratios present what the financial situation is, but do not consider what the financial situation should be from the perspective of investors. This deficiency is addressed through the use of economic value added (also known as residual income), which subtracts a capital charge from profits to account for the opportunity cost of debt and equity investors. Economic value added is not a ratio and it is not typically used by investors and analysts. However, economic value added is a very effective internal management tool since it describes in dollar amounts how much economic value is being created or destroyed by a business.

It is impossible for financial ratios to provide a complete picture of the financial situation and financial outlook of a company. That said, financial ratios are a good place to start these examinations. A thoughtful examination of financial ratios can highlight where a company is typical, where a company is not typical, and can help generate questions to be answered by more in-depth analyses.

6.9   STUDY QUESTIONS

1. What is the difference between gross profit margin, operating profit margin, and net profit margin?
2. Why are gross profit margins for typical utilities so much higher than operating profit margins?
3. Why are inventory turns for utilities often difficult to compare to other industries?
4. Why are days receivable for utilities often higher than other industries?
5. What is the difference between a debt-to-equity ratio and a debt ratio? How do utility values for these ratios compare to other industries? Explain the difference.
6. What values do investors generally prefer to see in a quick ratio? Why are utility quick ratios so much lower than this?
7. Why might an investor be interested in times interest earned? Explain why an investor might be more interested in times interest earned than the current ratio or quick ratio.
8. Why do some investors prefer market ratios to accounting ratios? What are some of the more popular market ratios?
9. What are the three factors that determine net profit margin in a basic DuPont analysis?
10. What is meant by the term economic profit? How is the concept of economic profit considered in economic value added?

7
Ratemaking

Ratemaking is a difficult topic to cover in a single chapter. Just in the United States, there are essentially fifty-one answers to every question. Each state plus the federal government has its own approach to utility regulation and ratemaking, not to mention different national approaches around the globe. Even within a state, ratemaking approaches may vary substantially for different types of utilities and for utilities with different ownership structures.

With ratemaking, the devil is in the details. Keeping in mind that each regulator and each utility will have many unique aspects, there are certain overarching principles that guide the ratemaking process. The focus of this chapter is on this high level view rather than on nuances and exceptions, with the goal of providing the reader with a sound basis for correct thinking about ratemaking and related topics.

In many ways, ratemaking is the most important activity for a regulated utility. Rates are the primary mechanism for revenue generation. Rates strongly influence how customers perceive utility value. Rates are how utilities are often compared. Rates are a major factor in attracting and retaining utility-intensive businesses to an area. Rate case proceedings can strongly affect the utility relationship between regulators and other stakeholders. The list goes on.

All else equal, utilities want rates to be set at levels that maximize revenue. Everyone else wants rates to be as low as possible. Since there is no free market to set utility prices, it is the job of regulators to balance these competing perspectives. The function of ratemaking is to have regulators ensure that rates proposed by utilities are fair to utilities, are fair to customers, and serve the greater interest of society. The remainder of this chapter discusses the theory, process, and common approaches to ratemaking and rate design.

7.1 REGULATORY GOALS

Since they are monopolies, utilities motivated strictly by profit maximization will wish to set rates higher than otherwise would occur in a competitive environment. This is not a criticism, just a truism. The monopoly status of utilities is the primary reason for rate regulation. Utilities have a financial incentive to desire high rates, and regulators try their best to ensure that rates are close to what they would otherwise be in a competitive free market.

The reader may recall from Chapter 3 that prices in a free market achieve equilibrium when industry supply equals customer demand. Businesses will continue to produce more units of their product or service until the marginal cost of producing an additional unit is equal to the market-clearing price of this additional unit. This supplier behavior is beneficial from a societal perspective since it maximizes economic surplus.

Chapter 3 also showed that a monopoly will desire to set prices where its marginal revenue for the next unit is equal to its marginal cost. Additional units will be produced until overall profits are maximized. This occurs at a higher price and at a lower quantity than where supply is equal to demand. The monopoly makes more profit, but society is left with less economic surplus.

A regulator striving for free market efficiencies will wish to set the price of utility services equal to the marginal cost of the utility producing an additional unit. In practice, this is difficult due to the capital-intensive nature of utilities. Once the initial infrastructure is built and customers are connected, it is very inexpensive to deliver incremental units of service to existing customers, up to a limit. It is also very inexpensive to connect new customers, up to a limit.

Eventually, an increase in demand will require infrastructure upgrades. For example, an increasing population will, at some point, lead to a need for an additional power plant, an additional telephone switching office, an additional water treatment plant, and an additional sewage treatment plant. Economists categorize these types of large investments – spending that must be periodically incurred as production volume increases – as *semi-variable*.

When a large percentage of total costs are fixed or semi-variable, as is the case for utilities, it is essentially impossible to determine a meaningful measure of marginal cost. Therefore, regulators strive for the next best alternative, average cost. Consider a utility that needs to invest in one billion dollars of infrastructure before a single customer can be served. Interest payments related to this infrastructure amount to one hundred million dollars annually. Next assume that the incremental cost to serve a typical customer is five hundred dollars per year. If the utility provides service to a single customer, the cost to serve this customer is one hundred million dollars for fixed costs plus five hundred dollars. If the utility provides service to two customers, the average cost per customer reduces dramatically to fifty million dollars for fixed costs plus five hundred dollars. If the utility provides service to one million customers, the average cost per customer becomes one hundred dollars for fixed costs plus five hundred dollars. The equation for cost per customer is the following:

$$\text{Cost per Customer} = \frac{C_{\text{fixed}}}{N} + C_{\text{incremental}} \tag{7.1}$$

C_{fixed} ; Fixed cost
$C_{\text{incremental}}$; Average incremental cost per customer
N ; Number of customers

The economic view of average cost is shown graphically in Figure 7.1. As the quantity of supplied utility services increases, average cost per customer decreases dramatically. Also shown in this figure are the demand and marginal cost curves. The demand curve shows that customers will purchase more as the price goes down. The marginal cost curve shows the cost to the utility for providing an incremental amount of services to customers. Under competition, market equilibrium will occur when marginal cost is equal to the price that customers are willing to pay at the current quantity demanded. This occurs where the demand curve intersects the marginal cost curve.

It is impossible for regulators to determine the hypothetical market equilibrium point since utilities have a large percentage of fixed and semi-variable costs. This means that ratemaking, for all practical purposes, cannot rely on marginal cost analysis. Instead, ratemaking must rely on average cost analysis, which is much easier to compute.

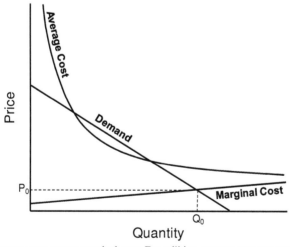

Figure 7.1. Average cost versus marginal cost. For utilities, average cost per customer decreases dramatically with quantity, whereas marginal cost gradually increases. Under competition, market equilibrium occurs at price P_0 and quantity Q_0, where the demand curve intersects the marginal cost curve. It is impossible for regulators to determine this point since utilities have a large percentage of fixed and semi-variable costs.

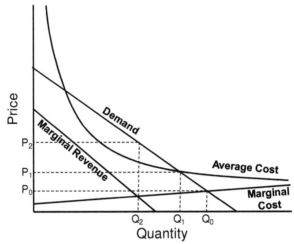

Figure 7.2. The economic basis of ratemaking. Societal benefit is maximized at price P_0 and quantity Q_0, where the marginal cost curve crosses the demand curve. Monopolistic profits are maximized at the higher price of P_2, where marginal cost is equal to marginal revenue. Since regulators cannot quantitatively determine P_0, they typically settle for a price that is equal to average cost, corresponding to a price of P_1 and a quantity of Q_1.

The economic basis of regulatory ratemaking is shown in Figure 7.2. Societal benefit is maximized at price P_0 and quantity Q_0, where the marginal cost curve crosses the demand curve. As a monopoly, the utility prefers that prices be set at the higher price of P_2, where marginal cost is equal to marginal revenue. This higher price results in a lower quantity demanded. Since regulators cannot quantitatively determine P_0, but do not want utilities to set a price of P_2, they will typically settle for a price that is equal to average cost. In the figure, this corresponds to a price of P_1 and a quantity of Q_1.

Average cost is a somewhat simple concept for use in ratemaking. Utilities compute their average cost per customer. Rates are then set so that the total amount paid by each customer is equal to this average cost. Each regulatory approach will have variations and complications, but the basic foundation remains. Regulatory rates are set so that price is equal to average cost, resulting in an approximation of what would otherwise occur in a competitive free market environment.

It is important to understand that setting rates based on average cost approximates free market prices, but does not simulate free market forces. In a competitive free market, companies are constantly striving to reduce cost so that they can remain cost competitive. If a utility reduces cost, rates reduce accordingly and customers receive the entire benefit. In a competitive free market, companies are constantly striving to improve quality so that they can attract and retain customers. A utility gains no revenue benefit for improving quality; they already serve all customers within their service territory. Other aspects of regulation besides ratemaking must deal with these issues. For example, regulators can impose financial penalties for utilities with continuing cost inefficiencies or continuing low quality of service.

A guiding principle in ratemaking is that rates must be "just and reasonable." These vague words are difficult to apply, but do set generally agreed upon bounds for utility rates. At the low end, rates must be sufficient to cover a utility's cost of providing service. It is not reasonable to expect utilities to operate at a loss by regulatory design. At the high end, rates must not exceed the value that the utility service provides to customers. Rates higher than this result in decreased usage and cause overall harm to society. The just and reasonable test also applies to rate design, which is discussed in Section 7.3.

When assessing whether rates are just and reasonable, it is tempting for regulators to make comparisons. Comparisons between similar utilities are common. If the proposed rates of a utility are significantly higher than similar utilities, the utility must provide compelling evidence why these higher rates are reasonable. Comparison with substitutes is also common. If proposed rates are significantly more expensive than the cost of a close substitute, the utility may also experience increased scrutiny. But comparisons are never perfect, and average cost remains the primary measure of whether rates are just and reasonable.

7.2   REVENUE REQUIREMENT AND THE RATE BASE

The total amount of revenue that regulators decide is appropriate for a utility to earn in a year is called the utility's *revenue requirement*. As discussed in the previous section, the revenue requirement of a utility is almost always based on the economic cost to serve its customers. Since the economic cost of a utility includes a fair return for investors, revenue requirement can be mathematically described in the following two ways:

$$\text{Revenue Requirement} = \text{Economic Cost of Service} \qquad (7.2)$$

$$\text{Revenue Requirement} = \text{Nominal Cost of Service} + \text{Fair Return} \qquad (7.3)$$

The nominal cost of service is easy to determine, at least from an historical perspective. This is simply the total amount of expenses that a utility actually incurred. Determining the fair return is a bit more complicated.

Everyone agrees that a well managed utility should be able to provide a fair return to its investors. Interest must be paid to lenders, and sufficient dividends must be paid to shareholders. If returns are not sufficient, a utility will not be able to raise the capital it needs to maintain and expand its infrastructure.

Recall from Section 6.7 that a fair return to investors is equal to the capital charge, which is defined as the product of capital employed and the weighted average cost of capital (WACC). Revenue requirement calculations are almost always based on this premise, but substitute the *rate base* for capital employed and substitute the allowed *rate of return* (ROR) for WACC. This approximate transformation is summarized as:

$$\text{Capital Charge} = \text{Capital Employed} \quad \times \quad \text{WACC} \qquad (7.4)$$

$$\downarrow \qquad\qquad\qquad \downarrow \qquad\qquad\qquad \downarrow$$

$$\text{Fair Return} = \text{Rate Base} \quad \times \quad \text{ROR} \qquad (7.5)$$

At most utilities, the rate base is a very important concept in both practical terms and in corporate culture. In practical terms, the rate base drives accounting, budgeting, and ratemaking practices. In cultural terms, employees associate the rate base with making money, and tend to view spending as either going into the rate base and earning profits or not going into the rate base and not earning profits. This mindset is an oversimplification and is discussed further in Section 7.5, but remains strong for most utility employees.

Spending that goes into the rate base is capitalized on the balance sheet rather than expensed. As such, rate base outlays are called *capital expenditures*, or *CAPEX*. Other operational cash outlays are treated as an expense and are charged against retained earnings. Therefore, operational cash outlays are called *operational expenditures*, or *OPEX*. Other expenses that utilities incur as a cost to provide service include depreciation on the rate base and income taxes. These result in the following equation commonly used for revenue requirement:

$$
\begin{array}{r}
\text{OPEX} \\
\text{Income Taxes} \\
\text{Depreciation} \\
+ \quad \underline{\text{Rate Base x ROR}} \\
\text{Revenue Requirement}
\end{array}
\qquad (7.6)
$$

In accordance with this formula, all prudently incurred OPEX and taxes are recovered at cost. They do not require any investment (aside from the working capital already included in the rate base) and therefore are not allowed a rate of return. Items in the rate base require invested capital, and therefore are allowed a rate of return. CAPEX typically goes into the rate base, with a few exceptions that make the rate base slightly different from capital employed. Typical items that are included in and excluded from the rate base are listed in Table 7.1.

Often the rate base is perceived as a listing of "real stuff" that a utility has purchased. Although real stuff is in the rate base, the purpose of the rate base is to approximate invested capital. Therefore, like capital employed, the rate base must compensate for working capital. Current assets like cash must be included since they are needed to run the business, and require the use of capital. Current liabilities, like customer advances and deferred income tax payments, reduce capital requirements and must therefore be subtracted.

Table 7.1. Items used in determining the rate base.

Included in Rate Base	May or may not be included in Rate Base
• Plant in service	• Construction work in progress
• Vehicles	• Plant held for future use
• Land being used	• Land held for future use
• Buildings	• Allowance for funds used in construction
• Inventory materials and supplies	
• Fuel stockpiles	
• Cash and cash equivalents	
Deducted from Rate Base	
• Accumulated depreciation	• Accumulated deferred income tax
• Contributions in aid of construction	• Investment tax credits
• Customer advances and deposits	

Since customers ultimately pay for everything allowed in the revenue requirement, there is a governing regulatory principle that all allowed costs must be for the benefit of customers. Utilities can spend money on anything they want, but spending that fails this test may be disallowed for recovery. For example, a utility may wish to spend money on general advertising to strengthen its brand recognition. Regulators may decide that customers should not have to pay for this expense since they do not receive any benefits.

For inclusion in the rate base, many regulators require that an asset be *used and useful*. This means that costs incurred for construction work in progress (CWIP) have typically not been included in the rate base. Regardless of whether CWIP is included in the rate base, it requires capital and the associated return on this capital. Some regulators allow interest on CWIP to be charged as OPEX. An argument against this practice is that current customers are paying interest on investments that will benefit future customers. Therefore, some other regulators allow interest on CWIP to be accrued into an account called *allowance for funds used during construction (AFUDC)*. When the project is completed and put into useful service, both the direct construction costs and the associated AFUDC are placed into the rate base.

An argument against AFUDC is that utilities will be more reluctant to fund expensive construction projects when there is no guarantee of interest expense recovery. To encourage capital investment, regulators may allow CWIP to be included in the rate base, either all the time or in special circumstances. Allowing CWIP in the rate base violates the used and useful principle, but satisfies the accounting matching principle since the utility is generating revenue from the CWIP, which is listed as an asset on the balance sheet.

Other capital expenditures may not strictly meet the requirements of used and useful. This includes completed projects that are not currently in service, but have definite plans to be put in service. Sometimes regulators allow this *plant held for future use* to be included in the rate base if they feel that the investment benefits customers. Similarly, a utility may decide to pre-purchase items such as land, which may not be available in the future when it is needed. If the land has definite plans of being used, regulators will often allow *land held for future use* to be included in the rate base.

Sometimes customers pay directly for a portion of infrastructure construction. For example, builders of new subdivisions may be required to install a certain amount of utility infrastructure that will later be transferred to utilities. Utilities may also require customers to pay for infrastructure costs that go beyond basic service. Direct customer payments for utility infrastructure are called *contributions in aid of construction*. Although these payments result in a utility asset, they are not included in the rate base since they are not funded by investors. Similarly, the depreciation expenses for contributions in aid of construction are not recoverable expenses since customers should not have to pay for the infrastructure twice.

In summary, a utility is allowed to recover its operating expenses and make a return on its rate base. The general regulatory approach is to set the ROR for a well run utility equal to WACC. However, the proper value of WACC can be debated since it includes an equity component. The required returns for bonds and loans are known. The required return for investors is not known with certainty. In any case, ROR must result in an allowed rate of return that satisfies three conditions. First, it must maintain the credit rating and financial integrity of the utility. Second, it must allow the utility to raise capital at a reasonable cost. Third, it must allow the utility to achieve equity returns comparable to other investments with similar risks.

When considering revenue requirement, it is easy to misinterpret the word "requirement" to mean that the utility is required to achieve a certain level of revenue. This is not the case. In any given year, the utility is given an opportunity but not a guarantee for achieving the revenue requirement target. Revenue variation is inevitable. In some years revenue may be lower and in some years revenue may be higher. Regulators are generally not allowed to set rates based on what happened in the past, called *retroactive ratemaking*, and are therefore prevented from compensating for historical revenue shortfalls and surpluses.

7.3 RATE DESIGN

Once revenue requirements are determined, rates must be established that will result in customer billings approximately equal to the revenue requirement. There are a vast number of ways to do this, and it is helpful to examine some of the simpler approaches before discussing more complicated approaches.

Consider a utility that expects to sell ten billion units of service over a year, has a revenue requirement of one billion dollars, and serves one million customers. The easiest rate structure is to divide the revenue requirement equally among all customers. In this case, each customer would annually pay $1 billion ÷ 1 million = $1,000 per year in utility charges. This approach is sometimes used when metering data is not available, such as street light electricity or wastewater disposal. A variation of this approach is to weight each customer based on estimated usage, such as making each customer pay in proportion to building size.

If metering data is available, revenue requirements can be allocated to units of service. In this case, each unit of service must result in $1 billion ÷ 10 billion = 10¢. Each month, meters are read and customers are billed ten cents per unit of utility services used. If customers consume ten billion units during a calendar year, the utility will earn its revenue requirement.

Though simple to administer and easy for customers to understand, the above methods of rate design are typically not acceptable to either customers or regulators. Not all customers are equal, and the volume of utility services consumed by a customer is typically not a good indication of the cost to serve this customer. Recall that a high percentage of utility cost is embedded in fixed infrastructure, and does not change with consumption. These infrastructure costs, normalized by consumption volume, are typically lower for larger customers. They are also lower for customers that tend to use a constant amount of service a large percentage of the time, as opposed to customers that use very small amounts of service most of the time and large amounts of service periodically.

Consider a grocery story that is open around the clock. This commercial customer uses a large amount of electricity compared to residential customers, but requires a single meter, a single meter reading per month, and a single bill. These costs are much less on a per-unit basis than for typical residential customers because the grocery store has lights, refrigeration equipment, and air handling equipment on twenty four hours per day. Electricity consumption will vary somewhat, but the peak usage is not too much higher than the average usage (i.e., high *load factor*). Compare this to a residential home. In the morning, people wake up, turn on the lights, turn on the air conditioners, take showers (electric water heater), and cook breakfast (electric stove). This results in a large spike in electricity usage. Everyone then leaves for the day and electricity usage decreases dramatically. People come home at dinnertime and usage peaks again until bedtime, when usage decreases again. For the residential customer, peak usage is much higher than average usage.

The utility must build its system to handle peak usage, but bills are commonly based on average usage over an entire month. Therefore, the per unit cost of fixed infrastructure for the commercial customer is lower than for the residential customer. This is because the infrastructure has a higher level of average utilization, and therefore a higher number of units which infrastructure costs can be spread across.

Regulators acknowledge that different customers have different per unit costs to serve. However, it is not practicable to assign costs to individual customers. Imagine a utility with two million customers, each with a different rate. Not only would this be an administrative nightmare, but it would be confusing, seem unfair to many customers, and likely be discriminatory in certain cases. Therefore, most rate structures group customers into billing categories such as residential, commercial, and industrial.

Table 7.2. Cost of service allocation. In this example, the total utility cost to serve is allocated across three billing classes. Separate rates are determined for each class by dividing the allocated cost to serve by the total units used by the class. These types of rate differences are common for utilities.

Billing Class	Customers	Units Used (millions)	Cost to Serve (millions)	Rate ($/unit)	Average Annual Bill
Residential	895,000	5,000	$550	0.110	$615
Commercial	100,000	3,000	$300	0.100	$3,000
Industrial	5,000	2,000	$150	0.075	$30,000
Total	**1,000,000**	**10,000**	**$1,000**		

Consider again the utility that expects to sell ten billion units of service over a year, has a revenue requirement of one billion dollars, and serves one million customers. This utility does a cost of service study and determines that (1) industrial customers account for fifteen percent of the total cost of service, (2) commercial customers account for thirty percent of the total cost of service, and (3) residential customers account for fifty-five percent of the total cost of service. The utility knows the amount of consumption for each of these classes, and can therefore compute separate rates for each class. These calculations are shown in Table 7.2.

The calculations shown in Table 7.2 result in a residential rate of $0.11 per unit, a commercial rate of $0.10 per unit, and an industrial rate of $0.075 per unit. These types of rate differences are common for utilities, and are called *differential rates*. Differential rates often cause people to assume that residential customers are subsidizing commercial and industrial businesses. Typically this is not the case. Commercial and industrial customers have lower rates since they have lower per unit costs to serve. Rates are set for different billing classes so that revenues from each billing class are equal to the cost to serve the billing class.

Dividing joint use cost categories into billing classes, such as the example in Table 7.2, is called *cost allocation*. Another approach is to identify specific costs incurred by specific customers, and embed the associated costs in these customer rates. This approach is called *direct cost assignment*. For example, an industrial facility may be served by several miles of dedicated facilities that are not used by any other customers. A cost assignment approach would make this industrial facility pay for all costs associated with these dedicated facilities. All unassigned costs are then allocated to billing classes as previously described.

It has been argued that increases in the consumption of utility services cost incrementally less due to the large amount of fixed costs involved. This logic has historically led to the justification of declining *block rates*, where the first block of units consumed in a month is charged at a high rate, the second block of units consumed is charged at a lower rate, the third block of units consumed is charged at a lower rate still, and so forth. In its simplest form, two blocks are used. The first block is at high rate to recover fixed costs. The second block is at a lower rate corresponding to variable costs. Two closely related rate approaches are a monthly minimum charge and a monthly fixed charge. A monthly minimum charge can be set to be the amount of the first full block. Alternatively, a single rate can be used and this same amount can simply appear as a fixed charge.

Consider a utility that needs to collect about $10 per customer per month to help cover fixed costs. Incremental costs for this utility are about $0.10 per unit. A declining block rate will design the first block so that, if fully consumed, it will contribute $10 to fixed costs. For example, the first block could charge $0.20 per unit for the first one hundred units, covering fixed costs and the incremental costs for this block. Rates beyond the first block are set to the incremental cost of $0.10 per unit. A minimum charge rate structure is identical except that the customer must always pay at least $10 per month, even if consumption is lower than 50 units. A monthly fixed charge simply adds $10 to every bill, and charges all consumption at the same rate of $0.10 per unit. A summary of these three similar rate approaches is:

Declining Block Rate
- First 100 units at $0.20 per unit
- Additional units at $0.10 per unit

Minimum Charge
- First 100 units at $0.20 per unit
- Additional units at $0.10 per unit
- Minimum monthly charge of $10

Monthly Fixed Charge
- All units at $0.10 per unit
- Monthly fixed charge of $10

Decreasing block rates have been criticized in several ways. Some consumer advocates argue that they discriminate against smaller customers since smaller customers will pay higher rates on average. This is because a higher percentage of their consumption is billed at the higher rate assigned to the initial block. Others argue that decreasing block rates encourage consumption, which is not good public policy. When these arguments prevail, regulators sometime use increasing block rates (sometimes called *inverted block rates*). For example, the first block for residential natural gas rates may be set at a low rate and sized for an amount necessary to heat a small home in the winter. The next block may be priced at a moderate rate and sized for normal usage in an upper middle class home. The third block is priced much higher to discourage high natural gas usage.

As can be seen, there are a variety of approaches to rate design. All rate designs should be based on revenue requirements, but other factors must also be considered. The most important considerations are the following:

Major Considerations in Rate Design
- Rates should reflect cost of service.
- Changes in rates should occur gradually.
- Rates should strive to retain most customers.
- Rates should be consistent with public policy goals.

As discussed earlier, rates should be allocated in a manner that correctly assigns costs to the rate classes responsible for these costs. In other words, rate design should avoid *cross-subsidies* from certain billing classes to other billing classes (as well as within a rate class). For example, rates should avoid charging artificially high residential rates so that commercial and industrial rates can be artificially low. It is fine for these rates to be lower if the proportional cost of service is lower, but not for public policy reasons such as encouraging new businesses to locate in the area with no corresponding benefit to the residential customers. Cross-subsidies are essentially a hidden tax on the customers paying artificially high rates, and a hidden tax subsidy for the customer customers paying artificially low rates. Artificially low rates for large customers may sometimes be acceptable if beneficial to all ratepayers, which is discussed shortly.

Most areas in developed nations are already served by utilities with established rate structures. As such, rates are rarely designed from scratch. Rather, rate adjustments are made to existing rates. When making rate adjustments, regulators must be careful to avoid changes that will result in much higher utility bills. Large rate increases can cause a variety of problems, collectively called *rate shock*. For example, after a seven year rate freeze, Baltimore Gas & Electric announced an upcoming 72% increase in rates to compensate for the increased cost of wholesale power. These rates would correctly reflect the cost of service, but the rate shock resulted in a massive political and regulatory maelstrom. Baltimore Gas & Electric was accused of profiteering. Regulators were accused of catering to the utility at the expense of consumers. According to *The Baltimore Sun*, the rate increases were the "dominant political issue" of the 2006 election campaign. In the end, the regulators allowed the new rates to be phased in over time, somewhat mitigating the effects of rate shock.

In certain special situations, it is beneficial for all ratepayers if certain rate classes have rates that are lower than those based strictly on cost of service. The reason lies with customer retention. Imagine a situation where large industrial customers are using natural gas supplied by the local utility. These customers can easily switch to propane, and will do so if the total cost of using propane falls below the total cost of using natural gas. If a large number of industrial customers leave the system, the fixed costs previously being paid by them must now be paid by the remaining customers, forcing a rate increase.

Table 7.3. The impact of artificially low rates to prevent switching. In the current situation, fixed costs and variable costs are allocated based on cost to serve. If the current rates remain, industrial customers will switch to a more cost-effective service and the associated fixed costs must be re-allocated. This results in a 1¢ rate increase for both residential and commercial customers. The industrial customers will remain if rates are lowered. To do this, fixed cost allocation is reduced for industrial customers and increased for other customers. The industrial customers now have an artificially low rate, but the increase in residential and commercial rates is only half the amount that would occur if the industrial customers left the utility.

Billing Class	Customers	Units Used (millions)	Fixed Cost to Serve (millions)	Variable Cost to Serve (millions)	Rate ($/unit)
Current Situation					
Residential	895,000	5,000	$350	$200	0.110
Commercial	100,000	3,000	$170	$130	0.100
Industrial	5,000	2,000	$80	$70	0.075
Total	**1,000,000**	**10,000**	**600**	**400**	
Rates if Industrial Customers Leave					
Residential	895,000	5,000	$400	$200	0.120
Commercial	100,000	3,000	$200	$130	0.110
Total	**995,000**	**8,000**	**600**	**330**	
Rates with Reduced Industrial Charges					
Residential	895,000	5,000	$375	$200	0.115
Commercial	100,000	3,000	$185	$130	0.105
Industrial	5,000	2,000	$40	$70	0.055
Total	**1,000,000**	**10,000**	**600**	**400**	

An example of the benefits of low rates for customer retention is shown in Table 7.3. The current situation is identical to the example previously discussed and shown in Table 7.2. There are residential, commercial, and industrial customers that have allocated costs to serve. In this example, costs have been broken out into fixed costs and variable costs. If a customer leaves, variable costs go away but fixed costs remain and must be absorbed by the remaining customers.

A differential rate analysis results in an industrial rate of $0.075 per unit. This is the rate that results in charges to industrial customers being equal to the cost to serve these customers. Now assume that a substitute product becomes available to the industrial customers for an effective rate of $0.06 per unit. If the utility rate remains unchanged, it makes economic sense for all of the industrial customers to quit using the utility service and start using the substitute. If this happens, the $70 million variable cost associated with the industrial customers goes away, but the $80 million in fixed cost remains.

If the utility allocates $50 million of additional fixed costs to residential customers and $30 million of additional fixed costs to commercial customers, residential rates increase from $0.11 to $0.12 per unit, and commercial rates increase from $0.10 to $0.11 per unit.

The utility finds that it can retain the industrial customers if rates are lowered from $0.075 to $0.055 per unit, making the utility more cost effective than the substitute. To do this, it reduces the fixed cost allocation of industrial customers from $80 million to $40 million. This is compensated for by increasing the fixed cost allocation of residential customers by $25 million and of commercial customers by $15 million. The industrial customers now have an artificially low rate, but the increase in residential and commercial rates is only half the amount than would occur if the industrial customers left the utility.

Artificially low rates designed to retain customers are called various terms including *anti-bypass rates* and *load retention rates*. The same logic that is used to justify customer retention can be applied to low rates designed to attract new customers. When low utility rates are designed to attract new businesses to the area, they are typically called *economic development rates* or *incentive rates*. If designed correctly, these low rates will benefit all customers and have the added benefit of improving the local economy. Some state regulators allow incentive rates if they meet certain criteria, but others reject incentive rates because they violate the ratemaking principle of nondiscrimination.

With or without incentive rates, customers may choose to use bypass utility services. When this happens, customers often want to retain the utility connection for backup services; consumption and peak demand are typically zero, but periodically may be large. In these cases, the utility has the right to recover the costs associated with retaining the ability of the customer to use the utility service connection, even though usage may be rare and of extremely low volume. These types of rates are typically called *standby rates*.

At this point, it should be clear that there are many rate design approaches that utilities and regulators can pursue to correctly allocate costs of service to the customers responsible for these costs. When deciding upon which approach is best, key considerations are rate complexity and bill complexity. Most customers will not appreciate overly complicated bills that make it difficult to determine how charges are being assessed and if these charges are correct. Customers want a simple bill that links understandable usage characteristics to charges. A simple and understandable monthly bill for a residential customer is shown in Figure 7.3. Customers receiving this bill can, at a glance, understand precisely how the bill was computed and how utility usage will affect future bills.

Compare this simply utility bill to the slightly more complicated bill shown in Figure 7.4. The more complicated bill uses block rates. It also breaks out certain fixed charges to increase billing transparency to customers. This utility also computes sales tax separately. Many customers will appreciate the additional information that the more complicated bill provides. Other customers will be more confused and will start to question how their monthly payments are being used, even though this bill is still very basic.

Simple Utility Bill	
Billing period:	Jan. 1st – Jan. 31st 2009
Monthly fixed charge:	$5.00
352 units consumed at $0.11 per unit:	$38.72
January total:	$43.72
Unpaid balance:	$0.00
Amount due:	**$43.72**

Figure 7.3. A very simple utility bill. Customers can examine simple bills like this quickly, and have a clear understanding of the charges and how consumption affects these charges. Most utility bills are more complicated than this simple example, but rate and billing simplicity remains an important goal of rate design.

Slightly Complicated Utility Bill	
Billing period:	Jan. 1st – Jan. 31st 2009
Monthly fixed charge:	$4.50
Amortized connection charge:	$0.50
Green initiative surcharge:	$0.35
Economic development surcharge:	$0.75
Deregulation transition charge:	$0.32
352 units consumed:	
First 50 units at $0.05 per unit:	$2.50
Next 50 units at $0.10 per unit:	$5.00
Remaining 252 units at $0.11 per unit:	$27.72
January subtotal:	$41.64
5% sales tax:	$2.08
January total:	$43.72
Unpaid balance:	$0.00
Amount due:	**$43.72**

Figure 7.4. A slightly complicated utility bill. This bill is based on increasing block rate charges, requiring the bill to separate consumption charges into three line items. This bill also chooses to break out fixed charges. This gives greater transparency to the customer, but complicates the bill and may raise questions or cause confusion with certain customers.

Sections 7.3.1 through 7.3.4 discuss approaches to rate design that can dramatically increase the complexity of bills. Certain large customers may have the expertise to examine and fully comprehend extreme bill complexity. These customers may prefer complicated rate designs if the added complexity provides opportunities for cost savings and better reflects the cost of service. However, most customers will not have the ability to fully understand very complicated utility bills, making rate and bill simplicity an important consideration in rate design.

7.3.1 Unbundled Rates

In the past, most utilities were vertically integrated and performed all activities required to produce and deliver services to customers. As such, the vertically integrated utility could simply base rates on its total cost of service.

Deregulation and industry restructuring has resulted in the vertical separation of many utilities. For example, a former vertically integrated utility could now consist of four separate companies: a production company, a transmission company, a distribution company, and a retail services company. Each of these companies contributes to the utility service and must be compensated for its services. One way to do this is to keep the bill the same as if the vertical utility existed and divide up the customer payments according to some formula. Another approach is to have separate rate structures for each company and have the customer bill list each charge separately. This approach is called *rate unbundling*.

Consider the following example. A retail electric service company is responsible for purchasing wholesale electricity and selling it to customers. It has an approved rate of $0.05 per unit of consumption based on the expected cost of energy purchases and the other associated costs of the business. It does not include the cost of transmission or distribution. Transmission rates are calculated separately, and consist of a rate of $0.01 per unit plus congestion charges that may occur if too much electricity is being routed through the same transmission lines. Distribution rates are calculated as a fixed monthly charge of $5.00 plus $0.02 per unit consumed. Last, the bill must include rates for *ancillary services*, which are required to maintain sufficient system reliability and service quality. A utility bill based on unbundled rates is shown in Figure 7.5.

With deregulation and vertical unbundling, utilities often expose themselves to increased cost uncertainty. This cost uncertainty leads to increased financial risk with traditional rate structures. For example, local gas distribution companies might have to purchase gas at market prices but sell it to customers at a fixed rate. If the market price of gas exceeds the rate that customers are charged, the utility must lose money, since it cannot choose to stop service. Similarly, a local electric distribution company might have to purchase electricity at market prices but sell it to customers at a fixed rate. The resulting financial risk exposure is similar to the gas case.

Unbundled Utility Bill	
Billing period:	Jan. 1st – Jan. 31st 2009
Consumption charges:	
350 units consumed at $0.07 per unit:	$24.50
Transmission charges:	
350 units consumed at $0.01 per unit:	$3.50
Congestion charge:	$0.00
Distribution Charges:	
Monthly fixed charge:	$5.00
350 units consumed at $0.2 per unit:	$7.00
Ancillary Services:	
Scheduling, System Control and Dispatch:	$1.17
Reactive Supply and Voltage Control:	$0.65
Regulation and Frequency Response:	$1.05
Energy Imbalance:	$0.25
Operating Reserve – Spinning:	$0.45
Operating Reserve – Supplemental:	$0.15
January Total:	**$43.72**

Figure 7.5. An unbundled utility bill. This bill breaks down service charges into four categories corresponding to unbundled utility operations. The consumption charges go to the retail service company, who must pay wholesale energy providers. Transmission and distribution charges go to those respective companies. Ancillary charges go to the regional transmission organization who must pay the companies that provide these various services.

In theory, utilities could file for rate changes every time the cost of service significantly changes. In practice, this is not feasible since rate case proceedings can take many months, and are expensive (this is discussed in detail in Section 7.4). Instead, many regulators allow for periodic rate adjustments based on changes in cost of service that the utility cannot control or influence. For example, many local gas distribution companies have purchased gas adjustment clauses that will automatically increase or decrease rates as the wellhead price of gas increases or decreases. Similarly, electric utilities are increasingly being allowed to have fuel adjustment clauses based on the wholesale price of natural gas, coal, and oil. Electric utilities that have divested their generation facilities are often allowed to have automatic rate adjustment clauses based on the wholesale price of electricity.

Critics of periodic rate adjustment clauses argue that they transfer financial risk from the utility to the customer. This is certainly true, and should be considered when setting the allowed rate of return and the corresponding revenue requirement. Utilities argue that automatic rate adjustment clauses correctly link rates to cost of service.

7.3.2   Rates and Peak Demand

As briefly discussed in Section 1.5, a large percentage of utility cost is based on peak demand requirements. A utility must build its system so that it can provide services during time of maximum consumption, even though this may only occur for a few hours over the course of a year. Therefore, it makes sense in certain cases to base a portion of rates on peak demand rather than purely on average consumption. This approach has the potential to more correctly allocate costs, to incentivize customers to reduce usage during peak periods, and to defer infrastructure expansion by lowering peak system demand.

Customers with peak demand charges require special meters that record peak demand. These meters are expensive and have historically only been cost effective to deploy at large customer sites. The cost of sophisticated meters is declining rapidly and most electronic meters now have the ability to compute peak demand. As such, utilities going forward will have increasing flexibility to include demand charges for a larger percentage of customers. Electronic meters with two-way communications are commonly called *smart meters*, and a system of such meters is referred to as *advanced metering infrastructure (AMI)*. For electric utilities, AMI is an important part of realizing the *smart grid*, which relies heavily on digital devices with communications capabilities.

Utility actions attempting to control or reduce peak usage are collectively referred to as *demand side management (DSM)*. The logic of DSM goes as follows. An increase in peak demand will require expensive investments in infrastructure expansion. Therefore, a reduction is peak demand is equal in value to infrastructure capacity expansion of the same amount. A utility should not spend money to expand infrastructure if DSM can free up peak capacity for less. When supply side planning and demand side planning are performed together, the process is called *integrated resource planning (IRP)*.

The two most common types of DSM are interruptible rates and direct load control. Customers with interruptible rates are charged less with the agreement that the utility can curtail service, with advance notice, a certain number of times per year. For example, an electric utility during a period of peak loading may experience a power plant outage. This utility can call all of its customers with interruptible rates and require them to curtail or greatly reduce electricity usage. The resulting reduction in peak loading will hopefully allow the utility system to supply the remaining load until demand goes down. Direct load control is similar in function, but typically targets residential appliances. For example, a utility might offer a reduced rate to residential customers that allow their electric water heater to be shut off by the utility for up to two hours per day. Participating customers have a special device installed that allows the utility to remotely connect and disconnect electricity to the affected devices. During peak loading conditions, the utility can remotely disconnect enough appliances to prevent system overloads.

Another rate approach that addresses both cost allocation and DSM issues is *time-of-use rates*. Time-of-use rates recognize that some time periods are more costly than others for the utility to provide service. A simple example is *seasonal rates*, where months of high demand have high rates and months of low demand have low rates. For example, a natural gas utility may set winter rates high since wellhead prices are highest in the winter and peak demand also occurs in the winter. Seasonal rates are attractive because they do not require special metering equipment.

A more granular approach to time-of-use rates will have different rates for different times of the day, for different days of the week, and for different months of the year. For example, a utility might have low rates at nights and on weekends; have moderate rates during the daytime in the fall and spring; and have high rates during the daytime in the winter and summer when demand is highest. This approach provides good pricing signals to customers, but requires expensive metering equipment and more complicated bills.

Yet another approach to time-of-use rates is the use of *real-time rates*. Special meters are constantly fed price information based on production costs, system loading, and potentially a variety of other components. The meter displays the current rate, and customers can modify usage accordingly. Rates can be truly real time, but more commonly are hour-ahead or day-ahead so that customers have a chance to plan their usage. An increasing number of vendors are developing *smart appliances* and home controllers that automatically manage utility service consumption based on real-time price signals.

A recent approach to peak demand reduction is *critical peak pricing*, sometimes called *dynamic pricing*. Customers with critical peak pricing pay standard rates most of the time. However, with advanced notice, the utility can significantly raise rates a certain number of times per year. For example, standard rates for an electric utility might be $0.12 per unit, but $0.10 per unit for customers with critical peak pricing. Up to twelve times per year, the utility can inform these customers that critical peak pricing of $1.50 per unit will be in effect the next day from 2:00 pm to 7:00 pm. These customers have a financial incentive to reduce consumption during these hours, which will result in reduced peak demand.

Critical peak rebates are similar to critical peak pricing, but pay customers for reduced consumption during peak. For example, a utility might expect a critical peak the next day from 3:00 pm to 6:00 pm. The utility has meter data that shows an industrial customer will typically have a peak demand of fifty thousand units during this time. The utility informs the customer that any peak below fifty thousand units will result in a proportional rebate on its next bill.

In summary, there are many approaches to peak demand reduction. These necessarily involve rates since customers must somehow be given financial incentives. Peak demand reduction is beneficial from a societal perspective, but often requires a significant investment, especially if sophisticated meters and communications systems are required. Proper rate design should ensure that the benefits of peak demand reduction exceed the associated costs.

7.3.3   Rates and Consumption

Most people agree that utility services should be used as efficiently as possible. This presents a problem with traditional utility rates based on consumption. If customers use less utility services, the utilities have a corresponding reduction in revenue. If customers with gas heat insulate their homes, gas utilities make less money. If customers replace old electric appliances with energy efficient appliances, electric utilities make less money. If customers install restricted flow shower heads, water utilities make less money.

Financially, utilities will typically prefer that customers use more of their services instead of less. For this reason, many utilities find declining block rates attractive. Not only do decreasing block rates do a good job of cost allocation, they also result in higher consumption from price sensitive customers. In fact, federal legislation in 1978 prevented the use of declining block rates by electric utilities in most situations, subject to state discretion, for this very reason.

Financial compensation is needed in order to gain the cooperation of utilities in efficiency programs. Many approaches have been tried, but there is growing consensus that major efficiency gains can made only if utility revenue is independent of consumption. Rates based on this approach are called *decoupled rates*. With decoupled rates, revenue requirements and rates are set based on expected efficiency gains. If efficiency gains exceed expectations, rates are automatically adjusted upwards to compensate for the revenue shortfall. If efficiency gains are less than expectations, rates are automatically adjusted downwards to compensate for the revenue excess.

Critics argue that decoupled rates only make utilities neutral to efficiency gains, and do not provide a positive incentive. Though true, efficiency gains and other aspects of business performance improvement can be incentivized in other ways, which is discussed in the next section.

7.3.4   Rates and Business Performance

Rates that allow utilities to recover their cost of service have a major problem. If a utility can recover its costs, it has no financial incentive to become more cost effective. Worse, if a utility earns a return on its rate base, it has a financial incentive to spend as much money as possible if it will increase the size of the rate base. Regulators can disallow costs that are not prudently incurred, but cannot disallow costs on a suspicion that the utility could be more cost effective.

There are several accepted ways for regulators to incentivize utilities to improve business performance. Perhaps the easiest is to simply freeze rates. When rates are approved, both the utility and the regulator agree that the rates cannot be changed for a certain number of years. During this time, any cost reductions made by the utility result in increased profits since revenues remain unchanged. After the rate freeze is over, the regulators have the options to reduce rates and pass on the cost reduction benefits to customers.

Rate freezes are attractive since they roughly mimic the dynamics of a competitive market. Under competition, a company that improves business performance is able to achieve higher profits. These higher profits are not permanent. They only continue until competitors achieve similar improvements in business performance. When this happens, the benefits get passed on to customers through lower prices.

There is an important difference between rate freezes and the free market. Cost reductions in the free market are only desirable if product quality is not unduly affected. Since utility customers have no alternatives, there is no competitive pressure for utilities to keep service quality high while cutting costs. For example, a utility beginning a rate freeze period may have historically spent $200 million on inspection and proactive maintenance activities. If the company reduces this amount to $150 million, the difference of $50 million will go straight to the bottom line. Ideally, the company will find innovative ways to spend the reduced budgets so that equipment and system conditions do not suffer. Unfortunately, budgets cuts often happen without fully quantifying the likely detrimental impacts to service quality.

Fundamentally, utilities can spend money in two ways. First, they can spend money now to save more money in the future. This type of spending can be justified by a net present value analysis. Second, utilities can spend money to improve system performance and customer service. There is no direct business case for this type of spending. If a utility under a rate freeze expenses an additional $10 million to improve its call center, profits for the year are $10 million lower. If a utility spends an additional $20 million to replace old and dilapidated equipment, cash flow for the year is $20 million lower.

Regulators understand that cost reduction runs the risk of underinvestment, worsening reliability, declining levels of customer service, and other deleterious effects. To prevent this, rate freezes are often accompanied by a set of performance metrics and targets. Typically, each metric for a utility will be assessed based on recent history at the utility and possibly industry benchmark data. Target levels are then set for each year of the rate freeze. Each year, actual performance is compared to the target, and financial penalties are assessed if the target is not reached. Penalties typically increase in proportion to the amount that the target was missed, and are capped at a maximum amount. There may also be financial rewards if the target is exceeded by a certain amount. The use of financial rewards and penalties based on performance metrics is called *performance-*

based regulation. An example of performance-based regulation has been previously shown in Figure 1.8.

A side effect of rate freezes is the incentive of utilities to time spending activities. Early in the rate freeze there is a strong incentive to keep costs low. Towards the end of the rate case there is a conflicting incentive. Since new rates will be based on current levels of spending, there is a financial incentive to temporarily increase spending so that new rates will be higher. This is especially true of capital spending. Throughout the rate freeze, reduced capital spending will result in a smaller rate base and the potential for lower profits. Therefore, utilities are financially incentivized to "shore up" the rate base through massive capital spending just before the rate freeze ends. This behavior is often reinforced through regulatory behavior that bases future rates on past spending decisions. These types of financial games are not in the best interest of customers and will never be the explicit policy of utilities. Regardless, the financial incentives for the timing of spending will always exist under rate freezes.

This chapter so far has been exclusively focused on rate regulation, which is based on cost of service. There is another approach for regulators to protect customers against the monopoly power of utilities – *price regulation.* As the name suggests, price regulation sets restrictions on the price that utilities can charge for services, and does not require that these prices be based on cost of service. Most commonly, maximum price schedules are generated that are similar to rate schedules. Utilities are free to offer customers lower prices if they so choose. They can also offer customers alternative pricing schemes as long as at least one option is clearly at or below the regulated price cap.

Price regulation has two immediate benefits. First, it allows utilities to deal more effectively with competitive threats. Under traditional ratemaking, it is difficult for utilities to use pricing to attract new customers and to retain existing customers. Price regulation provides this flexibility with the assurance that no customers will be charged unduly high rates. Second, it allows utilities to be innovative and design flexible rate structures that are valuable to customer niches that do not necessarily correspond to traditional rate categories.

Price regulation is often linked with business efficiency improvement expectations. For example, a utility may have a current regulated rate of \$0.15 per unit, but wishes to switch to price regulation. The regulators agree, but only if there is an immediate rate reduction to \$0.14 per unit, which is the initial price cap. The regulators allow this price to automatically increase annually based on an inflation factor, I, minus a productivity factor, X. Each year, the new price cap level is computed as follows:

$$\text{New Price} = \text{Old Price} \times (1 + I - X) \tag{7.7}$$

I ; inflation factor
X ; productivity factor

For example, regulators may set the inflation factor equal to the percent change in the consumer price index (CPI). It also sets a productivity factor of 2%. Recall that the first year price cap is $0.14 per unit. At the end of the year the CPI has risen by 3.5%. Therefore, the price cap increases by 3.5% – 2% = 1.5%, to a value of $0.1421 per unit. Assuming that the utility increases prices to the new cap level, any cost reductions beyond 2% result in an increase in profits.

The use of price regulation raises several concerns that are similar to those of rate freezes. Namely, price caps give utilities a financial incentive to reduce spending, which may degrade system performance and customer service. Therefore, price regulation is usually accompanied by performance-based regulation in a manner similar to what is done with rate freezes.

7.4   RATE CASES

Rate structures do not change automatically. Once in place, they stay the same until someone, typically the utility or the regulator, requests a change. Sometimes this change is minor, such as a utility adding a new customer class or changing the way that contributions in aid of construction are calculated. For minor changes like these, the utility prepares a suggested set of rate schedules, called a *tariff*, and submits it to the regulators for review. If the regulators find the tariff acceptable, they allow the utility to implement the new rates after affected customers have been notified.

If regulators have questions or concerns with a proposed tariff, they will enter into discussions and negotiations with the utility. Often this will result in a revised tariff that is acceptable to the utility. If agreement on a tariff does not occur, the utility may be required to justify its entire tariff, including revenue requirements and rate design, through a public hearing called a *general rate case*.

There may be times that a regulator feels that the rates of a utility are too high or that the rate design is no longer just and reasonable. When this happens, the regulator can issue a *show cause order*, forcing the utility to demonstrate why its rates should not be changed. If the utility cannot do this to the satisfaction of the regulators, a general rate case may be required. Some states require utilities to file general rate cases periodically, regardless of whether regulators have concerns about current rates.

It is most common for a general rate case to occur when a utility determines that current rates are insufficient to meet its revenue requirement. In other words, the allowed return on rate base is not being met. The shortfall between actual return and allowed return is called a *return deficiency*.

A return deficiency can result from many factors. The rate base could be much larger now than when existing rates were set. Operations and maintenance expenses could have increased. Taxes could be higher. Sales volume could be lower. Capital structure could have changed, increasing return requirements.

Inflation could be higher, also increasing investor return requirements. Typically, a combination of these factors will be involved.

After a utility identifies the need for a general rate case, it prepares its case. This includes supporting documentation and analysis related to consumption, cost of service, taxes, rate base, quality of service, the cost of capital, and a variety of other aspects. It also includes proposed rate designs including customer classes and rate structures for each of these classes.

A rate case is based on a *test year*. The simplest form of test year is based on the actual customer consumption and utility spending that occurred in the most recent fiscal year. Since both costs and consumption during the test year are known, the results that the proposed tariff would have produced in the test year can be calculated with a high level of confidence. Using historical values also assures that all revenues are matched against all historical costs.

In certain cases, the test year can be modified from actual values of a historical *base year* to account for unusual costs or revenues during the base year or to account for expected costs or revenues that are not reflected in the base year. For example, the base year could have experienced low consumption due to mild weather. Some regulators may allow assumed consumption to be modified based on what would occur during normal weather. A recent tax law change may result in higher future taxes that are not reflected in the base year. Some regulators may allow assumed taxes to be modified accordingly. The process of adjusting the base year so that a projected test year is more likely to be representative of future years is called *normalization*.

Some utilities argue that a historical base year, even if normalized, cannot reflect proposed changes in spending patterns. For example, a utility may wish to increase spending in an effort to replace aging equipment and improve overall reliability. If rates are always based on a base year, the utility will not be able to recover at least one year of spending increases; the spending increases must occur before they can be reflected in a base year, and rates cannot be set to recover past losses. Therefore, some regulators allow for partially forecasted test years and some allow for fully forecasted test years. The use of partial and full forecasting has advantages, but is time consuming and presents a large number of assumptions that can be challenged in the general rate case.

After the case is prepared, the utility petitions the regulators for a general rate case. This will include the submission of a proposed tariff, revenue requirements, test year documentation, accounting records, quality of service records, and a lot of additional information called the *minimum filing requirement (MFR)*. General rate cases can take a long time, often more than six months. Therefore, in its petition, the utility may request an interim rate increase to be effective during the rate case proceeding. If the regulators approve the interim rate increase, additional amounts paid by customers are collected under bond and subject to refund plus interest if the permanent rates turn out to be lower than the interim rates.

After the case is filed, the regulators will create a docket with key dates and deadlines. A period of time is allowed for the commission staff and interested third parties to register as *intervenors*. Common intervenors include the office of public council, ratepayer advocacy groups, business advocacy groups, environmental groups, and possibly large industrial customers. Intervenors are allowed to participate in the discovery and hearing processes.

Although the utility will have submitted a large amount of data and analysis with its initial filing, the commission staff and intervenors will inevitably have many questions and want additional information. This occurs in the *discovery* process. There are various mechanisms at work during the discovery process. Initially, this occurs through several rounds of data requests where intervenors ask for documents, supporting data, or justifications for the proposed tariff. This process also allows formal questions to be asked of the utility, called *interrogatories* or *data requests*. The commission staff and intervenors may also conduct formal interviews, called *depositions*, with key utility personnel and expert witnesses used by the utility to justify the proposed tariff. It is also common for regulators to audit the utility's financial statements during this time.

After the discovery process, the commission staff and intervenors prepare written testimony discussing their recommendations with regard to the proposed tariff and justifications for these recommendations. The utility then has an opportunity to review and respond to this testimony through a written *rebuttal*. This rebuttal will sometimes agree with certain points and make rate case adjustments accordingly. The rebuttal will often disagree with certain points and provide data and analysis that either supports the original tariff, or makes modifications different from the original tariff but not in accordance with the recommendation of the intervenor testimony.

At the end of the discovery process, the commission staff, the utility, and all intervenors prepare a list of contested issues and the preliminary position of each interested party on each issue. This document is called the *prehearing order*. Negotiations occur during this process in an attempt to settle as many issues as possible. Unsettled issues in the prehearing order are the focus of the formal commission hearing.

A rate case hearing in many ways resembles a civil trial. The utility presents its case by calling witnesses, who are cross examined by the commission staff and intervenors. Opposing sides also present their case by calling witnesses, which are cross examined by the utility. A hearing typically examines revenue requirements, rate design, and quality of service issues. At any time during the hearings, parties can enter into settlement agreements for any particular issue under dispute. After the hearings, all sides will prepare *briefs* of their final arguments. Transcripts of testimony, supporting exhibits, and final arguments are all introduced into the permanent hearing *record*.

After the hearing, the commission staff will perform a detailed review and analysis of the entire record including the final arguments of all interested parties. The commission staff will then present its recommendations to the commissioners, including all proposed adjustments to the utilities initial rate case filing. The commissioners review the staff's recommendations along with the entire record, make their final ruling, and issue this ruling with a commission order.

The order will typically include the ruling, the basis for the ruling, the approved new rates, and the date that these new rates will become effective.

It is likely that not all parties will be happy will all aspects to the commission's order. If this happens, any party can ask the commission to reconsider any of its positions on one or more issues in the order. If still not satisfied, any party can file an appeal of the decision with the courts.

7.5   RATE BASE MISCONCEPTIONS

The traditional ratemaking process has lead to a strong "rate base culture" at many utilities. The rate base is perceived as how a utility makes money and stays financially healthy. A rate base culture tends to group spending into two bins, capital spending that goes into the rate base and operational spending that does not go into the rate base. It is taken for granted that capital expenditure, CAPEX earns a return. Operational expenditures, OPEX, are only recovered and are somehow less desirable than CAPEX.

The rate base culture stems directly from the equation used in ratemaking: revenue requirements are equal to non-capital expenses plus the rate base multiplied by a rate of return. Based on a narrow view of this equation, the spending strategy of a utility seems clear: address problems with capital dollars whenever possible. The rate base culture not only biases spending towards CAPEX, it sometimes creates a preference towards more capital spending. When facing pushback about a proposed capital project, some engineers will think, "What is the problem? Let's just build it, put it in the rate base, and make money on it."

From an accounting perspective, the rate base culture has some merit. In the short term, rates are fixed and do not change based on the mix or level of CAPEX and OPEX. All OPEX is expensed and results in a one-to-one reduction in pretax profits. In contrast, CAPEX is amortized over time and has much less of an effect on short term profits. For example, a $30 million increase in OPEX will result in a $30 million reduction in IBT for the year. In contrast, a $30 million increase in CAPEX, if amortized equally over thirty years, will only result in a $1 million reduction in IBT for the year (although this expense will recur for the entire thirty-year period).

One must be extremely careful when using accounting nuances to justify spending behavior. This is especially true for utilities making CAPEX and OPEX decisions. At this point, it has been recognized that the revenue requirement equation allows for a return on the rate base, and that CAPEX results in higher short-term profits when compared to OPEX. Does this mean that CAPEX is truly more desirable than OPEX? The answer is "no" for many reasons.

Recall that investors value companies based on expected future free cash flow. From this perspective, OPEX is more desirable than CAPEX because it results in lower taxes in the short term. Imagine a utility that performed an extensive overhaul on a heavy vehicle, costing $1 million. It is not clear whether this cost should be CAPEX or OPEX; both can be justified. The rate base culture would immediately assume that the cost should be capitalized if possible to grow the rate base. Assume that the utility tax rate is 30% and that the cost is depreciated by $100,000 per year. The tax benefit this year is 0.3 x $100,000 = $30,000. If the cost is expensed, the tax benefit this year is 0.3 x $1,000,000 = $300,000. From an investor perspective, it is much better to expense the cost than to capitalize the cost.

Besides tax implications, there are other reasons why the distinction between CAPEX and OPEX is somewhat artificial. It is easy to view the rate base as representing stuff that the utility owns. Investment is needed to purchase this stuff, and investors must be fairly compensated. All true, but oversimplified. Consider the daily activities of a typical utility crew. This crew could spend a day inspecting equipment and making necessary repairs. In this case, the cost associated with the crew goes to OPEX. Alternatively, this same crew could be assigned to a "capital improvement" project intended to extend the life of equipment. It still spends a day inspecting equipment and making necessary repairs, but this time the cost goes to CAPEX. In both cases, no infrastructure is purchased. The utility pays workers for their time, but assigns these costs differently based on accounting rule interpretations.

A large percentage of utility cost is related to labor, and there are wide differences in accounting treatment. Should a utility that aggressively capitalizes everything it can make more money than a utility that is more conservative with capitalization practices? The rate base of the aggressive utility is higher, but common sense dictates that two utilities that spend money on the same things should achieve similar returns for investors, regardless of accounting practices.

Consider two utilities with different accounting practices. Both utilities are identical in terms of revenue and spending patterns. Annually, both spend about $2 billion in gross OPEX, $2 billion in gross CAPEX, and $2 billion in overhead expenses. Utility A allocates overhead expenses to all activities, resulting in net OPEX of $3 billion and net CAPEX of $3 billion. Utility B treats overhead as an expense, resulting in net OPEX of $4 billion and net CAPEX of $2 billion. Over time, the rate base of Utility A will grow significantly larger than Utility B. Does this mean that Utility A will make more money than Utility B, even though the utilities are essentially identical?

Since Utility A and Utility B have the same cash requirements for capital purchases and operational expenditures, their investment needs are identical and their revenue requirements should also be identical. Consider Table 7.4, which shows the CAPEX, OPEX, and rate base amounts for the two utilities. The revenue requirements of each are assumed to be $5.5 billion. This amount allows the utilities to pay all of their debt interest and provide an acceptable return to shareholders (tax effects are ignored in this example). To achieve this level of revenue requirement, Utility A must be allowed a 13.3% return on its rate base of $15 billion, while Utility B must be allowed an 11.7% return on its rate base of $10 billion. Using these numbers, each utility earns the same revenue. The

Table 7.4. Utility A and Utility B have the same cash requirements for capital purchases and operational expenditures, but Utility A allocates overhead across all activities while Utility B treats all overhead as OPEX. Since these two utilities are essentially identical, their revenue requirements should also be identical, in this case assumed to be $5.5 billion. To achieve this, Utility A must be allowed a 13.3% ROR, while Utility B must be allowed an 11.7% ROR. Dollar values in millions.

	Utility A	Utility B
Gross OPEX	$2,000	$2,000
Gross CAPEX	$2,000	$2,000
Overhead	$2,000	$2,000
Net OPEX	$3,000	$4,000
Net CAPEX	$3,000	$2,000
Net Rate Base	$15,000	$10,000
Depreciation period (years)	÷ 30	÷ 30
Depreciation Expense	$500	$333
Revenue Requirement	$5,500	$5,500
Net OPEX + Depreciation	− $3,500	− $4,333
Revenue Shortfall	$2,000	$1,167
Required ROR	13.3%	11.7%

Table 7.5. In 1980, a utility files for a rate case when inflation is high, resulting in a nominal rate of return of 20%. In 1985 the utility finds that revenues are the same (no customer growth), but OPEX and CAPEX spending have increased due to increases in labor costs and material costs. During this time, the rate base has also increased slightly. However, inflation has dramatically declined and a new rate case will reduce the nominal rate of return. Even though the rate base, operating expenses, and depreciation expenses have all increased, the revenue requirement has decreased significantly (all dollar amounts in millions).

	1980 Rate Case	1985 Rate Case
Inflation	13%	3%
Real ROR	7%	7%
Nominal ROR	20%	10%
OPEX	$3,000	$3,500
CAPEX	$3,000	$3,500
Net Rate Base	$15,000	$15,750
Depreciation period (year)	÷ 30	÷ 30
Depreciation Expense	$500	$525
Rate Base x Nominal ROR	$3,000	$1,575
OPEX + Depreciation	+ $3,500	+ $4,025
Revenue Requirement	$6,500	$5,600

allowed rate of return is different not because of differences in the cost of capital or cost structure. Rather, the rate of return is adjusted so that revenue requirements can be met. Revenue requirements are fundamental. The rate base and allowed returns on rate base are mechanisms to achieve the revenue requirements.

Revenue requirements, not rate base, should drive utility spending decisions. In fact, most utilities are obligated to make spending decisions that result in *minimum revenue requirements* while maintaining adequate service quality. From this perspective, should utilities avoid CAPEX spending, since it results in utilities getting a return? Again the answer is no. Shifting the same amount of OPEX spending to CAPEX spending will not necessarily impact revenue requirements. In a perfect world, returns on rate base could and should be ignored when making spending decisions.

In reality, the rate base culture remains strong and CAPEX is preferred over OPEX. Although utility executives and managers will not state this explicitly, CAPEX and OPEX budgets are made and tracked separately. CAPEX budgets strive to ensure that capital spending is at least much as the rate base depreciation expense. If not, the rate base will shrink, referred to as *rate base depletion*. Adjusting OPEX budgets is the primary mechanism of achieving earnings targets.

The rate base culture exists for a reason. There is a perception that regulators place a strong emphasis on setting the rate of return during a rate case. Instead of being incidental to revenue requirements, utility executives feel that revenue requirements and rate of return are determined in tandem, that past rates of return influence future rates of return, and that there is an aversion of regulators to increase rates of return. If any of these perceptions are real, a utility will benefit from biasing spending towards CAPEX, as long as there is not regulatory reprisal for imprudent spending decisions.

Like interest rates, the nominal allowed rate of return on the rate base has an inflation component and a real component. If inflation is high when rates are set, the nominal allowed rate of return will be relatively high. This can create a strong disincentive for utilities to file rate cases when inflation rates are low. Consider the example shown in Table 7.5. In 1980, the utility files for a rate case when inflation is 13%. The allowed real rate of return is set at 7%, for a nominal rate of return of 20%. Based on OPEX, CAPEX, and rate base, revenue requirements are set at $6.5 billion.

In 1985, the utility finds that revenues are the same (no customer growth), but OPEX and CAPEX spending have increased from $3.0 billion to $3.5 billion due to increases in labor costs and material costs. During this time, the rate base has increased slightly from $15 billion to $15.75 billion. However, inflation has dramatically declined from 13% to 3%. Therefore, regulators in a new rate case will reduce the nominal rate of return from 20% to 10%. Even though the rate base, operating expenses, and depreciation expenses have all increased, the revenue requirement has decreased from $6.5 billion to $5.6 billion due to the lower rate of return. This utility is not likely to file for a new rate case since rates will decrease rather than increase.

Of course, regulators at any time can instruct utilities to show cause as to why rates should not be reduced. In practice, this does not typically occur and utilities will often delay rate cases when inflation rates are low.

The rate base is an accounting artifact that attempts to estimate the amount of capital invested in the utility. From the perspective of an investor, a dollar is a dollar. All investments go into the treasury account, and this money is used to pay all cash expenses including CAPEX and OPEX. There are regulatory and accounting issues that, in certain cases, may result in advantages for rate base spending versus other types of spending. However, the culture of rate base is generally stronger than it should be as a result of many misconceptions.

7.6 SUMMARY

For some utility engineers, especially those involved with field operations, rates and ratemaking are far removed from daily job responsibilities. Rates are what they are, and do not affect how the utility system is built, operated, or maintained. Even so, rates and ratemaking are fundamental to utilities as businesses, and all engineers are encouraged to become familiar with the basics.

Other utility engineers will find many of their job responsibilities tied to rates and ratemaking. This might involve rate case preparation, demand side management, wholesale markets, performance-based regulation, and a variety of other possibilities. Although these engineers may not be involved in designing tariffs, the rate structures contained in these tariffs must be fully understood so that proper engineering decisions can be made.

The goal of ratemaking is to set rates for monopoly utilities that would otherwise occur in a competitive free market. Under competition, price is equal to the marginal cost of producing the next unit. Since marginal cost is virtually impossible to determine for utilities, regulators are typically satisfied with price being equal to average cost.

It is in everyone's best interest for utilities to be financially healthy. Therefore, rates must be set so that revenue is sufficient to cover all costs and provide a fair return to investors. The amount of revenue that achieves these goals is called the utility's revenue requirement. Typically, revenue requirement is formulated as expenses plus a rate of return on the utility's rate base.

After a utility's revenue requirement is determined, rates must be designed so that the revenue of a utility is likely to achieve its revenue requirement. A guiding principle of rate design is that rates should correspond to cost of service. Since different customer types have different cost of service characteristics, typical tariffs will have separate rate structures of different customer classes such as residential, commercial, and industrial.

Most rates are based on volume usage, but cost of service is highly impacted by peak demand. DSM attempts to reduce peak demand so that utility infrastructure expansion can be avoided or deferred. Most DSM approaches rely in some way on rates giving customers an economic incentive to reduce peak demand. This might include peak demand charges, critical peak pricing, reduced rates in exchange for load control, and a variety of other options.

There is tension between conservation of utility services and utility revenue. If customers use less of a utility service, utilities will make less money. Therefore, there is a financial disincentive for utilities to promote conservation. Many utilities try to work closely with regulators to effectively address conservation issues, and decoupled rates appear to be an effective approach.

Sometimes utilities go a long time between rate adjustments. During these times, there is a strong financial incentive for utilities to reduce costs, since this will result in increased profits. However, imprudent cost reductions may result in a worsening of system condition, reliability, customer service, safety, or other key performance areas. Therefore, rate regulation typically considers quality of service, and it is increasingly common for regulators to use performance-based regulation where quality targets are set and financial penalties are assessed if utilities do not meet these targets.

Rates are typically determined through a general rate case. A utility proposes a tariff and justifies this tariff in its initial filing. A lengthy process then begins where a variety of interested parties can request information, ask questions, and challenge the proposed tariff on a variety of grounds. This process culminates in a public hearing resembling a courtroom trial, where all sides present their arguments through witnesses and exhibits. After the hearing, the commission staff makes its recommendations, which are considered by the commissioners before they make their final decision.

7.7   STUDY QUESTIONS

1. What are some of the goals that regulators have when setting rates?
2. Do rates based on average cost mimic competitive free markets? Explain why rates are typically based on average cost.
3. What is the equation used to determine revenue requirements?
4. What does the rate base represent? What are some items that are sometimes allowed in the rate base and sometimes are not?
5. Explain how contributions in aid of construction are handled in terms of rate base, depreciation expense, and revenue requirements.
6. What are some of the main goals of rate design?
7. Why is peak demand important to consider during rate design? What are some approaches utilities can use to reduce peak demand?
8. Why is efficiency important to consider during rate design? What are some ways to encourage utilities to pursue efficiency programs?
9. What are some of the advantages and disadvantages of basing proposed rates on a historical test year?

10. What is meant by the rate base culture? Is it justifiable to prefer CAPEX to OPEX? Explain.

8
Budgeting

The process of corporate budgeting addresses the "who, what, when, where, and why" of corporate spending. Of course, budgeting determines the amount of money that will be spent. Once this is done, it identifies who is in charge of this spending, what the money will be spent on, when the money will be spent, where in the company the money will be spent, and why the money is being spent.

At the highest level, budgeting consists of identifying the total amount that a utility will spend over a fiscal year. At the lowest level, budgeting consists of identifying specific amounts of money for specific projects. There are many levels of budgeting in between these two extremes. Furthermore, there are often overlooked process requirements to ensure that low-level detailed budgets are consistent with high-level general budgets.

The budgeting process predicts future spending needs. Therefore, the only thing that is certain about a budget is that it is wrong. This said, budgets provide a baseline to which actual spending patterns can be tracked. In some cases, it may be desirable to manage spending so that budgets are met. For example, a utility may have a regulatory requirement to replace a certain percentage of meters each year. If spending on meter replacements is under budget, it may be desirable to increase the spending rate in this area. In other cases it may be desirable to deviate from the original budget. For example, the current price of meters may be higher than the price that was assumed during the budgeting process. In this case, it may be necessary to spend more than the original budgeted amount.

Budgets play a stronger role in utilities than in most other companies. This is because budgets are used to determine revenue requirements during the ratemaking process. When utilities deviate from budgets used during the ratemaking process, it is common for intervenors to demand justification. For example, a utility may have budgeted $20 million per year for equipment inspections in the last rate case, but only spent $15 million in the last several years. Intervenors will be reluctant to allow the utility to include more than $15 million per year for this activity when determining new rates, since the utility budgeted more last time and did not spend what they said they were going to spend.

Because of the nature of utility ratemaking, there are advantages for utilities to have spending correspond closely to budgets. There are many disadvantages as well. For example, it is common near the end of the year for spending in a specific area to be either over budget or under budget. If spending is over budget, utilities will often halt or severely curtail activities in this area so that budgets can be met. If spending is under budget, utilities will often increase activities in this area, even if the need is marginal, so that budgets can be used up. Although the utility can show that it is meeting its budgets, spending in this manner is often inefficient and is not how unregulated companies typically operate.

Utilities can become much more comfortable with budget deviations if they institute credible budgeting systems and processes. These systems and processes are able to justify all money that is spent in terms of benefits and costs. If spending is under budget in a specific area, a credible budgeting system will explain why money was either not spent (e.g., additional spending would not be cost-effective) or spent in other areas (e.g., spending was more cost effective in other areas). Not all utilities have this type of budgeting sophistication, but it is increasingly being recognized as desirable or even mandatory. Therefore, this chapter has a strong focus on *spending justification*, which goes well beyond the traditional budgeting function of *spending allocation*.

8.1 TOP-DOWN BUDGETING

Conceptually, high level budgeting at a utility is fairly simple. First, earnings targets are set. Second, revenue projections are made. Third, the total expense budget is assigned a value that is equal to the revenue projection minus the earnings target. In this case, the expense budget includes all items that will appear as an expense on the income statement. This relationship is expressed as follows:

$$\text{Expense Budget} = \text{Revenue Projection} - \text{Earnings Target} \qquad (8.1)$$

Since revenue projection is such a critical aspect of top down budgeting, many utilities have entire departments dedicated to this function. For most utilities, revenue is most strongly associated with consumption. Therefore, correctly projecting future revenue requires a utility to correctly project future consumption. A basic way to accomplish this is to estimate the consumption of a future year based on consumption that actually occurred in the most recent year. For example, consider a utility that experienced a total consumption in year y of

C(y). The weather in this year may not be typical. Therefore, this consumption is adjusted by a weather normalization factor, W. The utility expects that, on average, consumption per customer as a percentage of current levels will be CPC. The utility also expect that its total number of customers, as a percentage of current customers will be TC. Expected consumption for the upcoming year is calculated as follows:

$$C(y+1)=C(y)\times W\times CPC\times TC \qquad (8.2)$$

C   ; total consumption
y   ; year
W   ; weather normalization factor
CPC   ; projected consumption per customer as a percentage of current
TC   ; total projected customers as a percentage of current

Take for example a utility with current total consumption of 100 million units. This level of consumption occurred during a year of mild weather. The utility expects that consumption would have been 5% higher had normal weather occurred, resulting in W = 1.05. The utility also expects that conservation efforts will cause consumption per customer to decrease by 1%, resulting in CPC = 0.99. Last, the utility expects that the number of customers it serves will grow by 3%, resulting in TC = 1.03. Given these assumptions, the expected consumption for the upcoming year is 100 million x 1.05 x 0.99 x 1.03 = 107 million units. Assuming average rates remain the same, this utility can expect a 7% increase in revenue due to a 7% increase in consumption.

Consumption and revenue forecasts are typically much more complicated than the above example. Forecasts typically must be made, at a minimum, for each rate class and for each variable component of the rate structure within each of these rate classes. For example, consider a utility with separate rate structures for residential, commercial, and industrial customers. Residential and commercial rates are based on consumption. Industrial rates are based on both consumption and peak demand. To forecast revenue, consumption must be projected for all customer classes and peak demand must also be projected for industrials. Forecasting is further complicated if customers have the option to change rate structures, such as by participating in a demand response program. In these cases, expected participation must be forecasted so that the impact to consumption and revenue can be determined.

Once revenue projections are determined, the expense budget is set to a level equal to the revenue projection minus the earnings target. The budget will be smaller for operational outlays that are expensed (i.e., OPEX), requiring non-operational expenses such as taxes, interest payments, depreciation expenses, and potentially other items be subtracted. A general formula for determining the total corporate OPEX budget is the following:

$$OPEX = \text{Expense Budget} - T - I - D - N \qquad (8.3)$$

T   ; tax expenses
I   ; interest expenses
D   ; depreciation expenses
N   ; Other non-operational expenses, net

Operational outlays that are capitalized (i.e., CAPEX) require different treatment since they impact the income statement much less than OPEX. For example, consider an annual CAPEX level of $3 billion spent equally throughout the year with an average depreciation time of 30 years. The projected depreciation expense for this CAPEX in the year that it is spent will be approximately equal to the average amount that has been spent (in this case 50% of the total) divided by the depreciation time. In this case, the projected depreciation expense is $3 billion ÷ 30 ÷ 2 = $50 million. This is the incremental depreciation expense for new CAPEX, and does not include depreciation for existing plant, which will be much higher.

Because of the weak association of CAPEX to near-term earnings, capital spending must be justified based on a long-term economic analysis of capital needs, called *capital budgeting*. For most industries, capital budgeting assesses potential capital expenditures with a net present value analysis. For utilities, capital budgeting typically involves the identification of capital expenditures needed to build new infrastructure, replace old and failed infrastructure, and enhance existing infrastructure so that it performs better or is more cost-effective (e.g., through advanced technologies). This amount can be estimated as follows:

$$CAPEX = \text{Expansion} + \text{Replacement} + \text{Enhancement} + \text{Investment} \qquad (8.4)$$

A utility must connect new customers to its system and expand its system to accommodate expected growth in usage. The precise number and location of new customers cannot be known with certainty, but good planning will result in good estimates and accurate associated costs. Similarly, expansion planning will identify the most cost-effective way to add required system capacity. These costs together are the expansion portion of the CAPEX budget. Expansion costs will also include facility relocations, since the new plant is being constructed for reasons other than wear-out. Expansion costs allow a utility to meet its obligation to serve the customers in its service territory.

Depreciation represents the gradual wearing out of physical plant. To remain sustainable in the long run, a utility must replace its existing plant at least at the rate of depreciation, probably more since depreciation expenses are based on historical costs rather than current costs. Replacement costs can generally be categorized into reactive and proactive. Reactive replacements happen after equipment fails. The utility generally has no choice but to fix or replace the failed equipment, making reactive replacements an important CAPEX category.

Proactive replacements address old, worn-out, or technologically obsolete equipment before it fails, including capital maintenance activities. Proactive replacements can be very cost-effective and beneficial for reliability. But the utility can be challenged during a rate case as to why it is replacing equipment that is still functioning. Therefore, many utilities are cautious about proactively replacing equipment until there is strong evidence of imminent failure. Replacements allow a utility to continuously rejuvenate its existing infrastructure.

A utility might also wish to spend capital dollars to enhance various aspects of performance quality such as safety, reliability, customer service, or regulatory compliance. Sometimes this type of spending can be discretionary, such as replacing old but still functioning safety equipment. Other times this type of spending can be mandatory, such as replacing old safety equipment as required by a union contract. Enhancements allow a utility to meet regulatory requirements and to generally make things nicer for customers and employees.

It is often desirable to spend money now so that more money can be made in the future. Utilities typically justify these types of investment based on business cases showing a positive net present value. Financial benefits are typically future costs savings, but could also include future increases in revenue.

Although different utilities will categorize capital spending in various ways, the approach just presented is helpful in that it bases categories on the primary justification for why the money is being spent. Expansion CAPEX connects new customers and addresses capacity shortfalls. Replacement CAPEX takes care of failed and aging infrastructure. Enhancement CAPEX is nice to do if there is extra money to spend. Investment CAPEX will pay off over time. A summary of high-level categories for capital spending is provided below.

Broad Classifications of Capital Spending

Expansion:	new connections, system expansion, relocations
Replacement:	reactive replacements, proactive replacements
Enhancement:	safety, reliability, customer service, regulatory
Investment:	cost reduction, revenue enhancement

At this point, the reader may be wondering, "If next year's CAPEX does not have a strong impact on next year's earnings, how are CAPEX budgets determined?" The short answer is cash. Capital projects require cash, and the amount of capital spending is limited to available cash. All businesses would like to make more capital purchases than they are able to afford, for the same reason.

Recall from Chapter 2 that the statement of cash flow organizes cash transactions into three categories: cash from operating activities, cash from investment activities, and cash from financing activities. Once the OPEX budget is determined, cash from operating activities can be determined. The finance department will determine the cash expected to be generated by (or consumed by) financing activities. The CAPEX budget is equal to cash spent on investment activities. Assuming no change in cash balance, the capital budget is equal to the following:

Capital Budget = Cash from Operations + Cash from Financing   (8.5)

There will always be certain capital expenses that are mandatory. There will also always be additional capital expenses that are desirable. What happens if the capital budget, as estimated from the above equation, is more than mandatory plus desired CAPEX? If nothing else is done, the cash account of the utility will increase by the surplus. If the difference is substantial, the finance department will probably use the surplus to "strengthen the balance sheet" by buying back debt (which reduces cash from investing and therefore balances the capital budgeting equation). Buying back debt will reduce interest payments and result in higher future earnings.

More likely, there will be more mandatory and desirable capital projects than the utility has the ability to fund. In these cases there are three options. First, the utility can choose to only fund high priority capital projects that can be funded with the capital budget. All lower priority projects are rejected or deferred. Project prioritization is discussed further in Section 8.4. Second, the utility can choose to raise more cash through investment activities, such as issuing more equity or debt. Third, the utility can cannibalize its cash account, although this option is typically limited in extent.

To summarize so far, OPEX budgets are driven by revenue projections and earnings targets. CAPEX budgets are driven by cash flow projections. OPEX budgets are fairly constrained since they directly impact earnings. CAPEX budgets are more flexible and are developed iteratively with planned financing activities.

By itself, a high level *master budget* based on revenue and cash flow projections is not very useful. The master budget allows pro forma financial statements to be generated, and provides a baseline from which earnings and cash flow can be measured and controlled. However, the master budget does not identify how budgeted money will be spent. This requires a hierarchical budgeting allocation process that successively splits budgets into smaller amounts until they are associated with specific spending activities.

Figure 8.1 represents a typical top down budgeting process. At the top, organizationally-speaking, revenue and cash flow projections are made and used to create the master budget, including CAPEX and OPEX. These amounts are then allocated to various business units, including corporate overhead. Each business unit then allocates its portion of the master budget to various spending categories such as capital projects, inspection & maintenance, operations, and information systems. Each of these categories will have a certain amount of CAPEX and OPEX to spend. These amounts are allocated to specific budgets, each with an assigned budget number and an assigned budget owner. Budget owners can authorize spending from their budgets. Authorized spending is charged against its associated budget so that low level budgets can be tracked throughout the year.

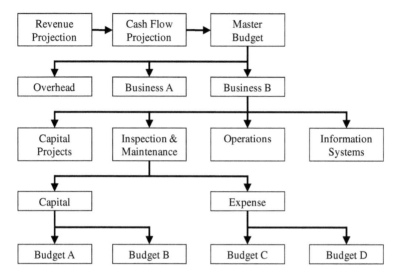

Figure 8.1. A typical top down budgeting process. At the top, revenue and cash flow projections are made and used to create the master budget. These amounts are then allocated to various business units. Each business unit then allocates its portion of the master budget to various spending categories. These amounts are allocated to specific budgets, each with an assigned budget number and an assigned budget owner.

Projections are never perfect, and budgets typically need to be updated throughout the year. For example, a utility may be experiencing lower levels of revenue than were assumed when developing the master budget. In order to meet earnings targets, the utility must reduce spending accordingly. Spending cuts can be made "across-the-board," such as by reducing all OPEX budgets by 10%. More commonly, budgets with significant discretionary spending and deferrable spending are identified and are assigned the majority of the budget reductions.

Revenues can be higher than projected, and projects can cost less than the budgeted amount. Both cases will lead to higher than projected earnings and/or higher than projected free cash flow. The utility can either maintain its current budget and have excellent financial performance for the year, or increase spending to bring financial performance more in line with projections. Typically the latter option is preferred due to the regulated nature of utilities. Higher-than-projected financial performance can lead to pressure from customers and regulators for rate reductions. Therefore, utilities will often strive to meet or slightly exceed their financial projections. To do this, budgets must be spent or slightly underspent. It is not uncommon for work activities to either increase substantially at the end of the year in an effort by managers to "burn up" their budgets, or to decrease substantially in an effort by managers to avoid budget overruns.

The process of projecting revenue, projecting cash flow, creating budgets, and spending budgets is called a *budget cycle*. Since the financial performance of a utility is assessed annually, budget cycles are almost always one year in length. The annual budget cycle presents difficulties for several reasons. First, many utility activities require substantial planning, engineering, and lead times for ordering equipment. Assume that budgets for the next calendar year are set in October of the current year. This only gives the utility two months (November and December) to prepare for the work that will be started in January. Practically, many major decisions must be made far earlier than two month in advance. Therefore, it is often necessary for money to be committed or spent before budgets are set. Some examples are the following:

Where Money Is Sometimes Spent before Budgets Are Set
- Ordering major pieces of equipment
- Purchasing land or land rights
- Construction permits
- Environmental assessments
- Engineering and design

Consider a construction project that is initially expected to start in January. For this to happen, equipment is ordered, land rights are acquired, construction permits are obtained, and engineering drawings are produced. When the budget is finalized, there may not be sufficient money for this construction project. If this happens, the ordered equipment will be cancelled if possible, probably with a financial penalty. Otherwise the ordered equipment will be (1) used in another project (hopefully), (2) placed into inventory, or (3) re-sold on the secondary market. Other expenditures related to this project may have value if the project is pursued in future budget cycles, or will be wasted if the project never happens.

Another drawback to annual budget cycles is that many utility activities should be planned over many years. For example, major construction projects can sometimes take a decade to plan, engineer, permit, and construct, with the construction activities alone spanning multiple budget cycles. Large enterprise software projects can take multiple years to implement. System-wide inspection and maintenance cycles can easily be ten years or longer. Hopefully, these activities are driven by a bottom-up planning process as described in Section 8.2. Too often, utilities do not have sufficient planning resources, resulting in spending activities that are primarily determined by the budgeting process. This short-term view can lead to sub-optimal decisions and, while saving money in the short term, can result in much higher long-term costs.

Some of the drawback to an annual budget cycle can be mitigated by budgeting farther out into the future. For example, an eighteen-month budgeting cycle will greatly aid in ordering equipment with long lead times. Typically, a budget that extends longer than one year is updated annually and is called a *roll-*

ing budget. For example a utility may utilize a three-year rolling budget. Each year, the second-year budget for the previous year is updated and set as the upcoming annual budget. In addition, budgets are estimated for the following two years so that spending can be better planned and coordinated from year to year. It takes more time and effort to create and manage a multi-year rolling budget, but many utilities find that the budgeting cycle becomes easier since much of the upcoming annual budget has already been created in prior years, and only requires updating.

Consider a utility that uses a three-year rolling budget and finishes its budgeting process in October. In October 2009, a detailed budget will be set for the upcoming year (2010). In addition, a less detailed budget will be created for the following year (2011). This budget will include many specific spending items, and will also include many generic spending allocations to account for spending that has not yet been identified with specificity. Last, a third budget will be created for 2012 that is even less specific. In October 2010, the somewhat detailed budget developed for 2011 in the previous year is updated and expanded into a detailed budget. Similarly, the not very detailed budget developed for 2012 is updated and refined to include specific spending items. Last, a new 2013 budget is created from scratch, but is not given much detail. This rolling budget process then repeats each year, as shown in Figure 8.2.

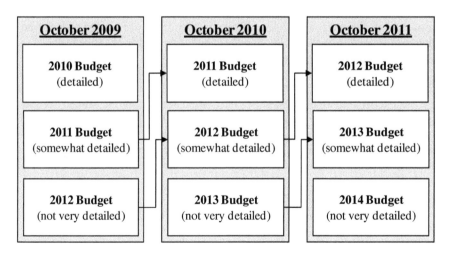

Figure 8.2. Example of a three-year rolling budget. In October 2009, the utility creates a detailed budget for 2010, a somewhat detailed budget for 2011, and a not very detailed budget for 2012. In October 2010, the somewhat detailed budget developed for 2011 in the previous year is updated and expanded into a detailed budget. Similarly, the not very detailed budget developed for 2012 is updated and refined to include specific spending items. Last, a new 2013 budget is created from scratch, but is not given much detail. This rolling budget process is repeated each year.

Rolling budgets should not be used in place of proper bottom up infrastructure planning. Rather, rolling budgets are a good mechanism to reflect the budgeting needs of the short-term portion of an infrastructure plan in a way that avoids the short-sightedness of a strictly annual budgeting process.

8.2   BOTTOM-UP PLANNING

Famous quotes about the benefits of planning abound. Two of the author's favorites are from World War II military generals. According to George S. Patton, "A good plan today is better than a perfect plan tomorrow." Dwight D. Eisenhower insists, "Plans are nothing; planning is everything." The author (who has been involved in the planning process at dozens of utilities) regularly reminds himself, "The only thing we know for certain about a plan is that it is wrong."

Why bother with planning if plans are not perfect? The answer is different depending upon whether one is referring to *short-term planning* or *long-term planning*. The goal of short-term planning is to anticipate upcoming system needs so that undesirable problems can be avoided. The goal of long-term planning is to ensure that actions performed today have lasting value.

Short-term planning exists because some problems take time to address. For example, a vertically integrated electric utility with growing demand will eventually need a new power plant. If the process of designing, permitting, and building a power plant requires five years, the short-term planning process must identify the need for the power plant five years ahead of time. Short-term planning is proactive in the sense that future problems and needs are recognized and addressed before they manifest themselves in an unacceptably negative way.

Short-term planning starts with the existing system in terms of equipment, customers, peak loading, average loading, recent performance, and so forth. Short-term planning should also consider committed system additions that are in the near-term construction pipeline. Last, short-term planning should forecast growth in terms of new customers, new customer locations, peak load growth, and other factors that have the potential to impact system performance. The short-term planning process uses this information to identify potential problems that will occur in the near future. Various alternatives are evaluated and the most cost-effective alternative is selected for implementation. The short-term planning process is shown in Figure 8.3.

Short-term planning is closely tied to budgeting. At a minimum, the short-term planning process identifies projects that consume budgets. For example, a utility may have a budget for system expansion to connect new customers. When planners identify new locations that need utility service, they will engineer a solution and create work orders that will eventually be charged to the system expansion budget. Tension occurs if there is insufficient budget to execute all of the projects identified in the short-term planning process.

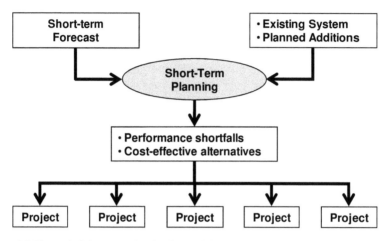

Figure 8.3. The goal of short-term planning is to anticipate upcoming system needs so that problems can be avoided. Short-term planning predicts the performance of the existing system after committed system additions and forecasted growth occurs. Potential problems are identified and cost-effective projects are selected for implementation.

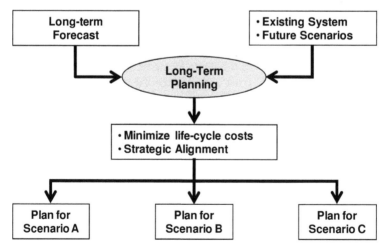

Figure 8.4. The goal of long-term planning is to ensure that actions performed today have lasting value in the future. A long-term plan looks at what the system should ideally look like in the future in terms of life-cycle cost minimization and strategic alignment with corporate goals. Short-term projects are then chosen so that the system can gradually transition the system from its current state to the desired future state.

Some projects, like new customer connections, must be done. Therefore, budget overruns in these "must do" situations will result in the transferring of money from budgets that have projects deemed to be less urgent. An engineer in charge of a budget could do a great job at identifying projects that fall within the budget. In the middle of the budget cycle, the great planning may all be for naught if another budget had poor planning, runs short of money, and ends up cannibalizing other budgets.

Budget shortfalls are not always due to poor planning. In fact, the budgeting process at many utilities is "top down" as described in the previous section and does not include significant input from the short-term planning process. Engineers can certainly present their case that budgets should be increased. The problem is that all budget managers typically request increases, and there is not enough money to fund all of the requests. If there is not a robust corporate process to prioritize spending, the budgets for next year are typically based on the budgets for this year.

The author firmly believes that, "the asset plan should drive the budget; the budget should not drive the asset plan." A utility system requires a certain amount of investment to provide adequate service to its customers. More spending than this required amount is wasteful, and less spending will result in either inadequate service, higher costs in the future, or both. A utility focused strictly on short-term planning will find it difficult to focus on the overall cost-effectiveness of its system, since short-term planning is similar to fighting fires as they appear throughout the service territory.

Many decisions made in the short-term planning process will impact the cost and performance of the utility system for decades. The goal of long-term planning is to ensure that spending decisions made today are cost effective over their useful life and that they result in a future system that is well designed and able to cost effectively provide adequate levels of service. Long-term planning strives to avoid "band-aid" fixes and incremental decision making. Band-aid fixes address symptoms without fixing the root cause of the problem. The root cause will have to be addressed eventually, and may be much more expensive to address in the future when compared to now. Incrementalism is similar, but is related to system expansion decisions made without looking at the system needs many years in the future.

The long-term planning process is similar in design to the short-term planning process except that it relies on a long-term growth forecast rather than a short-term growth forecast. Since a long-term forecast is very uncertain, various future scenarios are also typically considered. A long-term plan does not look at what needs to be done in the near term. Rather, it looks at what the system should ideally look like in the future in terms of life-cycle cost minimization and strategic alignment with corporate goals. Short-term projects are then chosen so that the system can, in addition to addressing short-term performance shortfalls, gradually transition the system from its current state to its desired future state. The long-term planning process is shown in Figure 8.4.

The not-so-near future is uncertain in many ways. Utility planners do not have a proverbial crystal ball, and cannot know with confidence what system needs will be in twenty or thirty years. However, utility planners can identify a range of reasonably possible future scenarios that the system will have to accommodate should it occur. For example, an area could experience low growth or high growth. An area could attract a major business, such as a theme park, that dramatically alters development patterns. A new interstate highway may or may not be built, similarly affecting development patterns.

Planners may feel that one future scenario is more likely than others to occur. It is a mistake to select a project simply because it is most cost-effective for this single scenario, especially if the project is not very cost-effective should other scenarios occur. It may make sense to spend a little bit more so that the project is *robust* across all reasonably possible future scenarios, not just the most likely. The amount of incremental money that can be justified depends upon the incremental value for each scenario and the likelihood of the scenario occurring.

Long-term planning does not result in spending decisions. Rather, long-term planning provides information so that short-term planning can result in better decisions over time. If long-term planning stops today, the negative effects are not seen immediately. Therefore, many utilities under budget pressures tend to cut back on long-term planning, sometimes resulting in a complete elimination of the function. Although this will reduce costs now, it can be a very expensive proposition over time.

To illustrate the typical results of an unplanned system, consider Figure 8.5. The shaded circles represent an undeveloped area in the utility service territory, with "Parcel A" being the source of utility services. Over twenty years, this area systematically develops in alphabetical order. In an unplanned system, the tendency is to connect the recently developed parcel to the nearest portion of the existing system. The system is first extended from Parcel A to Parcel B, from Parcel B to Parcel C, and so forth. Although each incremental expansion of the system seems cost-effective in the short run, the result over twenty years is an entangled and serpentine system that is inherently inefficient, unreliable, and expensive to operate.

Contrast the unplanned system to the planned system. In this situation, the planners understand that the area will likely be developed over twenty years, and have a vision for what the system should be at this time in the future. When a new parcel develops, it is not necessarily connected to the nearest location on the existing system. In Figure 8.5, the planned and unplanned systems develop identically up through Parcel D. In the unplanned system, Parcel E is connected by extending down from Parcel D. Although this represents the least-cost approach in the short term, it is a poor decision in the long term. Therefore, the planned system connects Parcel E by extending down from Parcel B and then over. Although this costs a bit more now, it results in a much better system as the remaining parcels develop.

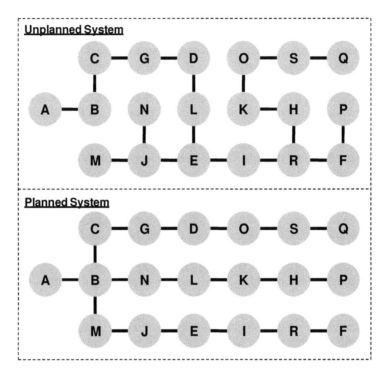

Figure 8.5. The difference between a planned and unplanned system. The shaded circles represent an undeveloped area. Over twenty years, this area systematically develops in alphabetical order. In an unplanned system, the tendency is to connect the recently developed parcel to the nearest portion of the existing system. The result over twenty years is an entangled and serpentine system that is inherently inefficient, unreliable, and expensive to operate. In a planned system, a newly developed parcel is not necessarily connected to the nearest location. Although this costs a bit more now, it results in a much better system as the remaining parcels develop.

Infrastructure planning is primarily an engineering function, and each type of utility will have highly technical issues that must be considered when planning the system. However, planning and budgeting are inextricably linked in both the short-term and in the long term. It is the burden of planners to identify both short-term and long-term infrastructure needs and ensure that these needs are properly reflected in budgets. A healthy process will be iterative, with an initial top-down budget followed by a bottom-up analysis of the implications of this budget. Planners might not always get all of the funding they feel is necessary, but at least the process will ensure that this funding shortfall is deliberate and well understood.

The effectiveness of planners and engineers in the budget process can increase significantly if funding requests are submitted with a rigorous and credible business case justification. This process is discussed in detail in Section 8.5.

The entire budgeting process will better align top-down budgeting constraints with bottom-up planning needs if it follows the principles of asset management. Asset management is discussed in detail in Chapter 9.

8.3   TYPES OF BUDGETS

The difference between CAPEX and OPEX budgets has already been discussed. Although important, there is another way to classify budgets that depends upon specificity. This includes, in order of most specific to least specific, project budgets, program budgets, and blanket budgets. Each of these budgets may include CAPEX, OPEX, or a combination of both.

8.3.1   Project Budgets

A project budget is typically prepared for a relatively large capital initiative with a well defined scope. Projects budgets are usually based on a detailed cost estimate. At a minimum, this cost estimate should be based on a detailed engineering analysis that identifies major cost items and verifies feasibility. Large and complicated project budgets are often based on detailed construction drawings.

A distinguishing characteristic of project budgets is that the work is self-contained. A typical project will result from a need that is identified through the planning process. The project will then typically proceed through conceptual design, engineering, construction documents, procurement, construction, and commissioning. After the project is commissioned, it is complete and no additional money should be charged to the project budget.

From a budgeting perspective, projects are desirable since both estimated and actual costs are transparent. The project budget is set when the project is approved. At the end of the project, actually incurred costs can be compared to the budgeted costs. Hopefully, many projects will come in under budget. Unfortunately, many projects exceed their initial budget. Cost overruns typically occur due to (1) underestimating the cost of materials, (2) underestimating the volume of labor, and (3) underestimating the scope of the project. Underestimating material and labor costs is understandable, especially if these estimates are based on a utility's cost accounting system. Underestimating the scope of a project is less forgivable, since it often results from insufficient attention to detail when setting the initial scope of work. Consider a project that involves the replacement of an old outdoor piece of equipment. The initial cost estimate includes the cost to purchase and install the new piece of equipment. When the project starts, it is found that the device does not fit on the old foundation, and a new foundation is required. It is also discovered that the control wiring for the new device is not compatible with the old control wiring. Both of these unanticipated costs will require money not in the original budget.

Project budget overruns are problematic in the overall corporate budgeting process since it is difficult and potentially expensive to stop projects once they exceed their initial budget estimate. Therefore, project budget overruns typically result in the cancellation of projects that have not yet been started or the reduction of non-project budgets that are perceived as discretionary. To minimize this problem, it is important for utilities to closely track ongoing projects against their initial budgets, to audit final project costs against initial estimates, and to continuously improve the scoping and estimating processes so that project budgets become more accurate over time.

8.3.2 Program Budgets

A program consists of a group of closely related activities that exist to achieve a specific outcome. Each activity within the group is typically small, but all of the activities grouped together result in a significant budget. Certain activities within a program may be known at the time of budgeting, but it is common for activities within programs to be specified throughout the year and well after the initial program budget is set.

A good example of a program budget is vegetation management. A vegetation management budget will typically include right-of-way branch trimming, hazard tree removal, and ad hoc trimming that addresses unanticipated vegetation problems. The utility will probably have a detailed list of upcoming rights-of-way that are scheduled for branch trimming in the next budget cycle. Therefore, this component of the budget can be managed in a detailed fashion similar to a project budget. Hazard tree removal may initially address a backlog of already identified hazard trees, but will also involve the removal of hazard trees that are newly identified throughout the year. Therefore, this portion of the budget will be based on a combination of identified specific tasks and unidentified but anticipated tasks. Ad hoc work, by definition, is not known with advanced notice and must therefore be budgeted based on anticipated volume.

Program budgets are manageable as long as the scope of included activities is specific. At the end of the year, the vegetation budget manager can communicate information such as the following: the miles of right-of-way trimming compared to budget, the dollars of right-of-way trimming compared to budget, the number of hazard trees identified and removed, the type and cost of various ad hoc activities, the actual unit costs compared to budgeted unit costs, and the total vegetation spend compared to budget. If a program is well designed, it is straightforward to keep the program budget manager accountable for the spending that occurs within the budget.

Programs tend to spend their budgets. If a program is spending less than budget, program managers will typically increase spending on unscheduled activities or accelerate work on backlog items. Similarly, if a program is spending more than budget, program managers will typically reduce spending on unscheduled activities or defer work, which increases the backlog. Because programs tended to spend according to their budget, it is often difficult to know whether program budgets are appropriate. For example, a program could be in charge of replacing a class of old and obsolete equipment. Initially, this program is designed to replace all of the equipment over five years. After the program begins, the program manager realizes that the cost for each replacement will be much more than the original estimate. After one year, the program has spent its full budget, but has only replaced half of original number of targeted devices. Was this program successful? Would it have been approved if the company had known that it would cost twice as much? Does it make sense to continue this program at the same rate and take ten years to complete the replacements? Often these questions are not investigated sufficiently. A good budgeting process will identify and address budget deviations early instead of waiting until after the end of the budget cycle.

An effective strategy for program budget management is to identify specific tasks within a program and separate these tasks out into a separate project. This allows identified activities to be examined in terms of scope and cost. Program management can then focus on how to identify good ways to spend the program budget. Projects are successful if they accomplish their scope within budget. Programs are successful if they can demonstrate cost-effectiveness for the actual work that is performed within the program.

8.3.3 Blanket Budgets

A blanket budget consists of an account number which "covers" a broad category of spending activities that are difficult to anticipate. When work is performed that is not associated with a project or a program, it will typically be charged to a blanket budget. Common examples of blanket budgets include new customer connections, replacement of failed equipment, unplanned maintenance, outage restoration, and major storms.

From a budgeting perspective, blankets are a necessary evil. They would not be a problem if blankets tracked costs in a meaningful and transparent manner. Unfortunately, costs get charged to incorrect blankets for a variety of reasons. There could be ambiguity, such as whether the replacement of a failed component on a piece of equipment should be charged to the replacement blanket or the unplanned maintenance blanket. Blankets can also be incorrectly used to perform work that was not approved as a project or program. Worse, costs can be charged to blankets simply because workers have memorized the blanket account number and it is therefore easy to enter this number.

Blanket budgets are typically very difficult to audit in terms of what was actually spent. At the end of a budget cycle, the total amount charged to a blanket will be known, but it will be unclear what the utility actually received in return for spending this money. In theory, a utility could gather all work orders, separate them based on blanket accounts, and attempt to glean information from these work orders. This would be an expensive and labor-intensive process, and it is questionable whether work orders will contain enough useful information for an examination of spending patterns. Practically, blankets budgets have little accountability since there is little transparency. Operational personnel will tend to use blanket budgets as they see fit, which may or may not agree with the intentions of management.

To make matters worse, blanket budgets often account for a majority of utility spending, especially for utilities that do not perform a large number of expensive capital projects. Projects and programs could be well managed, but a utility could still find that it does not have good controls on a large amount of spending due to blanket budgets.

There are several things that a utility can do to minimize the negative aspects of blanket budgets. The first is to have awareness and appreciation of the problem. The second is to keep the total value of blanket budgets as low as possible by forcing traditional blanket spending into programs and projects. The third is to keep blanket accounts specific. The fourth is to make it easy to enter appropriate blanket account numbers on work orders. The fifth is to have coded spending categories within each blanket budget. The sixth is to have a plan to audit each blanket to see if actual work is charged to the appropriate blanket and spending category. Last, any cost is that is assigned to the spending category "other" is immediately audited by management. These actions will provide much greater blanket budget transparency. Perhaps more importantly, workers will tend to be more thoughtful about charging work to blankets, knowing that they will be held accountable for these charges.

8.4   PROJECT PRIORITIZATION

Utilities will almost always have more requests for spending than they are able to fund. For this reason, it becomes necessary to prioritize spending. Sometimes prioritization is done very informally, such as spending that occurs within blanket budgets. Sometimes prioritization is forced to be explicit, such as when a proposed project is either approved or not approved. This section discusses how spending can be prioritized based on expected costs and benefits. It also discusses some of the difficulties that utilities typically encounter when attempting to prioritize projects.

8.4.1   Benefit-to-Cost Analysis

A benefit-to-cost analysis, as the name suggests, compares the expected benefits of a project to the expected net cost. The benefit-to-cost ratio (B/C) is generally not a pure economic measure since project benefits are typically measured in noneconomic terms, such as performance improvement or the elimination of an operational problem. To compare projects from a B/C perspective, the following formula is used:

$$B/C = \frac{\text{Expected Benefit}}{\text{PV of Costs - PV of Savings}} \tag{8.6}$$

The numerator of B/C is the expected benefit of the project or program. Ideally, this will be an engineering measure that can be both computed from historical data and predicted based on the expected impact of the project. Although benefit must be a single value to compute a ratio, multiple performance metrics can be combined through a formula into a single value if necessary.

The denominator of B/C is the present value of costs minus the present value of all savings (including expected increases in revenue). This is equal to minus the net present value of the project. If the denominator is negative, the project has a positive net present value and can be justified based on financial benefits alone.

It is important to measure costs and savings in terms of present value, and not in terms of impact to budget. For example, one project under consideration may have a low initial CAPEX but high yearly OPEX. Another project might have a high initial CAPEX but low yearly OPEX. Left alone, a CAPEX budget manager might well rank the two projects based in initial CAPEX to maximize the benefits gained by CAPEX spending. The correct approach is to minimize life-cycle costs, which requires the consideration of all impacted budgets.

Benefit-to-cost analysis is a reasonably effective way to prioritize projects, and many utilities are making use of B/C measures in their budgeting process. It is especially valuable when budgets are constrained and only a limited number of proposed projects can be approved. In such situations, all projects can be ranked based on their B/C scores and can be approved in order until budget limits are reached. The list of ranked projects can also be used throughout the budget cycle to approve new projects when more money becomes available and to defer projects when less money becomes available.

Benefit-to-cost analysis is not able to address a major problem associated with project ranking – the *gold plating* of projects. Consider a project that has a very high B/C score and will therefore be approved. Knowing this, the project manager could add additional scope to the project that will increase the cost without significantly increasing the benefits. Overall, the gold plated project still has a sufficiently high B/C score to get it approved. However, the project has embedded scope that is not cost-effective.

Although the use of B/C for project prioritization is a good start, it is inherently limiting since projects must either be approved or not approved. To reduce spending, the use of B/C requires entire projects to be rejected or deferred. A much better approach is to eliminate gold plating that may exist in proposed projects. More generally, a sophisticated prioritization process will "take budget out of projects rather than take projects out of the budget." To prevent gold plating and help ensure that the most value is being purchased for the lowest cost, benefit-to-cost analysis is insufficient and marginal benefit-to-cost analysis is required.

8.4.2   Marginal Benefit-to-Cost Analysis

How can a utility improve performance while cutting costs? Maybe it cannot, but it can assure that each dollar spent on the system is buying the most performance possible. A technique to help accomplish this goal is referred to as *marginal benefit-to-cost analysis* (MBCA).

MBCA is not magic. It simply states that dollars will be spent one at a time, with each dollar funding something that will result in the most incremental performance benefit, resulting in an optimal budget allocation that identifies the projects that should be funded and the level of funding for each. This process allows service quality to remain as high as possible for a given level of funding.

In the MBCA process, projects are not treated as a single set of activities that will either be completely approved or completely rejected. Rather, projects are defined by functional goals that may be achieved to different degrees and through different means. For the purposes of MBCA, a project is defined as follows:

> **Project –** A functional goal aimed at improving performance. A project may be effective to varying degrees (as measured by performance improvement), and can be achieved through different means.

A project addresses some issue of performance, but does not prescribe how to address the issue. For example, if a set of customers is experiencing poor reliability, a project might be identified that has the objective of improving the reliability of these customers.

Fundamental to MBCA is the concept of *project options* (also called *project alternatives*), which identify potential ways to achieve the functional goals of the associated project. These options should range from low-cost, low-benefit solutions to high-cost, high-benefit solutions. For the purposes of MBCA, a project option is defined as follows:

> **Project Option –** A method of potentially achieving the functional goal of an associated project. Each project option has an associated cost and an associated benefit.

Project options are an essential component of MBCA, but tend to be resisted at many utilities. This is due to the dominance of standard design practices. Engineers and planners are often reluctant to submit project options that are less expensive and less aggressive than what would have typically been proposed in the past. This reluctance is misguided. It may be better to address a project in a low-cost manner than to not address the project at all. For example, a utility may not have sufficient budget to improve the reliability of an underperforming area through the preferred engineering approach of a complete rebuild. If only the rebuild is proposed, it will probably be rejected. If many options from low cost to high cost are proposed, an appropriate level of funding can be selected based on value. The rebuild may even be selected. If it is, the MBCA process will ensure that there is no gold plating and that it is fully justified based on benefits and costs.

MBCA operates on the philosophy that each additional dollar spent must be justified based on the value it adds to the project. To do this, each project must have options ranging from inexpensive to expensive. If a project is approved, its cheapest option is approved first. A more expensive option will only be approved if its increased benefit compared to its increased cost is high compared to other projects and project options. More expensive options are allocated until constraints become binding, such as when performance objectives are reached or budgets are exhausted. The MBCA process optimally allocates funding to projects so that the net benefit of these projects is maximized. Some projects will be rejected and some projects will be approved. For those projects that are approved, a specific option associated with that project will be identified.

Each project option consists of a cost and a benefit. Each project starts with a "do nothing" option that will typically have zero cost and zero benefit. All additional options are sorted from the cheapest to the most expensive. An example of four projects, each with four options, is shown in Table 8.1.

The next step in the MBCA process assigns a default approved option to each project. The approved option of each project should initially be set to "do nothing." Exceptions occur when a project needs to be done for safety or legal reasons. In Table 8.1, Project 4 must be done for safety reasons. Instead of "Nothing," its initial set point is assigned to the next least expensive option. All other projects are initialized to a set point of "Nothing."

Table 8.1. Example of projects, project options, and marginal benefit-to-cost. Each project consists of a set of options ranging from inexpensive to expensive. Initially, each project is set to the cheapest possible option, typically "do nothing." The marginal benefit-to-cost ratio is then computed for all higher-cost options based on the currently approved options. Options are successively approved based on marginal benefit-to-cost until performance targets are achieved or no money is left.

	C	B	$\Delta B/\Delta C$		C	B	$\Delta B/\Delta C$
Project 1				**Project 3**			
→ 1.1 Do Nothing	0	0	---	→ 3.1 Nothing	0	0	---
1.2 Cheap	20	40	2.00	3.2 Cheap	10	25	2.50
1.3 Moderate	25	48	1.92	3.3 Moderate	12	40	3.33
1.4 Expensive	30	55	1.83	3.4 Expensive	20	45	2.25
Project 2				**Project 4**			
→ 2.1 Do Nothing	0	0	---	4.1 Nothing	0	0	---
2.2 Cheap	50	150	3.00	→ 4.2 Cheap	50	120	---
2.3 Moderate	60	165	2.75	4.3 Moderate	60	130	1.00
2.4 Expensive	70	175	2.50	4.4 Expensive	65	150	2.00

→ is the project set point, C is cost, B is benefit, and $\Delta B/\Delta C$ is the marginal benefit-to-cost ratio.

The next step in the MBCA process is to determine the marginal benefit-to-cost ratio ($\Delta B/\Delta C$) of each possible project upgrade. This is equal to the additional benefit of moving to this option (as compared to the currently approved option) divided by the additional cost of moving to this option. For example, the marginal cost for Project 1 to go from "do nothing" to "cheap" is 20, and the marginal benefit is 40. Therefore, the $\Delta B/\Delta C$ of this option is 2.0. The marginal cost for Project 1 to go from "do nothing" to "moderate" is 25, and the marginal benefit is 48. Therefore, the $\Delta B/\Delta C$ of this option is 1.92. Table 8.1 shows the marginal benefit-to-cost ratios corresponding to the initial project set points.

The MBCA process allocates additional money by selecting the project upgrade with the highest $\Delta B/\Delta C$. In this case, the best option is to upgrade Project 3 from "Nothing" to "Moderate." Notice that the "Cheap" option was skipped over. This illustrates that all possible upgrades must be considered – not just the next most expensive option. If a particular option has a low $\Delta B/\Delta C$ and is not allowed to be skipped over, it may block other possible project upgrades and prevent the optimal combination of project options from being identified.

After the option with the highest $\Delta B/\Delta C$ is identified, the MBCA checks to see if all performance objectives are satisfied, if budget constraints have been reached, or if any other constraints will prevent further project options from being approved. If not, the project containing the approved option must have its $\Delta B/\Delta C$ values for the more expensive options recomputed based on the newly-approved option. In this case, the new set point for Project 3 is "moderate." The incremental cost of moving from "moderate" to "expensive" is $20 - 12 = 8$. The incremental benefit of moving from "moderate" to "expensive" is $45 - 40 = 5$. Therefore, the $\Delta B/\Delta C$ of this Option 3.4 should be updated to a value of $5 \div 8 = 0.625$. After this, the budget is updated based on the marginal cost of the approved option and the process is repeated. A summary of the MBCA process is:

Marginal Benefit-to-Cost Analysis Process
1. Identify all projects and project options.
2. Identify the cost and benefit for all project options.
3. Initialize all projects to the lowest cost feasible option. This will be "Do Nothing" in most cases.
4. Determine the remaining budget.
5. Compute $\Delta B/\Delta C$ for all potential project upgrades.
6. Identify the project upgrade that has the highest $\Delta B/\Delta C$ without violating any constraints.
7. Upgrade this project and re-compute $\Delta B/\Delta C$ for potential future upgrades.
8. Update the budget.
9. Have performance goals been achieved or has the budget been exhausted? If yes, end.
10. Are there any additional project upgrades that do not violate constraints? If yes, go to step 6. If no, end.

The MBCA process is not complicated or difficult to understand. It is a pragmatic way for utilities to systematically address performance issues with constrained budgets. Large problems with many projects and project options cannot easily be solved by hand, but this algorithm can easily be implemented in a spreadsheet application or more sophisticated software.

The MBCA process can also be used to generate a benefit-versus-cost curve. As each incremental amount of spending is approved, the total spending, marginal value, and accrued benefit are recorded. This allows a utility to examine how much it will cost to achieve various levels of performance, and how quickly marginal benefit reduces with increased spending. An example of this approach is shown in Figure 8.6. This MBCA was performed by a US electric utility examining ways to cost-effectively improve customer reliability (as measured by expected reductions in interrupted energy, MVA-hr). The analysis consisted of approximately seventy-five projects and three hundred project options. For this utility, marginal value decreases sharply as total spending increases. Almost all of the reliability improvement is achieved with the first $10 million, and additional spending is essentially wasted.

MBCA is a useful tool for assisting utilities in prioritizing expenditures as well as optimizing performance within a given budget. MBCA does not in itself result in budget reductions, but does show how to best spend within a defined budget, and how to achieve performance targets for the lowest possible cost given a pool of projects and project options. In addition, the process can provide helpful insights on questions such as, "How much do I need to spend to achieve various levels of performance, what will the performance impact be of budget reductions, how should money be incrementally spent or saved if my budget is changed, and can I achieve my performance improvement targets for my allocated budget?"

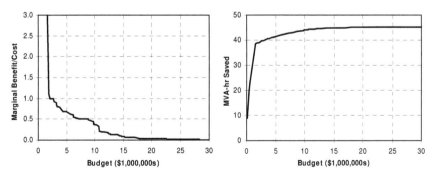

Figure 8.6. Marginal benefit-to-cost curves for an investor-owned US utility based on 75 project proposals containing approximately 300 project options (benefit is defined as MVA-hours of interruption saved and cost is defined as net present value). The left graph shows that marginal value drops off sharply as total expenditure increases. The right graph shows that almost all reliability improvements are obtained by the time $10 million dollars are spent.

8.4.3 Multiple Performance Targets

MBCA is straightforward if there is a single measure of benefit. However, most utilities have multiple performance measures that are of interest. These are commonly called *key performance indicators* (*KPIs*). Common non-financial categories of KPIs include safety, reliability, customer service, and customer satisfaction. Each category is likely to have multiple KPIs.

One approach to multiple performance targets is to have a separate MBCA analysis for each KPI. This is most effective if all KPIs are, for the most part, unrelated in the sense that projects impacting one KPI will not significantly impact other KPIs. For example, projects impacting a call center KPI are not likely to impact a safety KPI.

Once a set of mostly-unrelated KPIs is identified, several approaches can be taken. The most popular approach is to set a budget limit and a target for each KPI. Each MBCA analysis identifies the least-cost set of projects and project options required to meet its improvement target. If there is not a large enough budget to achieve the target, the required additional budget is determined. A separate process then looks at all of the KPI plans together. KPIs with spare budgets can shift money to KPIs with deficient budgets, targets can be reduced, and overall spending can be adjusted until each KPI has a budget, a target, and a spending plan to achieve the target for the least possible cost.

Another popular approach is to combine KPIs into a single measure using a mathematical function. Although this function can be as complicated as desired, most utilities use a simple weighted sum. The author recommends normalizing all KPIs based on their target values. This avoids the confusion that can arise from the use of KPIs with radically different units of measurement. For example, a call center KPI may be average wait time, measured in seconds. The existing call center KPI is 75 seconds and the target is 50 seconds. A reliability KPI may be average annual customer service interruptions. The existing reliability KPI is 1.2 and the target is 1.0. Without normalization, the weight of the reliability KPI will have to be sixty times that of the call center KPI, assuming equal importance. A less confusing approach is to normalize existing KPIs based on target values, so that the existing call center KPI becomes 150% and the existing reliability KPI becomes 120%. The equation for aggregate benefit with n KPIs is the following:

$$\text{Benefit} = \frac{w_1 \cdot \Delta KPI_1}{Target_1} + \frac{w_2 \cdot \Delta KPI_2}{Target_2} + \cdots + \frac{w_n \cdot \Delta KPI_n}{Target_n} \tag{8.7}$$

ΔKPI_i ; change in KPI_i
w_i ; weight of KPI_i
n ; number of KPIs

In the above equation, weights are proportional to the importance of the KPI. For example, a call center KPI with a weight of 1.0 would be half as important as a reliability KPI with a weight of 2.0.

A common approach similar to an aggregate measure as described above is called *economic mapping*. With this approach, each KPI is assigned a function that determines the economic value of the KPI at different levels. For example, a utility could determine that a 10% improvement in a KPI would be worth $2 million to the company, and a 10% reduction in the same KPI would be worth $3 million to avoid. Using these and a few more data points, a utility can generate the economic value associated with a range of KPI values. Economic mapping can seem appealing since, at some level, the motivation for all actions within a for-profit enterprise should be economic value. In practice, generating a credible economic map is difficult, and can lead to confusion, since the economic value of KPI benefit is closer to funny money than hard cash, and cannot typically be used to justify levels of spending (e.g., justifying a $2 million project since it results in $2.5 million in economic benefit). A safer approach is to avoid economic mapping and use weighted averages to combine multiple performance measures into aggregate measures of benefit.

Using an aggregate benefit measure has several advantages. First, it allows all projects to be examined by the same process, ensuring consistency. Second, it allows all projects to be examined at the same time and by the same people, greatly enhancing the ability of shifting budgets to balance KPI needs. Third, it allows projects with multiple KPI benefits to score better when compared to projects only impacting a single KPI. This encourages engineers to consider multiple KPIs when generating project options and can result in projects that "kill multiple birds with one stone."

Using an aggregate benefit measure also presents some problems. First, people who do not like the prioritized list of projects will tend to complain about the weights. Similarly, executives in charge of final project approval may start to use the prioritized list as a suggestion only, and approve many projects that they would have otherwise approved without the MBCA process. If this happens, cynicism will grow with the engineers in charge of project and option evaluation, resulting in an ineffective MCBA process.

A good approach to manage aggregate benefit measures is to develop a benefit-versus-cost curve similar to Figure 8.6, but showing the normalized value of all KPIs as spending increases. An example with three KPIs is shown in Figure 8.7. Currently, the KPIs are at 150%, 132%, and 120% of target values. After the initial project prioritization, it is shown that $15 million of the total $40 million budget is being spent on "must do" project options. Some of these options have KPI benefits, which are shown as a downward slope in the KPI curve. The "must do" project options of KPI_2 do not have any KPI benefits, resulting in an unchanging KPI value. After all of the "must do" projects are funded, project upgrades are successively approved based on their B/C ratios. At the budget level of $40 million, the KPI_1 target has been met, the KPI_2 target has not been achieved, and the KPI_3 target has been exceeded.

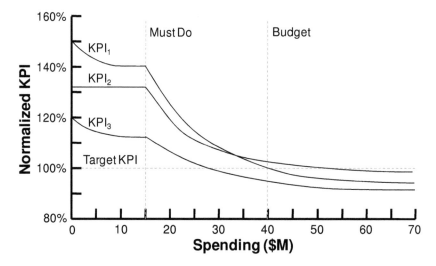

Figure 8.7. MBCA with multiple KPIs. In this situation, all project options are prioritized based on an aggregate benefit measure, but the results in the figure show the impact on each KPI. The first $15 million of spending is for "must do" project options, some of which have KPI benefits and others that do not. At the budget level of $40 million, KPI_1 is met, KPI_2 has not been achieved, and KPI_3 has been exceeded. This initial analysis should be used to refine "must do" requirements, targets, and budgets.

It is often helpful to create separate plots, similar to Figure 8.7, for each KPI. For these plots, spending only relates to spending that impacts the associated KPI, and may include a separate budget for improvement of this particular KPI. Plotting each individual KPI separately provides better information of benefit-versus-cost characteristics. This is because plots showing all spending will have flat portions for each KPI while spending is being allocated to other KPIs. It is also beneficial to rerun the MBCP without any project options being designated as "must do." The impact of "must do" project options can then be assessed by plotting the before and after curves for each KPI.

The initial prioritization list should typically not be the final list. Rather, it should be a starting point for budget refinement. Consider the initial MBCA analysis shown in Figure 8.7. KPI_3 exceeds its target in this scenario. A decision must be made whether to lower the target or to stop funding for this KPI when the initial target is reached. Assuming the latter, a second MBCA will be performed that shifts spending from KPI_3 to other KPIs after the target has been reached. Similarly, the target for KPI_2 is not reached in the initial analysis. It must be decided whether to raise this target, increase the overall budget, or increase the weighting of this KPI so that it is given a higher priority in the ranking process. The result of the MBCA process should be a set of KPI targets that are met but not exceeded when the allocated budget is spent.

It is important to distinguish between budget impact and present value when performing the MBCA process. All prioritization is based on the present value of project options so that life-cycle costs can be minimized. When a project option is approved, its impact to the upcoming budget is used to generate benefit-versus-cost curves. This distinction is described in further detail in Section 8.5.

8.4.4   Portfolio Effects

The importance of financial portfolios has been discussed in detail in Chapter 5. When performing project prioritization, it is similarly important to consider portfolio effects, but for projects rather than financial instruments.

As with financial portfolios, a benefit of project portfolios is diversification. A single project is relatively uncertain in terms of cost and benefit. It could cost more or less than expected, and benefits could be more or less than expected. Many projects allow these uncertainties to even out. Some projects will cost more and some projects will cost less, but the total cost of a diversified portfolio is likely to be close to the total initial estimate. Similarly, some projects will produce high benefits and some projects will produce less, but the total benefit of the portfolio is likely to be close to total initial estimate. Portfolio effects do not need to be considered when decisions are based strictly on expected value, but must be considered when evaluating the risk associated with not meeting performance targets. Section 8.4.5 discusses risk management in more detail.

Another important portfolio effect relates to project interactions. Some combinations of projects may result in higher total benefits than each project considered separately. For example, the addition of field automation equipment may produce modest benefits. Similarly, an infrastructure capacity upgrade may also produce modest results. If both projects are performed, the field automation can be used more extensively since it encounters fewer capacity constraints. Similarly, the capacity upgrades can be more fully utilized through the use of field automation. These synergies might make the combination of both projects cost-effective while each considered separately would not receive funding.

More commonly, projects experience diminishing benefits as the number of approved projects increases. Consider two reliability improvement projects designed to reduce a KPI with an initial value of 100. The first is able to reduce the KPI by 50 through a reduction in equipment failure rates. The second is able to reduce the KPI by 50 through improved response time. Clearly, the KPI will not be reduced to zero if both projects are approved. If the first project is approved, the second KPI will likely only produce an additional reduction of 25, since there are far fewer failures that will benefit from improved response times. If the second project is approved, the first project will likely only produce an additional benefit of 25, since each failure has less impact.

Project interaction effects such as those described in the previous paragraph pose a problem for project prioritization. Viewed separately, each project has a benefit of 50. Viewed together, the total benefit is only 75. When performing project prioritization, the B/C ratio of each project depends on which projects have previously been approved. There are several ways to address these interaction effects, each with difficulties.

First, all projects with interactions can be grouped as project options under a "meta project." Combinations of projects are listed as a separate option with a separate benefit. This approach only works if each project only interacts with a few others.

Second, a project interaction matrix can be devised that reduces benefits for project combinations. The default matrix has every element set to unity, indicating no interactions. Projects combinations with positive interactions have a value greater than one in the associated matrix element. Project combinations with negative interactions have a value less than one in the associated matrix element. This approach is flexible, but only explicitly captures the interaction effects of two projects.

Third, the total interaction effect for several levels of spending can be determined and used to develop a discount function based on total spending. A small amount of spending results in a small discount of total benefit. A large amount of spending results in a large discount of total benefit. An example is shown in Figure 8.8. The solid line shows how undiscounted benefits grow as prioritized projects are approved in order. For several spending levels, a detailed analysis determines the actual benefits considering the interactions of all projects. The ratio of the discounted benefit to the undiscounted benefit is the appropriate average discount factor to use for this spending level, and additional points can be used to generate an estimated discount factor for each level of spending. This approach is good for estimating total benefits but does not consider specific project interactions when creating the prioritized list.

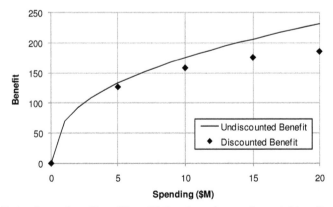

Figure 8.8. Project interaction effects. The solid line shows how undiscounted benefits grow as prioritized projects are approved in order. For several spending levels, a detailed analysis determines the actual benefits considering the interactions of all projects. The ratio of the discounted benefit to the undiscounted benefit is the appropriate average discount factor to use for this spending level.

After a group of projects and associated project options are identified for approval, the process is not complete. It is important at this stage to examine all of the tentative spending as an interdependent whole. This is important for two reasons. First, it allows all cost and benefit interactions to be explicitly identified so that overall costs and benefits are more accurate. Second, it allows the final mix of recommendations to be refined so that they work together as an integrated system. Taken together, the process of project prioritization, project interaction assessment, and portfolio refinement is called *project portfolio optimization*.

8.4.5   Risk-Based Spending

Project prioritization is based on average expected cost and average expected benefit. In terms of KPIs, utilities are often not interested in average performance. If average reliability is achieved, few people care. If better-than-average reliability is achieved, most people will not even notice. If much worse-than-average reliability occurs, there is trouble. The utility can be accused of not providing adequate service, not properly maintaining the system, and other aspersions. These dynamics are similar for other KPIs.

Utility analyses are typically based on expected performance, but utilities are typically interested in spending money to prevent unlikely but very bad outcomes. The author has heard this type of event called various things by various utilities including a high profile event, a negative media event, a big bad outcome (BBO), a high impact low probability event (HILPE), and a worst-case outcome of probability small (WOOPS). WOOPS will be used hereafter since the author finds it the most amusing.

It is often difficult to quantify the probability of occurrence of a WOOPS. Ideally, a WOOPS could be treated like other KPIs. For example, a KPI could correspond to the probability of a civilian death. A target could then be set such as no more than one death every five years. It is impossible to know from a few years of data what the actual probability of death is, unless many occur each year. Therefore, the current likelihood of a WOOPS must typically be estimated based on expert judgment.

It is somewhat easier to quantify the impact of a WOOPS should it occur since the impact is often embedded in the definition. If two similar events have significantly different outcomes, two different classifications of WOOPS are used. For example, a utility may be concerned about outdoor facility fires, but will create a separate WOOPS for rural facility fires and for urban facility fires.

Once all WOOPS are identified, they can be classified based on their probability of occurrence and severity of impact. An example is shown in Table 8.2, which categorizes the probability of occurrence of a WOOPS into four levels and categorizes the severity of impact of a WOOPS based on the expected media response. Combinations of probability and severity are designated as unacceptable, undesirable, or acceptable.

Table 8.2. Example risk matrix that classifies risks based on probability of occurrence and severity of outcome. Combinations of probability and severity are designated as unacceptable, undesirable, or acceptable. Risks classified as unacceptable are examined for cost-effective ways to reduce the probability of occurrence and/or the severity of outcome.

		Probability of Occurrence			
		Infrequent (1 year in 10)	Occasional (1 year in 5)	Frequent (1 year in 2)	Very Frequent (yearly)
Severity	Very Severe (national media)	OK	Undesirable	**Unacceptable**	**Unacceptable**
	Severe (state-wide media)	OK	OK	Undesirable	**Unacceptable**
	High (local media)	OK	OK	OK	Undesirable
	Moderate (lawsuits)	OK	OK	OK	OK

After all WOOPS are identified, they are evaluated and assigned a probability of occurrence and a severity of outcome. Based on the risk matrix, each WOOPS is classified as unacceptable, undesirable, or acceptable (i.e., OK). Other classifications systems can also be used. Unacceptable risks are then examined for cost-effective ways to either reduce the probability of occurrence and/or reduce the severity of impact so that the WOOPS is no longer unacceptable. If there are no cost-effective mitigation options, the utility should document the explicit decision that it understands and has decided to accept the risk.

WOOPS classified as undesirable can also be examined for inexpensive and easy ways to become acceptable. WOOPS with an acceptable risk classification do not require any mitigation actions.

Risk-based spending represents an important part of the budgeting process. However, it presents difficulties when combined with the project prioritization process. Mitigating unacceptable risks effectively results in "must do" projects. Often these risk-justified projects, combined with other "must do" projects, consume the entire budget and there is no spending left for project prioritization. This results in wasted effort and frustration for engineers creating projects and project options that have no chance of being funded. Risk-based spending on undesirable risks is also difficult to prioritize along with non-risk-based spending. Should a utility fund a project to move a risk from undesirable to acceptable, or fund a project with an attractive B/C ratio?

Risk-justified projects will often have KPI benefits. If so, these benefits should be considered in the overall budgeting process in the same way that all "must do" project benefits are considered. This ensures that all project benefits are considered when projecting future KPI values.

8.4.6   KPI Degradation

If a utility stops performing work on its system, most KPIs will deteriorate over time. Reliability will get worse; safety will get worse; and so forth. For this reason, it is helpful to include this natural degradation in the projection of future KPI values. If nothing is spent next year, each KPI can be expected to degrade. A certain amount of spending must then occur just to keep the KPI at its current level. Less spending will result in a worsening of the KPI. Additional spending will result in an improvement of the KPI.

Consider failures for a certain class of equipment. In the last five years, a utility has experienced a constant number of failures while keeping funding levels constant. An analysis shows that a cessation of inspection and maintenance would results in an increase in failures of ten percent per year. In the budgeting process, one option will be to keep the current inspection and maintenance program unchanged. This scenario can be examined in the budgeting process as follows:

Equipment Failures

2008 actual failures	1,000
10% annual increase	+ 100
Continue existing maintenance program	– 100
2009 projected failures	1,000

The benefits of this approach are twofold. First, it recognizes that the system will deteriorate over time if nothing is done. Second, it explicitly recognizes the benefit of existing activities. Often, benefit-to-cost analysis only looks at incremental spending. This approach allows new approaches to supplant existing spending if justified based on a higher B/C ratio. Assume that the existing maintenance program above costs $50 thousand per year, resulting in a B/C ratio of $100 \div 50 = 2.0$. For continued funding, this program must demonstrate that its B/C ratio is better than other projects considered in the project prioritization process.

Utilities may also have KPIs that are getting worse. This could be due to the deterioration of old equipment or a variety of other factors. Assume that a utility has experienced an average three percent increase in the number of equipment failures over the last five years, while keeping its maintenance program unchanged. Without maintenance, the utility estimates that the annual number of failures will increase by an additional ten percent. This scenario can be examined in the budgeting process as follows:

Equipment Failures

2008 actual failures	1,000
13% annual increase	+ 130
Continue existing maintenance program	– 100
2009 projected failures	1,030

If the utility wishes to keep equipment failures at 2008 levels, it must either keep its existing maintenance program and fund additional projects, or modify its existing maintenance program in a manner that reduces equipment failures by an additional three percent per year. When examining situations such as these, it is helpful to project KPIs out in time for at least five years. This allows KPI degradation to be proactively managed rather than reactively addressing KPIs that have already degraded to unacceptable levels.

8.4.7   Natural KPI Variation

Most KPIs will vary naturally from year to year. Some years will have good KPIs due to good luck and some years will have bad KPIs due to bad luck. This natural KPI variation has many similar characteristics to natural financial return variation. Therefore, it is helpful to think of KPI risk management as analogous to financial risk management.

KPI risk management is related to risk-based spending discussed in Section 8.4.5. However, KPI risk typically deals with the natural variation of broad performance measures while risk-based spending tends to deal with specific undesirable events. For example, it is common for utilities to measure average customer service availability with one or more KPIs. These measures will vary naturally from year to year, especially if service availability is impacted by weather patterns. This type of KPI is distinct from specific undesirable outcomes like the loss of a major facility that may have high visibility but not significantly impact a particular KPI.

KPI risk will not impact the project prioritization process if a utility is only interested in expected KPI results; that is, the average KPI that would result if a given year were repeated many times. However, many utilities have internal KPI targets linked to incentive compensation, have KPI targets set by regulators, and have additional reasons for wishing to achieve KPI targets a majority of the time. This poses a problem when KPI targets are set equal to expected KPI outcomes. Depending upon the probability distribution of KPI outcomes, this approach will result in utilities failing to achieve KPI targets about 50% of the time.

Consider the situation of a large US electric utility that is required by its regulator to track the average customer service unavailability each year (measured in minutes). This regulator has also set a KPI target that the utility is expected to meet. Results from 1998 to 2007 are shown in Figure 8.9, and break down the KPI into unavailable minutes that occurred during major storms, minor storms, and normal weather.

Clearly, there is large KPI variation from year to year based on major storm activity. This is typical across the industry, which is why it is common for regulators and utilities to exclude major storms when calculating service unavailability KPIs. Otherwise, utilities would typically meet KPI targets in years with mild weather and not meet KPI targets in years with severe weather. Also, large variations due to major storms will tend to mask underlying reliability trends such as gradual system deterioration or the benefits of reliability improvement initiatives. Regulators still expect a utility to efficiently restore service to its customers and repair its system after major storms, but these issues are examined apart from the unavailability KPI. As can be seen in Figure 8.9, elimination of major storms greatly reduces variations in the KPI.

The classification of a weather event as a major storm is somewhat arbitrary. Some weather events will meet the criteria and be excluded. Other weather events may not quite meet the criteria and, even though the weather was bad, will have to be included when computing the KPI. These not-quite major storms will be referred to as minor storms.

Similar to major events, minor storms can lead to large variations in service availability from year to year. Years with overall mild weather will tend to have a good KPI. Years with severe weather (but not severe enough to be classified as major storms) will tend to have a bad KPI. The example in Figure 8.9 shows that mild-weather years have historically experienced less than ten minutes of service unavailability due to minor storms. In contrast, severe-weather years have historically experienced more than forty minutes of SAIDI contribution due to minor storms. Therefore, service unavailability will regularly experience variations of thirty minutes or more purely due to the number and severity of minor storms. This variability presents difficulties when setting KPI targets. In many years, reliability index targets will be either achieved or not achieved based on the number of minor storms that occur, rather than on factors under the control of utility.

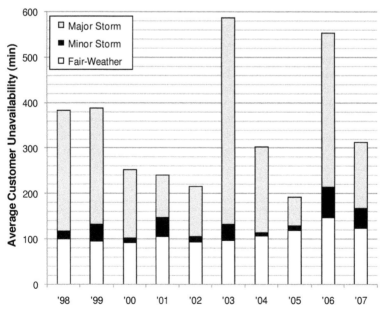

Figure 8.9. Example of KPI variability. For this utility, service unavailability varies dramatically with weather. Unavailability is low in years with mild weather and high in years with severe weather. This makes performance management based on unavailability problematic.

Service unavailability has been presented as an example, but natural variation will occur for all KPIs. Since this poses difficulties for performance tracking and performance management, it is worthwhile to examine how KPIs can be designed to minimize random variation. Major storm exclusion is an example of this approach for service unavailability. Another common approach is to associate upper and lower values for each KPI. As long as a KPI falls between its upper and lower value, performance is considered acceptable. A KPI falling outside of this range is investigated to determine whether the result was due to natural variation or a true degradation in the expected KPI value. In business operations, this approach of *statistical process control* is commonly applied to business processes with frequently updated quality metrics, but is equally valid for less frequently updated KPIs.

8.5   BUSINESS CASE JUSTIFICATION

As a thought experiment, think of two projects, Project A and Project B, that each are able to satisfactorily fix a problem that needs fixing. Project A requires an initial investment of $50 million and an annual cost of $1 million. Over the expected life of forty years, with a discount rate of 8%, the costs of this project have a present value of $62.9 million. Project B uses a cheaper approach. The initial cost is only $25 million, but requires replacement every ten years. Over the same forty year period, the costs of Project B have a present value of $69.2 million. Project A and Project B are summarized in Table 8.3.

Given these cost numbers, which project is more likely to be funded by senior management? Almost all utility engineers agree that, in their utility, Project B will be funded over Project A because it has a much lower initial cost. But the net present value of Project A is less, presumably making the project preferable from a shareholder perspective. Is senior management not looking out for the best interests of shareholders? The short answer is no. Senior managers are looking out for shareholders; they just do not trust the net present value calculation. They do trust the initial cost calculation, which is why they prefer Project B over Project A.

Table 8.3. Two projects that both satisfactorily fix a problem that needs fixing. Even though project A costs less from an NPV perspective, many senior managers will prefer Project B due to its lower initial capital cost. This outcome reveals that NPV calculations are not believed.

	Project A	Project B
Net Present Value ($M)	62.9	69.2
Benefit	fixes problem	fixes problem
Expected Life (years)	40	10
Discount rate	0.08	0.08
Initial Capital Cost ($M)	50	25*
O&M Cost ($M/yr)	1	1

* Replacement occurs in years 10, 20, and 30 with cost escalation of 3% per year.

There is a problem when projects are selected based on initial cost rather than on discounted life cycle cost as measured by NPV. Chapter 4 demonstrated that investors determine the value of a company based on an NPV analysis. Therefore, making decisions based on NPV maximizes shareholder value. The remainder of this chapter discusses how engineers can develop credible business cases using NPV that will avoid projects being approved primarily based on their impact to the current budget.

8.5.1 Cost Considerations

It has been repeated several times that the value of a business is equal to the net present value of expected future cash flows. This includes positive cash flows due primarily to customer payments, and negative cash flows due to the expenses of building, maintaining, and operating the system. These negative cash flows are typically called costs.

When examining the costs of a project, it is important to include both the costs directly associated with the project and all indirect cost impacts. Direct costs include items such as materials, labor, real estate, insurance, interest, and taxes. Indirect costs must consider issues such as *sunk costs*, *avoided costs*, and *opportunity costs*:

> **Sunk Cost.** A sunk cost is a cost that has already been incurred. From an economic perspective, sunk costs should not impact future decisions. In reality, sunk costs often influence later decisions due to noneconomic issues such as attempts to justify these prior expenditures.
>
> **Avoided Cost.** An avoided cost is a cost that would have normally been incurred but is not, as a result of a decision. Avoided costs are equivalent in value to cash costs and should be included in all economic decisions.
>
> **Opportunity Cost.** An opportunity cost is the cost of the next best economic decision. In business, opportunity cost is typically considered to be the cost of buying back bank loans, bonds, or stock. The weighted expected return of these cash sources is referred to as the weighted average cost of capital (WACC), and is the minimum opportunity cost hurdle rate of a company.

Costs can also be divided into *initial costs* and *recurring costs* (sometimes referred to as one-time costs and annual costs). Initial costs are typically incurred during the procurement and construction of a capital project, and do not repeat once they are paid. Recurring costs are the carrying charge of an asset and must be periodically paid as long as the asset is still owned. Table 8.4 provides a list of commonly considered initial and recurring costs.

Table 8.4. Common initial and recurring costs.

Initial Costs	Recurring Costs
Procurement Cost	Operating Cost
Material Cost	Maintenance Cost
Shipping Cost	Income Tax
Labor Cost	Property Tax
Equipment Rental Cost	Insurance Premiums
Commissioning Cost	(Depreciation)

Depreciation of an asset on a company's general ledger is treated as a negative recurring cost since it reduces a company's tax burden. Since depreciation lowers earnings by an equal amount, the negative recurring cost is computed by multiplying the amount of depreciation by the company's marginal tax rate.

The main difficulty in comparing the cost of different projects occurs when initial and recurring costs are different and/or occur at different points in time. If done correctly, an NPV analysis addresses this difficulty and is the preferred method to use. Annualized cost is equivalent to NPV and is also acceptable. Other common approaches include payback period and internal rate of return. Both have disadvantages and are not recommended, but still warrant an understanding.

8.5.2   Payback Period

The simplest method used to examine the economics of projects with both one-time and recurring costs is to examine actual cash payments. This is easily done by using cash flow diagrams such as those shown in Figure 8.10. These diagrams represent cash paid or cash received as arrows with length proportional to the amount of money for each cash transaction. The location of the arrow corresponds to when the payment takes place, with the start of the project corresponding to the leftmost point on the diagram and the project completion corresponding to the rightmost point. Multiple transactions occurring at the same time are represented by stacking arrows on top of each other, and a reduced cash flow diagram can be generated by summing all of the arrows corresponding to the same time period.

When making an investment, many simply want to know how long it will take for them to recover their initial outlay. This length of time is referred to as the *payback time* or *payback period*, and can be computed by summing values on a reduced cash flow diagram until the total becomes positive. If the project is characterized by an initial fixed cost (C_i), an annual recurring cost (C_a), and annual revenues (R_a), payback time can be computed as follows:

$$\text{Payback Period} = \frac{C_i}{R_a - C_a} \qquad (8.8)$$

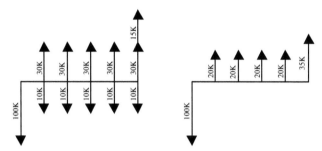

Figure 8.10. Cash flow diagrams. In the first year, $100K is spent constructing a project. In the next 5 years, the project generates $30K in revenue and consumes $10K in recurring costs. In the last year, the equipment is sold for $15K. The left figure shows all cash transactions and the right figure, a reduced cash flow diagram, shows the net cash generated or spent each year.

Payback period is useful in deciding whether or not to do a project (e.g., a company might choose not to do a project with a payback time of 10 years). Care must be taken, however, if payback time is used to compare projects. This is because payback time does not consider cash flow and the time value of money. For example, one project might cost $50 million and provide $10 million in revenue for 5 years. A second project might also cost $50 million and provide no revenue for 4 years but $50 million in revenue in the fifth year. Both of these projects have a payback time of five years. But the first project is more valuable than the second project because the utility will receive revenues earlier. These revenues can then be invested in other projects that have value.

8.5.3   Internal Rate of Return

NPV calculations require the use of a discount rate. The discount rate chosen can have a profound impact when comparing projects because a higher discount rate will favor options that defer investments until later years. This would not be a problem, except that discount rates are fairly "soft" values. Some will question the validity of NPV calculations by questioning the validity of the discount rate. This problem can be avoided by using the *internal rate of return*.

Internal rate of return (*IRR*) is defined as the discount rate that results in a zero NPV for a set of cash transactions in time. Stated another way, IRR is the discount rate that will make the present value of all costs equal to the present value of all revenue and/or savings.

IRR is computed iteratively. To begin, an arbitrary discount rate is chosen and the resulting NPV computed. Next, the discount rate is modified and a second NPV is computed. The next discount rate is chosen by either interpolating or extrapolating from these two values and the process repeats until a net present value of zero is obtained.

Table 8.5. Sample iterative calculation for internal rate of return.

Discount Rate (%)	PV of Costs ($ millions)	PV of Revenues ($ millions)	Net Present Value ($ millions)
0.0	150.0	200.0	50.0
10.0	130.7	122.9	− 7.8
8.0	133.6	134.2	0.7
8.2	133.3	133.0	− 0.3
8.1	133.4	133.6	0.2

For example, consider a project that has an initial cost of $100 million and annual costs of $5 million. This project is expected to produce revenues of $20 million per year and the life of the project is 10 years. The iterative IRR calculation for this example is shown in Table 8.5, with each row corresponding to a net present value calculation based on the stated discount rate.

The IRR calculation in Table 8.5 starts with a discount rate of zero, resulting in an NPV of $50 million. A discount rate of 10% is then chosen, and results in a net present value of minus $7.8 million. Since the value moves from positive to negative, the IRR must lie somewhere between 0% and 10%. Further computations show that the IRR for this example is 8.1%.

IRR can be used to rank projects in order of economic attractiveness, with better projects having a higher IRR. IRR can also be used to decide whether to perform a project or not. If the IRR is above a utility's weighted average cost of capital, the project should be considered for implementation since the economic return on the project is higher than the economic cost of the utility borrowing money to fund the project.

8.5.4   Developing a Business Case

Perhaps the most important aspect of a business case is credibility. If the managers in charge of the budget trust the business case process, they are much more likely to fund projects based on business cases that result from this process. A proven track record is invaluable. Imagine a budgeting process that has consistently demonstrated that (1) funded projects cost close to their initial estimates, (2) funded projects do not require extensive scope additions due to oversights, and (3) funded projects deliver all claimed benefits. With this track record, proposed spending plans will have a much higher probability of being taken seriously.

Most utilities do not have a strong track record of developing credible business cases, which is partially why it is difficult to justify projects in this way. The remainder of this section presents a recommended approach for business case development within utilities. Following these recommendations does not guarantee that projects will be funded, but increases the credibility of the business case itself and reduces the chance that a project will not be funded because the business case is incomplete, incorrect, not credible, or not compelling.

When approving spending, senior management typically asks, "Why do we need to spend this money?" This is a perfectly fair question, and always deserves a thoughtful and complete answer. If engineers cannot explain why spending is necessary or otherwise desirable, they should not expect the spending to be approved. With this in mind, it is helpful to think of the following four primary reasons why utilities spend money:

Primary Ways to Justify Spending
1. Positive NPV   (value creation)
2. Legal or regulatory compliance   (must do)
3. Improve performance   (good service)
4. Mitigate risk   (good image)

Positive NPV. It is always worth considering spending money today so that more money can be earned or saved in the future. More generally, positive NPV projects are of potential financial benefits to a utility since they represent the potential for lower revenue requirements and/or higher profits. An example of a positive NPV project is the replacement of old equipment that has high maintenance costs with new equipment that has low maintenance costs. A sufficient reduction in maintenance costs can justify the initial cost of the new equipment regardless of any other benefit that might result.

Legal or Regulatory Compliance. Often utilities must spend money to remain in compliance with regulatory or statutory obligations, regardless of whether the spending has a positive NPV. For example, utilities have an obligation to serve all customers in their service territory. If a new customer builds a home in a remote area within the service territory, the utility must extend service to the customer even if revenue from the customer is not expected to cover the costs of serving the customer. Through the rate case process, utilities will earn a fair return on prudently incurred investments performed to ensure legal and/or regulatory compliance.

Improve Performance. Sometimes a utility may wish to spend money so that performance levels exceed those that would result from minimum legal and regulatory compliance, or for areas of performance that do not have explicit legal or regulatory expectations. For example, a utility may decide to spend additional money to increase service availability, improve call center performance, or a variety of other possibilities. Ultimately, money spent in this category relates to the regulatory expectation of utilities providing "adequate service" in areas where specific regulations do not result in a "must do" situation.

Mitigate Risk. A utility may wish to spend money to mitigate an unacceptable risk by either (a) reducing its likelihood of occurrence, or (b) reducing its impact should the risk occur. Spending in this manner does not necessarily result in improved performance on average. Rather, the intent is to make performance more predictable and less subject to outcomes that deviate substantially from expected values.

Answering the question of "why do we need to spend this money" is a good start, but falls far short of a sufficient business case justification. A common mistake is for an engineer to simply (1) identify a problem, and (2) request approval to fund a project that solves a problem. Before spending significant amounts of money on projects, senior managers will want to delve deeper into the extent of the problem and why this particular project is the most appropriate course of action. The desired information can be classified into the following business case questions:

Business Case Questions
1. What is the need?
2. What is the project scope?
3. What will the project cost?
4. Will this project fully satisfy the need? For how long? Does it provide any additional benefits?
5. When will the project be complete?
6. What alternatives were considered? What were their costs and benefits?
7. Why is this recommendation the preferred alternative?

Knowing the answers to these questions is an important part of developing an effective business case. Having the ability to effectively communicate the answers to these questions to those holding the purse strings is even more important. The remainder of this section discusses each of the above business case questions in more detail and provides recommended approaches for developing comprehensive and credible answers.

What Is the Need?

The basic categories of need have already been presented and include good economics, compliance, improved performance, and risk mitigation. These are not mutually exclusive, and many projects will have benefits in more than one area. For example, replacing an old piece of critical equipment may result in economic benefits (reduced maintenance costs), improved system performance (the replacement uses better technology), and reduced risk (a catastrophic failure is less likely). On the other hand, some projects may have benefits in certain areas and detractors in other areas, for example, adding system redundancy to mitigate the impact of an unacceptable risk can result in expected performance degradation since there are more system components that can fail.

The following is a list of second level questions related to spending need. The first four items correspond to the basic categories described above. The remaining four items provide additional insight into the extent of the problem and the need to do something about it sooner rather than later.

Need – Second Level Questions
- Does the project have a positive NPV?
- Is this project required by regulation or statute?
- Will this project improve system performance?
- Does this project address an unacceptable risk?
- Are there strategic considerations?
- Is this an existing or anticipated need?
- How good are the assumptions and data?
- Is it worthwhile to improve assumptions and/or data?

Are there strategic considerations? It is not always possible to quantify benefits in specific terms. When this happens, it does not follow that benefits are non-existent. Sometimes decisions are made to support a high-level corporate strategy. For example, a utility may decide that part of its strategy is to transform its infrastructure into a "smart system" that has high levels of on-line monitoring and automated functions. A smart system requires an enabling two-way communications framework that is able to support a large number of devices spread across a wide service territory. When considered in isolation, many projects may not be able to justify the use of an expensive two-way communications system. However, these projects may specify two-way communications anyway so that they support the long-term vision of the company. If the strategy is sound, the benefits of the additional expense will materialize over time.

Justifying projects based on strategic alignment or strategic enablement should be done with caution. In the author's experience, the term "strategic project" is often a euphemism for "a project that will lose money." Strategy is important, and it is often necessary to invest money to pursue a strategy. But the goal of a strategic investment for a utility should always be to provide high levels of service quality for the least possible life-cycle cost.

Is this an existing or anticipated need? As discussed in the Section 8.2, the goal of short-term planning is to anticipate upcoming problems with enough lead time so that they can be prevented. Certainly, an existing problem with good documentation has a strong case for funding. The need for action is not as strong for an anticipated issue because there is uncertainty as to whether the problem will actually occur, when the problem will occur, and how bad the problem will be when it does occur. For example, demand forecasts may indicate the need for a new major facility that will take five years to design, permit, and construct. If this additional demand never materializes, money spent on the new facility will essentially be wasted.

How good are the assumptions and data? For both existing and anticipated needs, it is important to examine the basis for determination in terms of data and assumptions. Data is often not as reliable as one might think, and can lead to a false sense of confidence. Similarly, the determination that a problem exists or will materialize will often involve one or more assumptions, which may or may not be valid.

Is it worthwhile to improve assumptions and/or data? If there is not high confidence in the data and/or assumptions that have identified an existing or anticipated problem, it may be beneficial to improve the data and/or assumptions before spending money to address a situation that may not be as bad as initially thought. For example, an analysis may conclude that a problem probably exists, but with high uncertainty. Therefore, it recommends the deferral of any actions so that more and better data can be collected and so that assumptions can be verified and improved.

What Is the Scope?

To state the obvious, it should be clear what will be done if a project gets funded. In other words, the scope should be clear. In addition, a feasibility assessment should determine whether the proposed scope is sufficient and can be completed as proposed. The second level questions related to scope will vary depending upon whether the proposed spending is for a project, a program, or a blanket. Typical second level questions related to the scope of projects include the following:

Project Scope – Second Level Questions
- Is land available?
- Is there a conceptual design?
- Has there been a constructability review?

Is land available? Many utility projects require the use of new land. The land could be a self-contained parcel, a right-of-way for transmission or distribution facilities, or a combination of both. Without the ability to use the necessary land, the project as specified cannot proceed. Therefore, all project scopes should be clear and explicit about all land requirements. They should also be clear about the status of all related land purchases, leases, easements, permits, and other requirements for land use. Last, all project scopes should identify contingency plans should the preferred land options turn out to be infeasible.

Is there a conceptual design? At a minimum, the scope of a project should contain high-level schematics and drawings demonstrating the conceptual scope of the project. It may or may not be appropriate to perform a detailed engineering design prior to approval of the project, but it is difficult to have confidence in the scope of a project without the drawings and descriptions required to perform the detailed engineering design.

Has there been a constructability review? After looking at a conceptual design, an obvious question is whether a project can be completed as designed. It is not uncommon for a project to be approved, only to find upon further investigation that site conditions do not allow for completion of the project as original-ly specified. Therefore, it is often appropriate to perform *due diligence* to ensure that the project can, with reasonable certainty, be constructed as scoped. Often, this can be satisfactorily accomplished by having a person familiar with the con-struction of this type of project visit the intended site.

As discussed in Section 8.3, programs are different from projects in that they fund a large number of related activities, many of which may not be known at the time of the budget request. Since the development of a specific scope is often not possible for programs, the second level of scope-related questions is somewhat different than those for projects.

Program Scope – Second Level Questions
- What is covered in this program?
- What specifically will be done for the proposed budget?
- How much specific work has been identified?

What is covered in this program? For a program to have accountability, it must have clear rules about what types of activities can be funded through the program. It is not desirable for a program to become a type of money-laundering mechanism for projects that would otherwise not be funded. The simplest way to prevent unwanted spending is to limit programs to specific activities with specif-ic criteria. For example, a meter replacement program can replace meters if they are damaged or fail a calibration test. Some programs may be focused on solving problems, and may require a variety of possible activities to solve these prob-lems. In these situations, programs should have clear cost-to-benefit metrics and documented prioritization rules to ensure that all spending is justified. It is best if individual activities are not allowed to be too expensive. For example, a utility may rule that all activities costing more than \$30,000 cannot be funded within a program, and must be separately submitted as a project for approval.

What specifically will be done for the proposed budget? After a budget cycle, it should be possible to determine whether a program has accomplished what it claimed it would accomplish. To do this, the program must be specific in what it intends to do and in how it will measure its accomplishments against its goals. Spending its budget is not an acceptable program goal. A better approach can be demonstrated by using the meter replacement example. The program goal could be to replace a certain number of already identified meters and an addi-tional amount, not yet identified, based on projections. At the end of the year, this program can be examined based on volume (e.g., did fewer meters get re-placed due to fewer calibration failures?), and budget (e.g., did the budget run out early because of unanticipated increases in the price of meters?).

How much specific work has been identified? A program should be ex-pected to identify as much specific work that will be performed under the pro-gram as possible. Due to their nature, programs may not be able to identify in advance all work or even a majority of work. However, any program without at least 25% of work identified should be scrutinized closely and considered for budget reductions. It is not possible to credibly assess the value, in cost-to-benefit terms, for a nebulous program without significant identified work.

Compared to projects and programs, blanket budgets have poor transparency and poor accountability. Many field personnel like blanket budgets since there is relatively little bureaucracy, little paperwork, and little accountability for results. Many field personnel will make good spending decisions in a majority of cases. But this is of little help if the efficiency of spending is not demonstrable. At a minimum, it should be clear what can and cannot be charged to each blanket, so that overall spending on various activities can be tracked. The second level questions related to blanket scope are as follows:

Blanket Scope – Second Level Questions
- What is specifically covered in this blanket?
- What is expected to be done under this blanket?
- What is the basis for these estimates?
- What specifically was done under this blanket last year?
- Could anything in this blanket be treated as a project or program?

What is specifically covered in this blanket? By their nature, blankets have uncertainty in scope, and must maintain a certain amount of flexibility. This does not mean that anything can be charged to a blanket. Rather, a blanket should have clear guidelines about what can be charged to it and what cannot be charged to it. For example, a blanket may exist for the replacement of failed equipment. A utility is wise to define in detail what it means by failed equipment. Otherwise, at the end of the year, it will likely find a variety of additional activities being charged to this blanket if there is still budget left.

What is expected to be done under this blanket? Again, nobody expects perfect foresight regarding what will be performed under a budget. However, it is reasonable to ask the manager in charge of the budget to estimate the amount of money that will be spent on various activities. After the budget cycle, actual spending can be compared to these estimates, creating much more understanding and transparency than the traditional approach of only comparing the blanket spending to the blanket budget.

What is the basis for these estimates? Often, blanket budgets are increased as a matter of course, without scrutiny or justification. For example, there may be a blanket for new customer connections, which has traditionally been increased five percent per year to account for inflation. A better approach is to explicitly identify why a budget increase is needed, such as documented increases in labor costs, equipment costs, or an expected increase in the rate of new customer additions as estimated by economic forecasting services. The use of historical trends in spending should be avoided since blanket spending tends to closely match blanket budgets, rather than the amount of work that absolutely must be done under the budget.

What specifically was done under this blanket last year? Although related to the second question, this question is focused on the need for blanket budget managers to demonstrate that money spent under the blanket last year was appropriate, fully transparent, and cost effective. Answering this question can help to identify opportunities for blanket budget reductions. For example, an aggressive preventative maintenance program may result in reduced spending under the blanket for addressing failed equipment. If this reduction is not seen, the effectiveness of the preventative maintenance program can be examined. If the reduction is seen, the failed equipment blanket can be reduced accordingly so that extraneous money is not spent in lieu of the money no longer being spent in response to failures.

Could anything in this blanket be treated as a project or program? Since blankets are least able address spending based on cost-to-benefit ratios, and are the least transparent, they should represent as small a part of the overall budget as possible. A utility may have a brilliant spending prioritization process, but it will do little good if eighty percent of spending occurs in blankets (not uncommon). Therefore, there should be a gradual migration from blankets to programs and from programs to projects. Each budget cycle, each blanket should be examined for opportunities to split out part of the spending into programs or projects.

Once a project, program, or blanket budget is approved, there is a high risk of *scope creep* – doing more than was originally envisioned in the original approved scope. Scope creep is insidious for several reasons. First, it takes away money from other spending opportunities. For each dollar of scope creep, a corresponding dollar must not be spent somewhere else. Second, scope creep undermines the spending prioritization process. Engineers may try to get a project through the prioritization process on the cheap, and then have it done the "right way" through change orders during the construction process. Hopefully, at this point in the book, the reader understands the fallacy of this thinking. Last, scope creep destroys budgeting credibility. If spending consistently exceeds budgets, cost estimates will not be believed and, therefore, cost-to-benefit calculations will also not be believed.

Project scope creep generally takes two forms. First, a project is scoped without identifying all of the required work. For example, a project might specify the replacement of an outdoor control cabinet. When the project starts, it is found that the size of the new cabinet will not fit on the existing concrete foundation, requiring additional spending. It might also be found that the existing control wiring is not suitable for the control cabinet, requiring additional spending. Since these unanticipated costs can add up quickly, it is not a bad idea to add contingency funds to project budgets to cover these costs should they occur. It is also common for projects to simply grow after being initially approved. The thought process is often, "since we are already doing *this*, we may as well do *that* while we are here." Using the control cabinet example, a project manager may find that the existing concrete pad is usable but not in great shape. Therefore, the decision is made to replace the pad now since it will be much more expensive to do it later, after the new cabinet is installed. Perhaps this is a good idea, but additional spending like this should be prioritized along with all other spending requests, requiring an ongoing budgeting process that allows money to be shifted from one place to another.

Scope creep in programs and blankets tends to be a bit different from scope creep in projects. Typically, problems relate to either extraneous activities charged to the budget or the "burning" of unspent budget amounts towards the end of the budget cycle. Both issues can be minimized by ensuring correct charging, transparency of spending, accountability of people making the spending decisions, and minimizing the amount of money spent through programs and especially blankets. A sure sign that you have a problem is when workers have a memorized default blanket budget to which they, when in doubt, charge their activities (even worse if the number is inscribed under their hard hats!).

What Is the Cost?

For a business case to be credible in the eyes of the finance department, cost must be addressed in much more detail that a single number. Utility budgets are complicated and often interrelated. Prior spending and future spending should often both be considered. Project prioritization should be based on life-cycle costs rather than a single budget year. For business case justification, second level questions related to cost include the following:

> **Cost – Second Level Questions**
> * What budgets are impacted?
> * How good are these estimates?
> * Has previous money been spent that is related?
> * What discount rate is used?
> * What cost escalation is used?

What budgets are impacted? It is common for cost estimates to be limited to the impact on the budget that will be funding the proposed spending activity. For example, the addition of a piece of infrastructure will mostly impact the capital budget that funds its purchase and installation. Limiting the impact of spending to a single budget is insufficient for a proper life cycle cost analysis and is insufficient for a compelling business case. Instead, the expected impact to all budgets should be estimated. For example, adding the new piece of equipment will require an increase in the inspection and maintenance budget now and into the future. However, the new equipment may result in expected savings to other budgets. The author recommends that the expected impact to all budgets be projected out ten years. This is sufficient time to effectively compare CAPEX and OPEX projects, but is not so far into the future as to lose credibility, such as the

Table 8.6. A budget impact matrix for the proactive replacement of an old piece of equipment. The replacement will cost $50,000 in year zero, and avoid forced replacement costing the same amount in year four. Replacement avoids the current maintenance costs of $10,000 per year, but incurs new maintenance costs of $5,000 per year. The NPV of costs of this project over ten years is $12,900 (7% real discount rate).

Budget	Year									
	0	1	2	3	4	5	6	7	8	9
Proactive replacement	50									
Replace failed equip.					-50					
Maintenance (existing)	-10	-10	-10	-10	-5					
Maintenance (new)	5	5	5	5	5	10	5	5	5	5
Total	45	-5	-5	-5	-50	10	5	5	5	5
PV at 7% discount rate	45	-4.7	-4.4	-4.1	-38	7.1	3.3	3.1	2.9	2.7
NPV of Costs ($1000s)	**12.9**									

example described at the beginning of Section 8.5. The ten-year approach has multiple benefits. First, it allows NPV to be computed in a consistent and straightforward way for all projects. Second, it allows ten-year budgets to be easily examined. Last, it gives managers an advanced knowledge of spending in other budgets that might impact their budget, allowing them to act accordingly.

An example of a *budget impact matrix* is shown in Table 8.6. In this example, a proposed project will replace an old piece of equipment for a cost of $50,000, which will be charged to the "proactive replacement" budget in year zero. The expected budget impact does not end here. Based on an analysis of condition and historical experience, it is determined that the expected life of the old piece of equipment is five more years. It could fail earlier or later, but five years is the best guess. Therefore, the budget assumes that in five years it will have to be replaced due to failure, represented as an avoided cost in year four. From this perspective, the proactive replacement is advancing an unavoidable cost rather than generating a new cost. The existing equipment is expensive to maintain at $10,000 per year. These costs will be avoided after replacement (failure in year four is assumed to be mid-year). However, the new equipment will require maintenance costs of $5,000 per year, which is extended out ten years, except for a five-year overhaul in the fifth year costing $10,000.

The budget impact matrix can be used to easily compute the NPV of a project over the ten-year budgeting period. In Table 8.6, the NPV of costs is $12,900 using a real discount rate of 7% (having a positive NPV of costs shows that this project is not justified by pure economics, and must have other benefits to warrant approval). The set of budget impact matrices for approved spending activities can also be used to generate ten-year budgets.

How good are these estimates? It is relatively easy to develop a cost estimate. It is relatively difficult to develop cost estimates that accurately predict the actual expenses that will be incurred. Since the accuracy of cost estimates can vary widely, it is important to provide a basis for cost estimates and an indication of uncertainty. For example, a cost estimate may be based on a firm bid by a contractor, which can be assumed to have high credibility. A cost estimate could be derived from detailed design drawings and equipment quotes, which would similarly have high credibility. A cost estimate could also be based on unit costs from an accounting system, which would have less accuracy, or based on previous projects of similar scope, with similar implications. Developing accurate cost estimates can be time consuming and resource intensive, especially for complex projects that may not be approved. Therefore, it may sometimes make sense to have an initial cost estimate, say within plus-or-minus fifteen percent accuracy, which can be refined if the project is approved.

Has previous money been spent that is related? The well-defined budget cycle of a utility does not ensure that all projects will start and end neatly within this budget cycle. Provisions must be made for multi-year projects, multi-stage projects, and projects that, though less than one year in duration, span two budget cycles. Approval must still be explicit for each budget cycle, but approval of a project is nearly automatic when tagged with "in-progress, previously approved." Sometimes, up-front money will have been spent, but these sunk costs should not guarantee approval. For example, the engineering for a project may have been completed in the prior year, but the project no longer looks attractive compared to other projects, even with the engineering costs removed. Similarly, equipment may have been pre-purchased for a project. The project may no longer look attractive, even without the cost of pre-purchased equipment, especially if the equipment can be used in other planned projects.

In business case justification, care must be taken when previous money has been spent. A safe approach is to reduce the initial cost of the project by the already-incurred cost minus the value of anything that came from the expenditures, if any. For example, a project may be initially approved based on $10,000 in engineering costs in the first year and $90,000 in construction costs in the second year. After the first year, there is no value to the engineering work except for use in construction. Therefore, the second year spending must be justified based on the expected benefits compared to the $90,000 in construction costs. The $10,000 is a sunk cost and should not be factored into the analysis. Contrast this to a project where a specialized piece of equipment with a long lead time is ordered for $10,000. If the project does not go forward, the specialized piece of equipment can be sold in the secondary market for $6,000. Therefore, the project should be evaluated based on its original estimate (including the $10,000) minus the net loss of $4,000 from the specialized piece of equipment. In this case, only $4,000 should be considered a sunk cost since $6,000 can be recovered.

What discount rate is used? Since a business case justification is based on NPV, a discount rate is required. If the cash flow risk of the project is similar to the overall cash flow risk of the company, it is appropriate to use WACC as a discount rate. Stated differently, NPV should be based on WACC for typical utility activities where there is a high level of comfort in scoping, engineering, construction, commissioning, and operation. When the cash flow risk of a project is higher than the overall cash flow risk of the company, it is appropriate to use a risk-adjusted discount rate. Clear guidelines for discount rates should be developed in conjunction with the utility finance department so that NPV calculations and the resulting benefit-to-cost analyses will have maximum credibility. When asked why a discount rate was chosen, an engineer is well served by the response, "I used the discount rate guidelines developed by our finance department."

What cost escalation is used? Recall that nominal WACC consists of a real component plus an inflation component. Therefore, inflation must be applied to all future costs that are subject to inflation. Otherwise, the NPV calculation will be incorrect. In addition to inflation, certain cost may be expected to additionally rise in real terms due to increases in commodity prices, labor costs, and other factors. For these reasons, cost escalation assumptions must be clearly documented and consistently applied. It would be unfortunate for a project to be approved because the sponsoring engineer forgot to escalate prices, while a competing project was rejected because it correctly escalated prices and therefore seemed more expensive.

What Is the Benefit?

Section 8.5 started out by describing the basic categories of benefit including positive NPV, legal/regulatory compliance, performance improvement, and risk mitigation. When addressing the question of benefit, a business case must clearly identify each benefit, the category of each benefit, and if possible, quantify each expected benefit with clearly defined metrics. There will typically be one primary benefit, but all likely benefits should be listed.

If the primary benefit can be measured by a meaningful metric, a benefit-to-cost ratio should be calculated, which is equal to the expected change in the metric divided by the NPV (see Section 8.4.3 on how to address multiple benefits). The benefit-to-cost ratio allows all projects that impact this metric to be compared based on spending efficiency.

If the primary benefit is risk mitigation, the expected mitigation should be clearly stated in terms of reduced probability and/or reduced impact. It may be necessary to use a large amount of "engineering judgment" when estimating these benefits – an acceptable and necessary approach so that that risk-related projects can be better compared to other spending alternatives.

Last, any strategic considerations should be discussed. Sometimes the primary goal of projects or programs is to satisfy a strategic objective of the company. In these cases, the benefit discussion should focus specifically on how the spending will advance the corporate strategy, rather than assessing the merits of the strategy itself. More often, a project or program focused primarily on non-strategic benefits will have additional strategic implications. In these situations, it is important to discuss any strategic reasons for not pursuing any lower cost options or any higher benefit-to-cost options. It is also recommended to discuss strategic benefits even if the spending can be otherwise justified, thereby allowing a utility to have a better understanding of its strategic direction.

What Is the Timeline?

A rough timeline based on multi-year budgets will have been developed in response to the question "What is the cost?" In addition to a multi-year budget such as the one shown in Table 8.6, a good business case will describe the expected start date, the expected accomplishments for each budget cycle, and the expected completion date. The timeline discussion should also present the level of certainty of the estimated schedule. For example, the schedule of a project related to a city roadway widening project may be constrained by when the city undertakes the project.

A list, including status, of *critical path* activities should be included in the timeline. If a work item is on the critical path of a project, a delay of the work item will result in a corresponding delay of the overall project. The amount of time that a work item can be delayed before it becomes part of the critical path is called *slack*. Common critical path activities include land acquisition, permitting, engineering/design, and procurement of equipment with long lead times.

Last, each timeline should include a discussion of other projects that depend upon the completion of this project and/or other projects that must be completed before this project can be initiated or move past certain milestones.

What Alternatives Were Considered?

The next question that should be addressed in a business case relates to the alternatives that were considered. More than any other, this question demonstrates to senior management the richness and deepness of the thought process that resulted in the recommendation of specific spending activities.

The first alternative that should be considered is always "do nothing," even if doing nothing is not a realistic option. The "do nothing" option should describe what would likely happen if the recommendation were not funded. An equally important point to address is the impact of deferring this spending by one year. Everyone may agree that the proposed spending should eventually take place. A good business case should go beyond this conclusion by justifying why the spending should take place sooner rather than later. A robust discussion of deferral also allows the budgeting process to better prioritize spending when there is not enough money to pursue all desirable spending requests in the current budget cycle.

Of course, a discussion of alternatives should also address different approaches, including dissimilar alternatives and similar alternatives. Dissimilar alternatives try to solve the same functional problem in an entirely different way. For example, a recommended project may consist of building new facilities to address demand growth. Dissimilar alternatives might include customer consumption control, dynamic ratings for existing equipment, and tariff increases to reduce consumption. Similar alternatives might include a different choice of materials, overhead versus underground construction, different routes, different technologies, and so forth. The objective is to first discuss why the general approach is preferred (as opposed to dissimilar alternatives), and to next discuss why the specific choices within the general approach are preferred (as opposed to other similar alternatives).

Dissimilar alternatives should explicitly discuss whether it is possible to address the problem with CAPEX and/or OPEX. For example, if a project recommends the replacement of an old piece of equipment using CAPEX, the business case justification should discuss whether it is possible to make the equipment serviceable through OPEX activities such as increased inspection and maintenance. Similarly, if a project recommends an increase in OPEX to perform labor-intensive activities, the business case justification should discuss whether it is possible to achieve the same benefits through CAPEX related to process automation.

Similar alternatives should address the issue of marginal benefit-to-cost by discussing what would be done if the project were to spend a little bit more and what would be done if the project were to spend a little bit less. For example, a little more spending might allow a project to include some "nice-to-have" features. In contrast, a little less spending might force a project to exclude a few features that were included since they were nice-to-have, but not absolutely necessary. A well-structured prioritization system will automatically include these "spend a little bit more" and "spend a little bit less" options in the prioritization process so that spending recommendations can be based on project alternatives with the best marginal benefit-to-cost ratios.

In addition to discussing dissimilar and similar alternatives that essentially accomplish the same thing, a business case should present opportunistic alternatives that address other issues. For example, a capacity-related project might be able to address a reliability issue by spending a little bit more. Even though the capacity planner does not recommend this spending, it is mentioned in the business plan so that it is visible to those addressing reliability issues. Similarly, a project might be able to add future flexibility and options by spending a little bit more. The capacity planner has opted to keep the project as inexpensive as possible, but mentions this as a possibility to be considered in the budgeting process.

Last, additional spending could be spent in anticipation of future issues. It might only cost a little bit more to significantly increase the capacity of a new facility. Although the capacity planner does not recommend this incremental spending, it is mentioned as a possibility for consideration. A well-structured prioritization system and associated processes will automatically consider opportunistic alternatives in a manner similar to the way it considers dissimilar and similar alternatives.

Why Is This Recommendation the Preferred Alternative?

The last question to be answered in a business case justification is the equivalent of, "What is your closing argument?" The answer to this question typically takes the form of a summary of all prior questions organized in a clear and concise fashion, highlighting major points and avoiding details. The answer will typically take the form of a short narrative that can be quickly read and understood by individuals who will not necessarily scrutinize the entire business case. An effective answer as to why this recommendation should be pursued, and why it is the preferred alternative, will typically cover the following points:

> **Points to Cover in a Closing Argument**
> - It makes sense to do something now.
> - Spending a little bit less is less desirable.
> - Spending a little bit more is less desirable.
> - Different approaches are less desirable.
> - The scope is complete and credible.
> - The scope is fully coordinated will other activities.
> - The cost estimate is complete and credible.
> - The spending will delivered the promised benefit.

It is often helpful to group parts of the closing argument with other key pieces of information into a business case summary. Ideally, this summary is no more than a single page and presents all of the key information required for a reader to understand the problem, the recommendation, the cost, and the justification. Many individuals will not wish to read through the entire business case, and will greatly appreciate a single-page distillation of the important facts and recommendations. Supporting analysis for any of the summary information can be examined in the full business case if desired.

An example of a business case summary is shown in Figure 8.11, which recommends the purchase of an integrated voice response system (IVR). The summary starts out by stating all prior costs, the cost for the upcoming budget cycle, and the total cost. It then describes the primary benefit category of the project, in this case performance improvement. The business summary then describes the problem that this project is attempting to solve. A description of the recommendation is then provided, which is essentially a summary of the scope and estimated benefits. The last paragraph describes why this recommendation is being made, including other considered alternatives, the basis for the cost estimate, and any other key points. This type of business summary is invaluable during the budgeting process when dozens or hundreds of projects are being considered. Of course, each business case summary is backed up with detailed data and analysis which can be referenced, if necessary, during the budgeting process.

Business Case Summary			
Name:	Purchase IVR System	Prior Cost:	$0
ID:	P2010-046-03	2010 Cost:	$2.5 million
Prepared by:	John Doe, PE	Total Cost:	$2.5 million
Type:	Performance improvement		
Problem:	The current call center cannot adequately handle call volumes during large storms, resulting in busy signals, hold times exceeding ten minutes, extensive customer frustration, and complaints to the commission.		
Recommendation:	Install an Integrated Voice Response (IVR) system, allowing computers to answer calls if customer service representatives are not available. This system will triple the capacity of the call center during storms. The system can automatically cross-reference phone numbers to customer accounts and can automatically give region-specific information about storm damage and the status of restoration efforts.		
Justification:	The commission expects action due to many customer complaints after our last major storm. Other viable options include (1) expansion of our call center, which is much more expensive, and (2) routing overflow calls to an independent call center, which is problematic during a storm due to the difficulty of providing customers up-to-date information.		
	Pricing represents the lowest budgetary estimate of the standard package of three major vendors. More could be spent on added features, but these are not essential. More or less could be spent on call volume capacity, but tripling existing capacity is recommended based on a queuing analysis indicating a maximum two minute hold time during a major storm.		

Figure 8.11. A business case summary. In one page, this summary presents all of the key information required for a reader to understand the problem, the recommendation, the cost, and the justification. The points presented in the business case summary are backed up by the full business case analysis.

In addition to a business case summary, it is beneficial to have a short-form version as a stand-alone executive summary. Anyone reading the executive summary should have a good feeling about why the spending is needed, and what the spending will accomplish in terms of scope, benefits, and budget. For example, the executive summary in Figure 8.11 could be as simple as the following:

Install an Integrated Voice Response (IVR) system. This project will allow computers to answer calls if customer service representatives are not available, and automatically give region-specific information about restoration efforts during storms. This will solve the problems experienced during our last major storm. It will cost $2.5 million, based on vendor quotes, and will be done entirely within the 2010 budget cycle.

And so, a complete business case justification will typically consist of a short executive summary, a one-page business case summary, and an extensive business case analysis addressing all of the first-level and second-level questions presented in Section 8.5.4. The executive summary is used primarily by people wanting an overview of approved or tentatively approved spending. The business case summary is used primarily by people managing the budgeting and spending prioritization process. The detailed business case analysis is used primarily by the engineers making and defending the recommendation, and to give credibility to the executive summary and business case summary.

8.6   SUMMARY

The planning, budgeting, project prioritization, and business case justification process described in this chapter are potentially time consuming and expensive. As discussed earlier, it does not make sense to expend this amount of effort for every small spending decision. The point is to have robust and credible processes and systems so that the individuals responsible for funding the recommended spending decisions have a reasonable assurance that these spending recommendations are cost effective.

Without transparent and credible systems and processes, it is often difficult to justify any spending decisions that represent deviations from minimizing initial cost. Gaining credibility often takes time, but is well worth the effort. Utilities are encouraged to develop a multi-year plan that allows for a reasonable transition from the existing budgeting process to the desired budgeting process. This transition plan will address issues such as data, systems, methodologies, assumptions, scope, training, and culture change.

The budgeting approach laid out in this chapter represents a desired future state. Once this desired future state is reached, ambitious utilities can easily increase the sophistication of systems, processes, and analytics. However, the author recommends that each utility focus on the main aspects of this chapter first, which represent the football equivalent of "basic blocking and tackling." For example, it is risky for a utility to implement and customize a large enterprise software system, complete with business process redesign, before engineers have demonstrated the ability to develop and present a credible business case. Similarly, it is premature to implement a mathematically complicated budget optimization algorithm when there is no evidence that the utility can provide believable input data. Garbage in will result in garbage out, even if the "budgeting black box" is smart, sophisticated, and expensive.

The approach to budgeting described in this chapter involves a large number of details. Although these details are important for a budgeting process to have maximum credibility, they should be taken in context. If a utility wishes to significantly improve the way it spends money, there are several important issues to emphasize. First, the utility must be committed to an open and transparent process, where all spending is approved in the same way and executive "pet projects" are not allowed unless they go through the same process as every other spending request. Second, the utility must allow, at least in part, all problems to be addressed by CAPEX, OPEX, or a combination, as determined by minimum life cycle cost. Third, the utility must be willing to lower spending on projects, based on marginal benefit-to-cost, in a manner that may result in project scopes different from historical approaches. Last, utilities must attempt to reign in blanket budget spending in terms of lowering the total amount of blanket spending, and increasing the accountability of all blanket spending. Some utility executives may not desire one or more of these things, for various reasons. If this is the case, dramatic improvements in spending efficiency are not likely to occur through budgeting process improvements.

8.7   STUDY QUESTIONS

1. What are the primary factors that influence the expense budget?
2. What are the primary factors that influence the capital budget? Explain why these are different from those of the expense budget.
3. Explain the difference between short-term planning and long-term planning. Why are both important?
4. What are the three main types of budgets? Why is each type necessary?
5. Does a project with a good benefit-to-cost ratio always ensure that money is being spent efficiently? Explain.
6. What is meant by marginal benefit-to-cost analysis? Explain how a utility might look at a project proposal in terms of marginal benefit-to-cost.
7. What two factors must be considered when assessing an unacceptable risk and options for mitigating this risk?
8. What factors can cause a key performance indicator to vary from year to year? Why are these natural variations important to understand?
9. When developing a business case, explain the separate roles of net present value and impact to budget.
10. What are some of the important changes to which a utility must commit in order to make significant improvements in spending efficiency?

9
Asset Management

Asset management is becoming a frequently encountered term in utilities around the world. Although a few utilities have moved away from asset management, many more have committed themselves to asset management and expect it to be around for a long time. But when asked about asset management, utility engineers often have more questions than answers. Is asset management just another buzzword that will go away if it is ignored for long enough? If not, why is the company pursuing asset management and how will it affect my job? Do I need any new skills to be successful in an asset management environment? If so, what are these skills and how should I go about developing them? This chapter addresses these and other questions, and will begin with a sneak preview.

Asset management is not just a buzzword or a "flavor of the month" management initiative. It is a business approach that balances the financial aspects of a utility with the engineering and infrastructure aspects. As such, the content of this book provides a solid foundation for asset management. Not all utility engineers need to be business savvy beyond the essentials, and not all utility executives need to be engineering savvy beyond the simple basics. But effective asset managers at utilities need to be versed in both business skills and engineering skills.

As will be discussed later, the business environment for many utilities is undergoing a fundamental change. In many cases, old business models are not able to easily adapt to these changes. A new approach is needed, and many utilities feel that asset management is the answer. Business skills will help all utility engineers, but are particularly important for those in asset management organizations and for those in asset management positions.

9.1   WHAT IS ASSET MANAGEMENT?

Asset management is a business approach designed to align the management of asset-related spending to corporate goals. The objective is to make all infrastructure-related decisions according to a single set of stakeholder-driven criteria. The outcome is a set of spending decisions capable of delivering the greatest stakeholder value from the investment dollars available. For a utility, asset management allows a utility to provide acceptable service quality for the minimum revenue requirement in a credible and demonstrable way.

When asked, many utility employees are not able to provide a good definition of asset management. Perhaps the most common response is "it means different things to different people." In fact, asset management is a well-defined term that has a long history in the financial community. When an investment banker is asked about asset management services, the answer is very specific and will be similar to the following:

> **Financial Asset Management –** Making financial investment decisions so that returns are maximized while satisfying risk tolerance and other investor requirements.

Financial asset management is much more than a definition. The financial industry has a complete set of processes, data bases, theories, methodologies, and tools to help make financial investment decisions within an asset management framework. Many of these fundamental theories used in financial asset management, as discussed in Chapter 4 and Chapter 5, have had Nobel prizes awarded to the people who developed them.

Tangible infrastructure assets are different from financial assets. They deteriorate with age. They require periodic inspection and maintenance. They are part of an integrated system. Once installed, they cannot easily be taken out of service and sold. The list goes on. With all of these differences, it is expected that utility infrastructure asset management will not be identical to financial asset management, and will likely be more complicated.

By definition, for-profit businesses should be attempting to maximize profits and profit growth. Businesses are expected to pursue profits legally and ethically, but anything other than profit maximization within these guidelines violates the will of the business owners (i.e., common stockholders). Profit maximization is fundamental to a competitive free market, and businesses should proudly and aggressively pursue profits for their owners.

Profit maximization is nuanced for public utilities that operate as regulated monopolies. The regulated part of public utilities cannot strongly influence revenues, since rates are set by regulators. Therefore, profit maximization for the regulated portion of public utilities is virtually the same as expense minimization. With this point in mind, the definition of infrastructure asset management becomes the following:

Infrastructure Asset Management – Making data-driven infrastructure investment decisions so that life cycle costs are minimized while satisfying performance, risk tolerance, budget, and other operational requirements.

The easiest way to minimize life cycle costs is to stop spending money. Utilities clearly cannot take this to extremes since they must provide adequate service to their customers. Examples of performance requirements include connecting new customers, providing adequate service reliability, and achieving reasonable levels of customer service.

Utilities also have to work within a budget, and may have to levelize work over a multi-year period to keep within these budgets. Closely related to budgets is the use of craft labor. There are only a fixed number of employees to do the work, often requiring work to be spread out over time.

The above definition of infrastructure asset management makes sense from a regulatory perspective. Regulators expect utilities to provide adequate levels of service to customers for the lowest possible rates. The goals of infrastructure asset management are identical. Regulators also expect spending decisions to be backed up by good data and sound analysis so that budget and performance projections have high credibility. Asset management also has an expectation of rigor based on asset-level data.

Many utilities have applied the concepts of asset management to targeted aspects of their business. Examples include the following: equipment inspection and maintenance programs, computerized maintenance management systems, equipment condition monitoring, equipment utilization, risk reviews for cancelled projects, and software that prioritizes spending requests. These are all useful initiatives that can benefit from the concepts of asset management. However, they do not in themselves constitute asset management. Rather, they should be considered potential aspects of an overall corporate asset management framework. Asset management is a broad concept and, by definition, must consider all spending related to utility infrastructure.

Stated simply, asset management is a corporate strategy that seeks to balance performance, cost, and risk. Achieving this balance requires the alignment of corporate goals, management decisions, and engineering decisions. It also requires the corporate culture, business processes, and information systems capable of making rigorous and consistent spending decisions based on asset-level data. The result is a multi-year investment plan that maximizes shareholder value while meeting all performance, cost, and risk constraints.

Utility engineers will probably feel somewhat comfortable with the definition of infrastructure asset management since it represents an optimization problem. There is a single and clear objective: to minimize life cycle cost. However, the ability of a utility to minimize life cycle cost is a strong function of performance targets and risk exposure. A higher level of performance requires higher cost. A lower level of risk also requires higher cost. Therefore, inherent to asset management are the tradeoffs between cost, performance, and risk.

These tradeoffs must ultimately be aligned with overall corporate objectives set by senior management. With these points in mind, the goals of asset management become the following:

Goals of Infrastructure Asset Management
- Balance cost, performance, and risk.
- Align spending decisions corporate objectives.
- Base spending decisions on asset-level data.

At this point, asset management hopefully seems like a promising idea and a sound approach for utilities. If asset management makes so much sense, why have utilities only recently started to embrace the concept? To understand the answer to this question, it is necessary to understand the history of utilities in terms of where they have come from and where they are today.

9.2   HISTORY OF UTILITIES

As discussed in Chapter 1, public utilities are considered basic services that should be available to everyone. Further, public utilities require expensive investments such that it is generally not desirable or economical to have multiple companies with redundant infrastructure competing against each other.

For the better part of a century, many public utilities experienced rapid growth due to the connection of previously unserved customers and general population growth. Typical growth rates for US utilities from the 1940s through the 1970s were seven to ten percent, with even larger growth occurring after World War II in the 1950s and 1960s.

With high growth rates, utilities could easily satisfy investor expectations for annual increases in revenue. In addition, utilities were able to achieve significant economies of scale which allowed for lower and lower rates. In this business environment, everyone is happy. Profits are strong and growing. Rates are low and decreasing.

A seven to ten percent annual growth rate implies that the amount of utility infrastructure is doubling every seven to ten years. A necessary result of this situation was that most utility businesses were structured around new construction. As some would say, utilities were construction companies that happened to sell a service.

An efficient construction company is almost military-like in its organizational design. A command-and-control hierarchical structure allows orders from management to flow down the chain of command for implementation. The focus and value system related to this structure is on efficient infrastructure construction that can be built and forgotten about so as to not distract from even more construction activities in the future.

The combined effects of being a regulated utility, having high growth, and having an emphasis on new construction produced a distinct utility employee culture and mind-set. Cost becomes almost incidental. Utilities build new infrastructure according to their construction standards and it "costs what it costs." Less than perfect performance is not desirable since equipment failures or customer complaints distract from new construction activities. Risk is particularly undesirable and is to be avoided unconditionally. "Risk phobia" is further reinforced by risk-averse engineers being attracted to stable jobs within regulated utilities. Some have referred to a person with the typical utility mind-set as a DOUG – a darn old utility guy. A summary of the DOUG mentality is:

Traditional Utility Mind-Set
- It costs what it costs; just put it in the rate base.
- Failures are bad. Better performance is always better.
- Risk is to be avoided at all costs. Risk exposure results in stomach aches and insomnia.

Many utilities changed substantially during the oil embargo and resulting energy crisis of the mid 1970s. For energy utilities, the embargo resulted in a focus on energy reduction, greatly reduced growth rates, and even negative growth in many areas. Large construction projects (such as power plants) generally kept work forces busy through the mid 1980s followed by a need for dramatic staff reductions. In the telephone industry, the mid 1980s was characterized by the breakup of AT&T and the potential threat of competition from first generation mobile phones.

An important result of the 1980s was a change in leadership at the top levels of utilities. Until this point, most utilities were led by people with engineering backgrounds who had typically worked their way up through the organizational chart. These people understood engineering issues and engineering justifications for infrastructure spending.

With revenue and profit growth uncertainty, investors began giving increased scrutiny to utilities. This pressure led to widespread replacement of utility top leadership with financial professionals without engineering backgrounds. These new executives had a strong focus on cost reductions, initially through staff reductions, and later through reductions in expense and capital budgets. Often, these cost reduction programs were done in conjunction with rate freezes so that the utilities could keep the realized savings.

Reduced spending caught the attention of regulators due to the potential of reduced service quality. This led to an increased focus on performance measurement, performance management, and performance targets. This period also coincided with utility deregulation.

Recall that until the mid 1970s, utility infrastructure was typically doubling every seven to ten years. This means that half of all utility equipment was typically less than ten years old. In addition, most spending was driven by new customers or increased consumption, and was therefore associated with increased revenue.

About this time, many utility growth rates lowered to two percent or less. As a result, the average age of equipment began to rise dramatically. Utilities have always had to manage very old equipment, but this was typically a very small percentage. Today, a relatively small amount of infrastructure has been built over the last thirty years, and this equipment has typically been maintained less due to budget cuts. Further, the massive amount of equipment installed after World War II is now over forty years in age and is approaching the end of its useful life.

In short, many utilities are transitioning from a focus on building new stuff to a focus on managing old stuff. Managing an aging infrastructure requires a different skill set and a different mind-set. It also requires an increased need for spending justification since investments directed at aging equipment typically are not associated with increases in revenue. In the past, infrastructure was built to connect new customers, and the money automatically came when these new customers started paying their bills. Today, additional money to maintain and replace old equipment has to be requested from regulators, often through increases in rates.

Fortunately, asset management is a good fit for all of the issues facing utilities today. There are many details about asset management that are beyond the scope of this book, but the important point is that asset management requires a shift in mind-set towards the following:

Asset Management Mind-Set
- Primary focus on life cycle cost minimization.
- Exceeding performance targets is bad since less money could have been spent.
- Risk is an opportunity for increased profits. It is sought out, understood, and actively managed.

Notice that the asset management mind-set is the polar opposite of the traditional utility mind-set. In the old mind-set, cost was incidental. In the new mind-set, cost is the most important thing. In the old mind-set, better performance is always preferred. In the new mind-set, exceeding performance targets is a waste of money. In the old mind-set, risk is avoided. In the new mind-set, risk is actively managed.

In the past, utilities were run by engineers who made decisions because they made good engineering sense. Today, utilities are run by business people that expect decisions to be made because they make good business sense. Utility infrastructure is complicated and will always require sound engineering. However, engineering arguments are insufficient from an asset management perspective; engineering discussions are often not even possible with nontechnical people responsible for approving spending plans. Engineers with business knowledge can help bridge this gap. Asset managers represent an organizational function that is designed to bridge this gap.

Utility asset managers are a people with both engineering expertise and business expertise. They understand the technical needs of the system and the business needs of the company. They serve as a translator between the front line engineers and corporate officers. They have the ability to connect the "clean suits" with the "dirty boots."

9.3   HISTORY OF ASSET MANAGEMENT

In order to fully understand a subject, it is always helpful to be aware of its history and development. History helps to explain why things are the way they are, and whether these things are appropriate for your needs or are simply a historical artifact. For this reason, a brief history of asset management is now provided.

As mentioned before, asset management started in the financial industry. First to appear was diversification theory in the early 1950s. Diversification theory explains how risk and return are optimized on the efficient frontier. Next to appear was the capital asset pricing model (CAPM) in the early 1960s. CAPM built upon diversification theory by assuming that investors have a fully diversified portfolio. The last major milestone in financial asset management was options pricing in the early 1970s, which was a major breakthrough for risk management. These topics are discussed in detail in Chapter 5.

The first major application of financial asset management techniques to physical assets was the US nuclear arsenal. This is documented in the 1960 report *The Economics of Defense in the Nuclear Age*. It was written by Charles Hitch and Roland McKean, both financial economists, and provides an analysis of military policy planning in the nuclear age in terms of the most efficient allocation of available resources.

Asset management was gradually adopted by some of the more heavily regulated industries such as petroleum refining and chemical processing. The focus was quantititative risk management. Manufacturing industries also began adopting asset management with a focus of optimizing process uptime and system availability. The application of asset management to physical processes gradually became known as *physical asset management* (*PAM*).

Gradually, asset management started to be applied to public infrastructure such as roadways and water supply systems. An early example was the 1974 New Zealand Local Government Act which required local governments to set infrastructure performance measures and targets, report actual performance against targets, and add rigor to financial systems and policies. Asset management requirements were expanded in 1996 to require ten year infrastructure plans that consider the costs and benefits of various infrastructure options. To ensure consistency, the *Infrastructure Asset Management Manual* was published in 1996 by the Association of Local Government Engineers of New Zealand.

This document, now maintained by the National Asset Management Steering Group (NAMS), was most recently updated in 2006 and is now called the *International Infrastructure Management Manual*.

On the other side of the world, asset management gained momentum in the early 1990s with utility privatization initiatives in the United Kingdom. Privatization ultimately led to the 2002 merger of the Office of Electricity Regulation (OFFER) and the Office of Gas Supply (OFGAS) into the Office of Gas and Electricity Markets (OFGEM). OFGEM quickly initiated an Asset Risk Management Survey in 2002. After this survey, OFGEM began requiring the submission of annual asset management plans from its regulated entities. OFGEM, in coordination with other UK regulators agencies,[*] collaborated with the Institute of Asset Management (IAM) to write *Publicly Available Specification 55: Asset Management* (PAS-55). There are independent audit companies that check whether a company's policies and procedures are in compliance with PAS-55.

PAS-55 is intended to apply to "the management of physical infrastructure assets ... and in particular the assets that form the main element of our built environment such as utility networks, power stations, railway systems, oil and gas installations, manufacturing and process plants, buildings, airports, etc." Its stated objective for asset management is to have "the assets deliver the required function and level of performance in terms of service or production (output), in a sustainable manner, at an optimum whole-life cost without compromising health, safety, environmental performance, or the organization's reputation."

In the US, proliferation of infrastructure asset management has been steady yet less formal. Asset management conferences targeted to specific utility industries are common, as are consulting service offerings. A large focus has been on the deployment of information systems that enable effective asset management including executive dashboards, asset databases, maintenance management systems, and project ranking tools.

Perhaps the most organized activity has been done by the Electric Power Research Institute (EPRI) with its Nuclear Asset Management (NAM) program and its Asset Management Toolkit (AMT) project, which has published several useful reports including its *Guidelines for Power Delivery Asset Management*, initially published in 2004, revised in 2005, and re-written in 2008 (by the author of this book) to more closely align with PAS-55. In addition to applying the elements of PAS-55 to electric utilities, this document also adopts the separation of asset management functions into asset owner, asset manager, and asset service provider. Functional separation was first made popular by Yorkshire Electricity in the early 1990s after becoming an independent distribution company.

---

[*] The review panel for PAS-55 included: Chartered Institute of Public Finance and Accountancy, Department for Transport, Halliburton KBR, Health & Safety Executive, Interbrew, London Underground, the Ministry of Defence, Network Rail, OFGEM, Office of Water Services, Office of the Rail Regulator, UK National Air Traffic Services, and Yorkshire Electricity Group.

9.4 ORGANIZATIONAL FUNCTIONS

It is becoming more common for asset management companies to separate the asset management function from asset ownership and asset operations. The asset owner is responsible for setting financial, technical, and risk criteria. The asset manager is responsible for translating these criteria into a multi-year asset plan. The asset service provider is responsible for executing these decisions and providing feedback on actual cost and benefits. This decoupled structure allows each asset function to have focus: owners on corporate strategy, asset managers on planning and budgeting, and service providers on operational excellence.

In some cases, the separation of functions is dramatic. For example, in 1998, Long Island Lighting Company (LILCO) was taken over by Long Island Power Authority (LIPA). LILCO was a traditional electric and gas utility. LIPA is a pure asset owner that subcontracts all asset management and asset operation functions. Another example is Puget Sound Energy, which outsources almost all of its electric operations and gas operations.

Asset management is also about process. Instead of a hierarchical organization where decisions and budgets follow the chain of command into functional silos, asset management is a single process that links asset owners, asset managers, and asset service providers in a manner that allows all spending decisions to be aligned with corporate objectives and supported by asset data.

A conceptual diagram of an asset management organization is shown in Figure 9.1. The center circle represents primary functions. The outer ring represents the asset management process that links the primary functions. The primary inputs for the asset manager are corporate objectives from the asset owner and data from the asset service provider. The asset manager is then responsible for developing a multi-year asset plan that is able to achieve all corporate objectives for the least life cycle cost. Once a plan is developed, the asset manager translates the short-term portion of the asset plan into a budget and work packages that are contracted out to the asset service provider.

It is extremely difficult to perform effective asset management within the traditional utility top-down budgeting process described in Section 8.1. This is because managers tend to be very protective of their budgets, and are extremely reluctant to recommend a shift in spending from their budget to other budgets. Every year, every manager tends to ask for budget increases. In part, this is because the power and influence of managers are related to the size of their budgets. A reduction in budget results in fewer resources, less influence within the company, and a reduced ability to affect system performance.

The goal of asset management is to reduce spending wherever possible, but the personal desire of traditional managers is to increase their budgets wherever possible. Since managers are employed by the company, they are supposed to act in the best interest of the company. However, the personal budget goals of managers are often incompatible with the budget goals of the utility. This is called an *agency problem* since the managers have difficulty in acting as loyal agents for the utility.

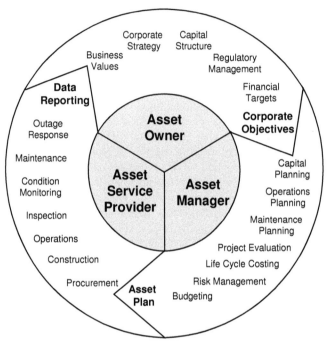

Figure 9.1. Asset management is based on three functions (asset owner, asset manager, asset service provider) surrounded by a single process. Functional separation allows asset owners to focus on corporate objectives, asset managers to focus on developing asset plans, and asset service providers to focus on efficient implementation of asset plans.

Functional separation goes a long way in solving the agency problem associated with traditional top-down budgeting. Since it separates the people who plan the work (asset managers) from the people who perform the work (asset service providers), there is no incentive for asset managers to protect budgets. In an ideal organization, asset managers make all infrastructure-related spending decisions at the same time and at the same place and with the same processes. If this process results in efficient shifts away from historical spending patterns, the asset managers are rewarded rather than penalized.

Even with a strong asset management organization, many spending decisions end up being made at an operational level. Problems arise every day that need to be addressed, and it is not practicable to require all operational actions to have prior approval by the asset management group. There is opportunity at most utilities to reduce blanket budgets for anticipated but unidentified spending. However, it is likely that these budgets will always be a large part of overall spending. For this reason, it is important to create an asset management mind-set for all employees who influence how money is spent. As stated previously, this means a shift away from "spending because it makes good engineering sense" to "spending because it makes good business sense."

9.5 ASSET MANAGEMENT PROCESS

Figure 9.1 helps to visualize asset management in terms of organizational function and scope. Figure 9.2 shows a typical asset management system from an organizational process perspective. Like Figure 9.1, the asset manager is given high level goals and objectives from the asset owner. The asset manager then examines current and potential performance gaps, identifies an optimal long-term asset plan, creates budgets based on the short-term component of the long-term plan, and contracts with the service provider to perform the work.

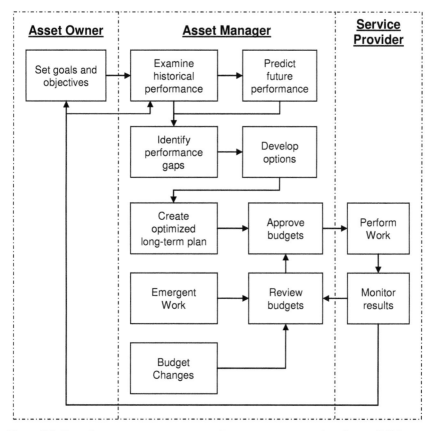

Figure 9.2. Example asset management system from a process perspective. Responsibilities are divided between the asset owner, asset manager, and asset service provider. The asset owner sets high level goals and objectives for the asset manager. The asset manager then develops a long-term plan that achieves all goals for the least life cycle cost. The short-term portion of the plan is translated into work items and budgets, which are executed by the service provider.

Although Figure 9.1 concisely shows the process flow for a typical asset management system, it does not provide details on the specific elements that constitute the process. Keeping in mind that different utilities will approach asset management in different ways, representative descriptions of the elements in Figure 9.2 are now provided.

Set Goals and Objectives. The asset owners must clearly communicate the priorities of the utility to the asset manager. Typically, this will involve goals and objectives related to OPEX, CAPEX, service quality, risk tolerance, and strategic direction. Sometimes these goals and objectives will involve specific metrics with specific targets. Just as common, they may involve high-level KPIs that will have to be mapped by the asset into more specific metrics. In any case, this step will typically involve a back-and-forth dialogue between the asset owners and the asset managers so that the asset management metrics sufficiently represent the goals and objectives of the utility.

Examine Historical Performance. After performance metrics and targets are set, the first job of the asset manager is to examine historical performance and determine all situations that are currently deficient vis-á-vis these metrics and targets. If these situations are expected to persist, they must be addressed in the asset plan.

Predict Future Performance. There are two aspects to this function, short-term performance and long-term performance.

Short-term performance. Many problems are not desirable to occur. This could be for a variety of reasons such as unacceptable safety, high cost to fix, lengthy time to fix, high impact to customers, and so forth. For these types of problems, it is the job of the asset manager to anticipate them before they occur. Problems should be identified in time to preempt or mitigate, considering the time required to plan and execute a solution.

Long-term performance. In the long-run, it is more cost-effective to design and engineer performance into the system than to reactively address performance shortfalls. Therefore, it is the responsibility of the asset manager to predict the long-term performance of the system so that strategic system approaches can be selected and pursued. Essentially, this function identifies long-term performance objectives, identifies a future system design that is able to achieve these long-term performance objectives for the least life cycle cost, and ensures that short-term spending decisions are compatible with the future system design.

Identify Performance Gaps. As mentioned in the previous two paragraphs, the asset manager must compile a list of all existing performance gaps, all emerging performance gaps, and all expected medium-term and long-term performance gaps. Addressing these gaps in a cost-effective way is the primary focus of the asset plan.

Develop Options. This process develops a range of options capable of addressing performance gaps in a way that achieves the overall goals and objectives of the utility. Many options will address a single performance gap, but others may address multiple performance gaps. It is helpful in this step to have a range of options including low cost, medium cost, and high cost.

Create an Optimized Long-Term Plan. Ideally, an asset plan will be based on a desired future state in terms of goals, objectives, and strategic direction. It is the job of the asset manager to have a roadmap. A roadmap is a high level, long-term asset plan that is expected to achieve a desired future state for the minimum life cycle cost. At a minimum, this long-term plan should contain a demand forecast twenty years into the future, a description of system expansion required to meet this demand forecast, a discussion of alternative scenarios, adoptions of different technologies and design standards, and high-level budgets extending at least five years. As discussed in Chapter 8, the long-term plan is a best guess, is likely wrong, and will have to be periodically updated. Its function is to ensure that short-term spending has long-term value, that strategic direction is systematically pursued, and that decisions are robust across a variety of future possibilities.

Approve Budgets. A large part of the asset management function involves budgeting, which is described in detail in Chapter 8. Once a long-term plan is established, the short-term component of this plan must be translated into budgets and work packages. These budgets should address performance gaps in a cost effective manner while balancing short-term needs with the long-term plan. The result will be budgets for approved projects, programs, and blankets. These can then be scheduled and gradually released to the asset service provider for implementation.

Perform Work. Once budgets and work packages are finalized, the asset manager releases them to the asset service provider, who is expected to perform the work on schedule and on budget. Typically, the asset service provider will consist of a variety of internal resources and a variety of external contractors. Work should be scheduled for both cost effectiveness and flexibility. For example, the initiation of large projects should be staged throughout the budget cycle to retain the ability to defer should unexpected costs or budget cuts emerge throughout the years (see "emerging work" and "budget changes" below).

Monitor Results. During normal operations, costs are tracked, schedules are tracked, asset performance is tracked, and system performance is tracked. All of this information is regularly fed back into the budget review and update process. This data is also provided to the asset owner for consideration when reviewing and updating goals and objectives. When performing work, the asset service provider should monitor progress in terms of schedule and budget. Any significant variances must be communicated to the asset management group for consideration. In addition, it is the responsibility of the asset service provider to collect field data about system performance and equipment condition. Data collection should be performed with the guidance of the asset management group so that it can support the computation of KPIs, help to predict future KPIs, determine whether projects deliver their expected benefits, help to develop and validate predictive reliability models, support condition-based decisions, and so forth.

Emerging Work. Budgets should anticipate a certain amount of emerging work. As the budget cycle progresses, the asset management group must identify variances between actual emerging work and the planned amount. Budgets and expected performance calculations should then be updated, potentially deferring formerly-approved projects if emerging work is more expensive than originally budgeted and potentially approving formerly-deferred projects if emerging work is less expensive than originally budgeted.

Budget Changes. Budgets often change during the course of a year. Changes could be from the top down, such as a directive from the CEO to cut O&M spending to meet earnings targets. Changes could also be from the bottom-up, such as a large unexpected expense (e.g., a major storm) that results in the need to reduce spending across a variety of budgets. When creating an asset plan and setting budgets, asset managers should anticipate budget changes and make sure that their budgets are able to reasonably accommodate these changes should they occur.

Review Budgets. Budgets should be regularly reviewed. These reviews should examine planned spending versus actual spending, planned schedule versus actual schedule, emerging work, budget changes, KPIs, changing priorities, and a variety of other factors that may require changes to budgets and approved projects. Ideally, asset managers should perform budget reviews at least monthly.

Communicate with Asset Owner. It is possible that resource or budget constraints preclude the possibility of addressing all goals, objectives, and performance gaps. In these cases, the asset manager should work with the asset owner to either increase resources or relax goals and objectives. If this is not possible, the asset managers should carefully document how work will be prioritized and what risks are involved for the organization.

As stated earlier, an asset management system can take many forms. The system described in this section is similar to the approach taken by many utilities, but does not necessarily represent all effective approaches.

9.6   SUMMARY

Many regulated utilities are transitioning from traditional organizations to asset management organizations. This change is appropriate since today's business challenges are different than the challenges in the past. Historically, most utilities primarily built new stuff. Today, many utilities primarily manage old stuff. In the past, an increase in spending was typically accompanied by an increase in revenue. Today, an increase in spending often requires an increase in rates. In the past, utilities were run by engineers that expected spending decisions to make good engineering sense. Today, utilities are run by business professionals that expect spending decisions to make good business sense. A hierarchical planning and budgeting process was appropriate in the past. An asset management planning and budgeting process is appropriate for many utilities today.

In many ways, an asset manager is a hybrid engineer and business person. An asset manager is able to understand the business objectives of the utility and communicate with senior management using business language and business concepts. An asset manager is also able to understand the engineering issues of the utility and communicate with operational staff using engineering language and engineering concepts. An asset manager takes pride in spending as little as possible so that profits can be maximized. An asset manager also takes pride in good engineering solutions that result in the achievement of all performance objectives.

Asset management is the art of balancing cost, performance, and risk. For an effective balancing act, utility engineers must become proficient in business essentials such as the material presented in the first eight chapters of this book. In addition, asset managers must embody a new mind-set that differs greatly from the traditional utility mind-set. A traditional utility engineer views cost reduction as undesirable and a threat to a well-engineered system. An asset manager views cost reduction as desirable as long as performance targets can be met. When performance targets are exceeded, the traditional engineer is happy because the system performed well. The asset manager is not happy, since less money could have been spent and the company could have been more profitable while still meeting all performance targets. Risk exposure gives a traditional engineer an upset stomach. An asset manager thrives on understanding and managing risk and views it as a potential means to achieve higher profits.

Although a change in mind-set is critical for asset managers, a large amount of spending decisions will always be made during everyday operations. It is therefore important for all utility engineers to shift their mind-set and become familiar with basic business concepts and jargon. Asset managers will need to thoroughly learn these business skills and will apply them extensively. But in order to achieve the next level in business performance, all utility engineers should learn the essentials of business and supplement their engineering prowess with business acumen.

9.7 STUDY QUESTIONS

1. What is the difference between financial asset management and infrastructure asset management?
2. Why might a utility choose to transition from a traditional business model to an asset management model?
3. What are the three basic functions of an asset management organization? What is the primary responsibility of each function?
4. What are some of the characteristics of a traditional utility engineering mind-set? What are some of the things that led to this mind-set? Are these still relevant today?
5. What are some of the characteristics of an asset management mind-set? How do these differ from a traditional utility engineering mind-set?

6. Describe the relationship between an asset manager and an asset owner. Is asset management the only concern of asset owner? Explain.
7. Explain how the asset service provider function is different in an asset management organization as compared to a traditional organization.
8. Explain some of the important skills for asset owners. Repeat this for asset managers, and asset service providers.
9. Once budgets are determined by the asset manager, is it ever appropriate for them to change? Explain why or why not?
10. Why is it important for all utility engineers, including asset managers, to be familiar with the essentials of business?

Index

A

Accelerated depreciation, 50
Accounting, *see also* Finance
  accounts, 38–41
  amortization, 49–51
  assets, 45–46
  balance sheet, 62–68
  basic accounting equation,
    33–37
  capital employed, 49
  conservatism, 43–44
  consistency, 43
  cost accounting, 77–78
  depreciation, 49–51
  financial statements, 55–74
  full disclosure, 42–43
  fundamentals, *xiii,* 33, 82
  historical cost, 42
  income statement, 55–62
  income taxes, 53–54
  inventory, 51–53
  journals, 38–41
  ledgers, 38–41
  liabilities, 46–47
  management accounting, 80–81
  matching, 42
  owner's equity, 47
  principles, 41–54
  project accounting, 77–78
  regulatory accounting, 75–77
  software impact on, 74–75
  statement of cash flows, 68–74
  study questions, 82–83
  tax accounting, 75
  types other than financial,
    74–81
  working capital, 48
Accounts, 38–41
Accounts payable, 35, 66
Accounts receivable, 35
Accounts receivable subsidiary
  ledger, 41
Accrual accounting, 42
Acid test ratio, 205
Acquisition, non-hostile, 131
Active investors, 174
Activity ratios, 200–203
Advanced metering infrastructure
  (AMI), 232
AFUDC, *see* Allowance for funds
  used during construction
  (AFUDC)
Agency problems, 311
Allowance for funds used during
  construction (AFUDC), 61, 221
Alpha, CAPM, 174
Alpha investor, CAPM, 174

Alternatives, business case
  development, 296–300
American Electric Power, 207
American financial options, 184
AMI, *see* Advanced metering infra-
  structure (AMI)
Amortization, 49–51
AMT, *see* Asset Management
  Toolkit (AMT)
Ancillary services, unbundled rates,
  230
Annual cost, 281–282
Annualized costs, 114–116
Annual revenues, 282
Annuities, 113–114
Anti-bypass rate, 228
Arbitrage opportunity, 187
Asset management
  financial asset management,
    304, 306
  fundamentals, *xiv,* 303,
    316–317
  historical developments,
    309–310
  infrastructure asset
    management, 305–306
  mind set, 308–309
  organizational functions,
    311–312
  process, 313–316
  study questions, 317–318
  utility historical developments,
    306–309
Asset Management Toolkit (AMT),
  310
Asset owner, communication, 316
Asset risk, 176, 179
Asset Risk Management Survey,
  310
Assets
  accounting equation, 33–36
  current, 46
  fundamentals, 33
  noncurrent, 46
  portfolio theory, 165

turns, 200–201
Association of Local Government
  Engineers of New Zealand, 309
AT&T
  deregulation, 29
  traditional mind set, 307
Average cost, ratemaking, 218
Avoided costs, 281

B

Balance sheet
  accounting equation, 34–36
  financial statements, 55, 62–68
  fundamentals, 34
  statement of cash flows, 69, 70
Baltimore Gas & Electric, 226
Baltimore Sun, The, 226
Bank loans, 131
Bankruptcy, 141–144
Base year, 238
Basic accounting equation, 33–37
Basic earnings per common share,
  61
BBO, *see* Big bad outcome (BBO)
B/C, *see* Benefit-to-cost (B/C)
  analysis
Bell curve, 154, 162
Benefits, business case
  development, 295–296
Benefit-to-cost (B/C) analysis
  business case development,
    295–296
  project prioritization, 264–265
Beta
  capital asset pricing model,
    176, 180
  stock price risk, 173–174
Beta investors, CAPM, 174
Big bad outcome (BBO), 275
Bills, utility, 228–230, 231,
  see also Consumption
Bimodal data set, 149
Biogas, 6

Black, Fischer, 187
Black–Scholes Option Pricing
    Model, 187–191
Black Swan, The, 164
Blanket budgets, 263–264
Block rates, 224–225
Book value, 36, 50
Boom, 102
Borrowed construction funds, 60
Bottom line, 61
Bottom up planning, 256–261
Brave New World, 102
Brief, rate case, 239
Brownian motion, 158
Bubbles, speculative, 135
Budget cycle, 254
Budget impact matrix, 293
Budgeting
    alternatives, 296–300
    asset management process,
        315, 316
    benefit, 295–296
    benefit-to-cost analysis,
        264–265
    blanket budgets, 263–264
    bottom up planning, 256–261
    business case development,
        280–300
    costs, 281–282, 292–295
    developing, 284–300
    fundamentals, 247–248,
        300–301
    internal rate of return, 283–284
    KPI degradation, 276–278
    marginal benefit-to-cost
        analysis, 266–269
    multiple performance targets,
        270–273
    natural KPI variation, 278–280
    payback period, 282–283
    portfolio effects, 273–275
    program budgets, 262–263
    program scope, 289
    project budgets, 261–262
    project prioritization, 264–280

risk-based spending, 275–276
    scope, 288–292
    spending justification, 285–286
    spending need, 286–288
    study questions, 301
    timeline, 296
    top down budgeting, 248–256
    types, 261–264
Buffet, Warren, 135
Business case development
    alternatives, 296–300
    benefit, 295–296
    cost, 292–295
    cost considerations, 281–282
    developing, 284–300
    fundamentals, 280–281, 284
    internal rate of return, 283–284
    payback period, 282–283
    scope, 288–292
    spending, 285–288
    timeline, 296
Business cycles, 101–106
Business performance, 234–237
Bust, 104

C

Call option
    Black-Scholes Option Pricing
        Model, 188–190
    capital structure, 138
    financial options, 181, 185
Canada, 140
Cannibalizing assets, 48, 72
CAPEX, *see* Capital expenditures
    (CAPEX)
Capital
    cash, 119
    defined, 48
    management, 51
Capital asset pricing model
    (CAPM), 173–180, 309
Capital budgeting, 250
Capital charge, 211–212

Capital employed, 49, 213
Capital expenditures (CAPEX)
    business case development, 297
    costs, 292
    goals and objectives, 314
    rate base misconceptions,
        240–244
    reduced, 130
    revenue requirement and
        rate base, 220
    semi-strong market efficiency,
        130, 131
    top-down budgeting, 250–252
    working capital, 48
Capital market line, 171
Capital markets, 17
Capital structure, 120, 136–140
Capital structure irrelevance
    principle, 140
CAPM, *see* Capital asset pricing
    model (CAPM)
Cash account, 34–36
Cash and cash equivalents, 65
Cash flow diagram, 110
Cash instruments, 180
Catastrophic events, 76
Central switching office, 9
Cesspit, 13
Cesspool, 13
Chapter 11 bankruptcy, 141–142,
    see also Bankruptcy
Civil trial, *see* Rate cases
Closing, 40
Closing entries, 40
Collection period, 201
Collector pipes, 12
Combined utilities, 14–15
Common mode root cause, 161
Common stock
    balance sheet, 67–68
    company valuation, 121–124
    privately-owned utilities, 17
    splits, 131
Commonwealth Edison, 207
Compact fluorescent light bulbs, 94

Company, defined, 165
Company valuation, 121–124
Compliance, 285
Compound interest, 110–111
Conceptual design, 288
Conservatism, 43–44
Consistency, 43
Consolidated Edison, 207
Consolidated financial
    statements, 44
Constructability review, 288–289
Construction vehicles and
    equipment, 78–79
Construction work in progress
    (CWIP)
    balance sheet, 64–65
    fundamentals, 52
    inventory turns, 202
    revenue requirement and
        rate base, 221
Consumer confidence, 102
Consumer Confidence Index
    (CCI), 106
Consumer Price Index (CPI),
    105, 237
Consumption
    fresh water, 9
    gas purchases, 7
    private water wells, 7
    rate design, 234
    retail electricity purchases, 4
    telephone service, 11
    top-down budgeting, 249
    utility rates, 23–24
Continuing operations, 60
Continuity of service, 26
Contraction phase, 104
Contribution margins, 80–81
Contributions in aid of
    construction, 222
Controlling account, 41
Cooperatives (co-ops), 19
Corporate finance, 109
Corporation commission, 21
Correlation coefficient, 150–151

Cost accounting, 77–78
Cost of capital, 119
Cost of goods sold, 52
Costs
    allocation, 224
    annual, 281
    avoided, 281
    business case development,
        281–282, 292–295
    disposal, 50
    escalation, 295
    initial, 281
    marginal, producer surplus, 95
    one-time, 281
    opportunity, 281
    prudently incurred, 24
    recurring, 281
    service standards, 26
    sunk, 281
Covariance, 150
Credit (cr.), 38
Creep, scope, 291–292
Critical path activities, 296
Critical peak pricing, 233
Critical peak rebates, 233
Cross-subsidies, rate design, 225
Crown corporations, 18
Cumulative density function,
    152–153
Current assets, 46, *see also* assets
Current liabilities, 46, *see also*
    Liabilities
Current portion of long-term debt,
    47, *see also* Long-term debt
Current ratio, 204
Curtailable rates, 24
Customer deposits, balance
    sheet, 66
Customer service, 26–27
CWIP, *see* Construction work in
    progress (CWIP)

D

Darn old utility guy (DOUG), 307
Data requests, 239
Days inventory ratio, 202
Days payable ratio, 201
Days receivable ratio, 201
Dead band, 27
Debentures, 138
Debit (dr.), 38
Debtor-in-possession, 142
Debt ratio, 203
Debts, *see also* Long-term debt;
    Short-term debt
    capital structure, 136–140
    current portion of long-term
        debt, 47
Debt-to-equity ratio, 203
Declining balance depreciation, 50
Declining block rate, 225
Decoupled rates, 234
Deferred tax liability, 53
Deflation, 105
Demand
    economics, 86–90
    market pricing, 92–94
Demand curve, 86
Demand side management
    (DSM), 232
Depletion, rate base, 243
Deposition, 239
Depreciation
    cost considerations, 282
    expense, 49
    fundamentals, 36, 49–51
    Iowa Curves, 51
    studies, 51
    top-down budgeting, 250–251
Depression, 104
Deregulation, 28–30
Derivative instruments, 180

Desalination facilities, 9
Design, project scope, 288
Development, business case
    budgeting justification, 284–300
Differential rates, 224
Diluted earnings per common
    share, 61
Diluted shares, 61
Direct cost assignment, 224
Direct load control, 232–233
Discontinued operations, 60
Discount rates, 116–119, 294–295
Discovery process, 239
Discrimination, *see* Nondiscrimina-
    tory service standards
Diseconomies of scale, 95
Disposal cost, 50
Distress, financial, *see* Bankruptcy
District heating system, 14
Diversification, risk, 160–164
Dividends
    balance sheet, 66
    declared, balance sheet, 66
    fundamentals, 37
    income statement, 362
    increase, 130
    omission, 129
    semi-strong market
        efficiency, 129–130
    yield, 207
Docket, 20–21
Double entry accounting, 38
DOUG (a darn old utility guy), 307
Dow Jones Industrial Average, 172
dr., *see* Debit (dr.)
DSM, *see* Demand side
    management (DSM)
Dual water systems, 9
Due diligence, 289
DuPont analysis, 209–211
Dynamic pricing, 233

E

Earnings, 35, 130, 131
Earnings before interest, taxes,
    depreciation, and amortization
    (EBITDA)
    income statement, 56
    times interest earned ratio, 205
Earnings before interest and taxes
    (EBIT)
    bankruptcy, 144
    company valuation, 124
    income statement, 55
    profit margin, 210
    tax shields, 140
    times interest earned ratio, 205
Earnings per share (EPS), 207
Earthquakes, 76
EBIT, *see* Earnings before interest
    and taxes (EBIT)
EBITDA, *see* Earnings before in-
    terest, taxes, depreciation, and
    amortization (EBITDA)
Economic development rates, 228
Economic dispatch, 2
Economic mapping, 271
Economic recovery, 104
Economics, business
    business cycles, 101–106
    demand, 86–90
    fundamentals, *xiii*, 85–86,
        106–107
    market pricing, 90–95
    monopolistic pricing, 97–101
    producer surplus, 95–96
    retail electricity purchases,
        amount, 4
    study questions, 107
    supply, 86–90
*Economics of Defense in the
    Nuclear Age, The,* 309

Economic value added (EVA), 209–211
Economies of scale, 95, 97
Edison International, 207
Effective annual interest rate, 111
Efficiency, service standards, 26
Efficient frontier, 168
Efficient market theory, 124
Efficient portfolios, 168
Einhorn, David, 164
Elastic demand, 86–88
Elastic supply, 88–90
Electric membership
    corporations, 19
Electric Operating Revenue, 76
Electric Power Research Institute
    (EPRI), 310
Electric transmission systems, 3
Electric utilities, 2–4, 14
Emergency Natural Gas Act
    (1977), 29
Employee stock ownership plans
    (ESOP), 68
Engineering economics, 85
Entergy New Orleans, 142–143,
    207
Environmental Protection Agency
    (EPA), 22
EPA, *see* Environmental Protection
    Agency (EPA)
Equation, basic accounting, 33–37
Equilibrium, 90
Equipment, 78–79, 277
Equity issues, 130
Escalation, costs, 295
ESOP, *see* Employee stock
    ownership plans (ESOP)
Estimates, 80, 293–294
European option, 184
EVA, *see* Economic value added
    (EVA)
Event study, 128–129
Excel STDEV/STDEP
    formulas, 150
Exelon, 207

Exercising options, 182
Expansion phase, 102
Expected value, 148
Expenditures, 130, *see also* Capital
    expenditures (CAPEX); Opera-
    tional expenditures (OPEX);
    Spending
Expiration date, options
    Black-Scholes Option Pricing
        Model, 190
    financial options, 184, 185–186

F

Fannie Mae, 164
FASB, *see* Financial Accounting
    Standards Board (FASB)
FCC, *see* Federal Communications
    Commission (FCC)
Federal Communications
    Commission (FCC), 22–23
Federal Energy Regulatory
    Commission (FERC), 22, 76
Federal Power Commission (FPC),
    22, *see also* Federal Energy
    Regulatory Commission
    (FERC)
Federal regulation, 21–23
FIFO, *see* First-in, first-out (FIFO)
    method
Finance, *see also* Accounting
    annualized cost, 114–116
    bankruptcy, 141–144
    capital structure, 136–140
    company valuation, 121–124
    compound interest, 110–111
    discount rates, 116–119
    distress, 141–144
    fundamentals, *xiii*, 109, 145
    market efficiency, 124–136
    net present value, 112–114
    semi-strong efficiency, 128–131
    strong efficiency, 132

study questions, 145–146
tax shields, 140–141
time value of money, 110–121
weak efficiency, 126–128
weighted average cost of
      capital, 119–121
Financial Accounting Standards
      Board (FASB), 44, 76
Financial asset management, 304,
      306, *see also* Asset management
Financial capital, 48, 119
Financial derivatives, 158
Financial engineering, 181
Financial instruments, 180
Financial leverage, increasing, 211
Financial leverage ratio, 203–204,
      see also Leverage ratios
Financial options, 180–191
Financial ratios
      activity ratios, 200–203
      DuPont analysis, 209–211
      economic value added, 209–211
      fundamentals, *xiv,* 195–196,
          213–214
      leverage ratios, 203–204
      liquidity ratios, 204–205
      market ratios, 205–208
      profitability ratios, 196–199
      residual income, 209–211
Financial statements, *see also*
      Accounting
      balance sheet, 62–68
      fundamentals, 55
      income statement, 55–62
      statement of cash flows, 68–74
Firm, defined, 165
FirstEnergy, 207
First-in, first-out (FIFO)
      method, 53
Fixed assets, 46
Fixed investment cycle, 102
Florida Power & Light, 207
Footnotes, 42
Forward contract, 181

FPC, *see* Federal Power Commis-
      sion (FPC)
FPL Group, 207
Free cash flow, 69–70
Free markets, 15, 235
Freezes, rates, 235–236
Fuel cost, 65, 92
Full depreciation, 50
Full disclosure, 42–43
Full diversification, 164, 173, 175
Functional separation, 312
Future value, money, 110, *see also*
      Time value of money

G

GAAP, *see* Generally accepted
      accounting principles (GAAP)
Gain on sale of investment, 59
Gas purchases, *see* Consumption
Gas utilities, 4–7, 14
GDP, *see* Gross domestic product
      (GDP)
GDP deflator, 105
General ledger, 41
Generally accepted accounting
      principles (GAAP)
      fundamentals, 44
      income taxes, 53
      tax accounting, 75
General Motors, 209
General rate case, 237
Germany, 140
GNP, *see* Gross national product
      (GNP)
Goals, asset management process,
      314
Going long/short, 184
Gold plating, projects, 265
Goodwill, 45, 65
Government-owned utilities, 18–19
Greater fool theory, 135
Grocery stores comparison, 208

Gross domestic product
(GDP), 101
Gross margin, 196
Gross national product
(GNP), 102
Gross profit, 196
Gross profit margin, 196
Group depreciation, 51
Growing annuity, 114
*Guidelines for Power Delivery
Asset Management,* 310

H

Heating utilities, combined, 14
High impact low probability event
(HILPE), 275
HILPE, *see* High impact low prob-
ability event (HILPE)
Historical developments
asset management, 309–310
costs, 42
utilities, 306–309
Hitch, Charles, 309
Housing starts, 104
Hurricanes, 76
Huxley, Aldous, 102
Hypothetical market equilibrium,
217–218

I

IASB, *see* International Accounting
Standard Board (IASB)
IBT, *see* Income before taxes (IBT)
Ice storms, 76
Idiosyncratic risks, 161–162
Impairing asset value, 45, 59
Incentive rates, 228
Income before taxes (IBT), 56
Income from continuing
operations, 60

Income statement, 55–62, 69, 70
Income taxes, 53–54, 67, *see also*
Tax accounting
Increasing economies of scale, 15
Industry restructuring, 28
Inelastic demand, 86
Inelastic supply, 88
Inflation, 105, 118
Infrastructural investment
cycle, 104
Infrastructure asset management,
305–306, *see also* Asset
management
*Infrastructure Asset Management
Manual,* 309–310
Infrastructure utilities, 1
Initial cost, 281–282
Initial public offering (IPO), 17
Intangible assets, 45
Integrated resource planning
(IRP), 232
Integrated voice response (IVR)
system, 298–299
Interceptor pipes, 12
Interest burden, 210
Interest payments, 36, 59–60
Interest rate swap, 181
Interest tax shield, 141
Interexchange carriers (IXCs), 11
Internal rate of return (IRR),
283–284
International Accounting Standard
Board (IASB), 44
*International Infrastructure
Management Manual,* 310
Interrogatories, 239
Interruptible rates, 24, 232
Interveners, 21
Intervenors, 239
Interview, *see* Deposition
"In the money," 182
Inventory
balance sheet, 65
fundamentals, 51–53
turns, 202–203

Inverted block rates, 225
Inverted yield curve, 118
Investments, *see also specific type*
    discount rates, 118
    malinvestments, 104
    return, 165
    statement of cash flows, 72
Investor-owned utilities (IOUs)
    fundamentals, 17
    government-owned
        comparison, 19
    member-owned
        comparison, 19–20
Investors, 174
Invisible hand theory, 117
IOU, *see* Investor-owned utilities
    (IOUs)
Iowa Curves, 51
IPO, *see* Initial public offering
    (IPO)
IRP, *see* Integrated resource
    planning (IRP)
IRR, *see* Internal rate of return
    (IRR)
IVR, *see* Integrated voice response
    (IVR) system
IXC, *see* Interexchange carriers
    (IXCs)

J

Journals, 38–41
Just and reasonableness
    characteristics, 219

K

Key performance indicators (KPIs)
    asset management process, 315
    degradation, 276–278

    multiple performance targets,
        270–273
    natural variation, 278–280

L

Labor costs, 78
Land, project scope, 288
Land held for future use, 222
Large cap, company valuation, 121
Last-in, first-out (LIFO)
    method, 53
Lateral pipes, 12
LEC, *see* Local exchange carriers
    (LECs)
Ledgers, 38–41
Legal compliance, 285
Lenders, bankruptcy impact, 143
Levelized cost, 115
Leverage ratios, 203–204
Levered beta, 176, 179
Liabilities
    accounting equation, 33–36
    balance sheet, 65–67
    current, 46
    deferred tax, 53
    fundamentals, 33
    noncurrent, 46
Liberalization, 28
Life cycle cost, 112, 314
LIFO, *see* Last-in, first-out (LIFO)
    method
LILOC, *see* Long Island Lighting
    Company (LILOC)
Linear regression, 174
LIPA, *see* Long Island Power Au-
    thority (LIPA)
Liquefied natural gas (LNG), 5
Liquidated assets, 48
Liquidity, 48
Liquidity ratios, 204–205
LNG, *see* Liquefied natural gas
    (LNG)

Load factor, 223
Load retention rates, 228
Loans, bank, 131
Local access and transport area, 11
Local exchange, 10
Local exchange carriers (LECs), 11
Lognormal distribution, 156–157
Long call, 182
Long Island Lighting Company
    (LILOC), 311
Long Island Power Authority
    (LIPA), 311
Long put, 184, 185
Long-term assets, 46
Long-term debt, *see also* Debts
    balance sheet, 65–66
    capital employed, 49
    current portion, 47
    statement of cash flows, 72, 74
Long-term liabilities, 46
Long-term performance, 314,
    see also Performance
Long-term plan, optimized, 315
Long-term planning, 256, 258
Loss on sale of investment, 59

M

Macroeconomics, 85
Malinvestments, 104
Management accounting, 80–81
Marginal benefit-to-cost analysis
    (MBCA), 266–269
Marginal cost, 95
Marketable securities, 181
Market cap, 121
Market capitalization, 121
Market crash, 135
Market efficiency
    fundamentals, 124–126,
        132–136
    semi-strong efficiency, 128–131
    strong efficiency, 132

weak efficiency, 126–128
Market equilibrium price, 90
Market portfolio, 171
Market pricing, 90–95
Market ratios, 205–208
Market risk premium, 173, 180
Master budget, 252
Matching, 42
Material costs, 78
Mathematical analysis, 128
MBCA, *see* Marginal benefit-to-
    cost analysis (MBCA)
MCI, deregulation, 29
McKean, Roland, 309
Mean value
    lognormal distribution, 156, 157
    normal distribution, 154
    statistics, 148–149
Median value, 149, 154
Member-owned utilities, 19–20
Merchant generation, 29
Methane, 5, 6
Mezzanine capital, 138
MFR, *see* Minimum filing re-
    quirements (MFR)
Microeconomics, 85
Microsoft company, 122
Microsoft Excel formulas, 150
Mid cap, company valuation, 121
Minimum charge, rate design, 225
Minimum filing requirements
    (MFR), 238
Minimum revenue requirement,
    26, 243
Minimum variance portfolio, 168
Mode value, 149, 154
Modigliani–Miller theorem,
    140–141
Money supply, 106
Monitoring, asset management
    process, 315
Monopolies
    fundamentals, 15, 97
    natural, 15–16
    regulatory goals, 216

Monopolistic pricing, 97–101
Monthly fixed charges, 225
Mortgage-based debt, 138
Multimodal data set, 149
Multiple performance targets, 270–273
Municipalization, 19
Municipal utilities, 18
Munis, 18

N

NAIRU, *see* Non-accelerating inflation rate of unemployment (NAIRU)
NAMS, *see* National Asset Management Steering Group (NAMS)
NARUC, *see* National Association of Regulatory Commissioners (NARUC)
National Asset Management Steering Group (NAMS), 310
National Association of Regulatory Commissioners (NARUC), 21
Natural Gas Policy Act (1978), 29
Natural monopolies, 15–16, 97
Natural unemployment rate, 105
NEC, *see* Nuclear Regulatory Commission (NEC)
Negotiable cash instruments, 180
Net income, 56
Net margin, 197
Net operating profit after taxes (NOPAT)
 company valuation, 124
 income statement, 56
 residual income and EVA, 211–212
Net present value (NPV)
 company valuation, 123

discount rates, 118, 294–295
 fundamentals, 112–114
 spending justification, 285
Net profit margin, 197
New Zealand Local Government, 309
Node, 153
Nominal interest rate, 117
Non-accelerating inflation rate of unemployment (NAIRU), 105
Noncurrent assets, 46
Noncurrent liabilities, 46
Nondiscriminatory service standards, 27–28
Nondiversifiable characteristic, 161
Non-hostile acquisition, 131
Non-marketable securities, 181
Nonnegotiable cash instruments, 180
Nontransferable securities, 181
NOPAT, *see* Net operating profit after taxes (NOPAT)
Normal distribution, 154–156
Normalization, rate case, 238
NPV, *see* Net present value (NPV)
Nuclear Asset Management (NAM) program, 310
Nuclear decommissioning trust, 65
Nuclear Regulatory Commission (NEC), 22

O

Objectives, asset management process, 314
Obligation to serve, 25
Odor, addition to gas, 5
OFFER, *see* Office of Electricity Regulation (OFFER)
Office of Electricity Regulation (OFFER), 310

Office of Gas and Electricity Markets (OFGEM), 310
Office of Gas Supply (OFGAS), 310
OFGAS, *see* Office of Gas Supply (OFGAS)
OFGEM, *see* Office of Gas and Electricity Markets (OFGEM)
One-time cost, 281
Open access, 29
Operating activities, 72
Operating income, 55
Operating margin, 196–197
Operating profit margin, 196–197
Operational expenditures (OPEX)
    business case development, 297
    costs, 292
    goals and objectives, 314
    rate base misconceptions, 240–244
    revenue requirement and rate base, 220–221
    top-down budgeting, 249, 252
Opportunity cost, 281
Optimal risky portfolio, 171
Optimized long-term plan, 315
Organizational functions, 311–312
Other income (income statement), 59
Outdoor water use, 9
Overbuilding, 104
Overhead, cost accounting, 79
Owner's equity
    accounting equation, 33–36
    balance sheet, 67
    company valuation, 33–34, 121
    fundamentals, 33, 47
Ownership
    fundamentals, 16–17
    government-owned utilities, 18–19
    member-owned utilities, 19–20
    privately-owned utilities, 17–18

P

Pacific Gas & Electric (PG&E), 142, 207
Paid-in capital, 34, 47
PAM, *see* Physical asset management (PAM)
Pari passu debts, 138
Partially inverted yield curve, 118
PAS-55, *see Publicly Available Specification 55: Asset Management (PAS-55)*
Passive investors, 174
Payable-to-sales ratio, 201
Payback period, 282–283
Payback time, *see* Payback period
Payment period ratio, 201–202
Payout ratio, 199
P/E, *see* Price-to-earnings (P/E) ratio
Peak demand, 24, 232–234
PECO Energy, 207
Performance, 285, 314
Performance-based regulation, 235–236
Perpetuity, 113
PG&E, *see* Pacific Gas & Electric (PG&E)
Physical asset management (PAM), 309
Plain old telephone service (POTS), 9, *see also* Telephone utilities
Plant held for future use, 221
Plowback, 123
Portfolio effects, 273–275
Portfolio interaction effects, 273–275
Portfolio theory risk, 165–172
Posting, journals, 40
POTS, *see* Plain old telephone service (POTS)
Preferred stock, 67–68

Prehearing order, 239
Present value, money, 110, *see also*
    Time value of money
Pretax margin, 197
Pretax profit margin, 197
Price discrimination, 99
Price momentum, 135
Price regulation, 236–237
Price-to-book (P/B) ratio, 206–207
Price-to-earnings (P/E) ratio,
    135, 206
Pricing, 90–95, 97–101
Primary treatment, wastewater, 12
Privately-owned utilities, 17–18
Private water wells, 7
Probability, risk, 148, 152–154
Probability density function,
    152–153, 154
Probability distribution function,
    152–153
Producer surplus, 95–96
Profitability ratios, 196–199
Profit and loss statement, 55
Profit margin, EBIT, 210
Program budgets, 262–263, 291
Progress Energy, 207
Project prioritization
    benefit-to-cost analysis,
        264–265
    fundamentals, 264
    KPI degradation, 276–278
    marginal benefit-to-cost
        analysis, 266–269
    multiple performance targets,
        270–273
    natural KPI variation, 278–280
    portfolio effects, 273–275
    risk-based spending, 275–276
Projects
    accounting, 77–78
    alternatives, 266
    budgets, 261–262, 291
    options, 266

portfolio optimization, 275
Project scope, *see* Scope
Provider of last resort, 30
Provisional liabilities, 46
Prudency reviews, 24
Prudently incurred costs, 24
PSC, *see* Public service commis-
    sion (PSC)
PSTN, *see* Public switched tele-
    phone network (PSTN)
Publicly Available Specification
    55: Asset Management
    (PAS-55), 310
Public offering, 17
Public service commission
    (PSC), 21
Public switched telephone network
    (PSTN), 11, *see also*
    Telephone utilities
Public utilities, *xiii,* 1
Public utility district (PUD), 18
Public Utility Holding Company
    Act (PUHCA), 22
PUD, *see* Public utility district
    (PUD)
PUHCA, *see* Public Utility Holding
    Company Act (PUHCA)
Put-call parity, 185–186, 190
Put option
    Black-Scholes Option Pricing
        Model, 190
    capital structure, 138
    financial options, 181, 185

Q

Quality of service, 26
Quants, 162
Questions for study, *see* Study
    questions
Quick ratio, 204–205

R

Random variables, 152
Rate base depletion, 243
Rate cases, 180
Rate design
    business performance, 234–237
    consumption, 234
    fundamentals, 222–230
    peak demand, 232–234
    unbundled rates, 230–232
Ratemaking
    business performance, 234–237
    consumption, 234
    fundamentals, *xiv,* 215,
        244–245
    peak demand, 232–234
    rate base, 219–222
    rate base misconceptions,
        240–244
    rate cases, 237–240
    rate design, 222–237
    regulatory goals, 216–219
    revenue requirement, 219–222
    study questions, 245–246
    unbundled rates, 230–232
Rate of return (ROR), 220–222
Rates, 23–25
Rate shock, 225
Rate unbundling, *see*
    Unbundled rates
REA, *see* Rural Electrification
    Administration (REA)
Real interest rate, 117
Real options, 191–192
Real-time rates, 233
Reasonableness and just
    characteristics, 219
Rebuttal, rate cases, 239
Recession, 104
Recombinant binary tree, 158
Record, rate cases, 239
Recurring cost, 281, 282

*Regulated Enterprises-Accounting
    for Abandonments and Disal-
    lowances of Plant Costs,* 44
Regulatory accounting, 75–77
Regulatory assets, 65
Regulatory compliance, 285
Regulatory goals, 216–219
Regulatory liabilities, 67
Re-regulation, 28
Research and development (R&D)
    expenditures, 130
Residual income, 209–211
Residual value, 122
Restrictive covenants, 143
Retail choice, 30
Retail competition, 30
Retail electric purchases, U.S., *see*
    Consumption
Retained earnings
    balance sheet, 67
    distribution, 37
    fundamentals, 35, 46–47
Retroactive ratemaking, 222
Return deficiency, 237
Return on assets (ROA), 198–199
Return on capital employed
    (ROCE), 198–199
Return on equity (ROE)
    capital asset pricing model, 180
    fundamentals, 199
    owner's equity, 47
Return on sales (ROS), 198
Revenue
    annual, 282
    forecasts, 249
    requirement, 219–222
Reversed transaction, 43
Risk
    capital asset pricing model,
        173–180
    diversification, 160–164

financial options, 180–191
fundamentals, *xiv,* 147–148,
    192–193
goals and objectives, 314
investments, 165
lognormal distribution, 156–157
normal distribution, 154–156
portfolio theory, 165–172
probability, 148, 152–154
real options, 191–192
spending justification, 285
statistics, 148–151
stock price movement, 158–160
study questions, 194
Risk-adjusted discount rates, 121
Risk-based spending, 275–276
Risk-free security, 171
Risk premium, 121
ROA, *see* Return on assets (ROA)
Robustness, bottom up planning,
    259
ROCE, *see* Return on capital
    employed (ROCE)
ROE, *see* Return on equity (ROE)
Rolling budget, 254–255
ROR, *see* Rate of return (ROR)
ROS, *see* Return on sales (ROS)
Rural Electrification
    Administration (REA), 19
Rural Utility Service (RUS), 19
RUS, *see* Rural Utility Service
    (RUS)
Russell 3000, 172

S

Safety, 26
Sales volume, 210–211
Salvage value, 50
San Diego Gas & Electric, 207
Schapiro, Mary, 44
Scholes, Myron, 187

Scope
    blanket, 290–292
    creep, 291–292
    program, 289–290
    project, 288–289
Scum, 12
Seasonal rates, 233
SEC, *see* Security and Exchange
    Commission (SEC)
Secondary treatment,
    wastewater, 13
Sectors, stock market, 172
Secured debt, 138
Security, negotiable contract, 165
Security and Exchange
    Commission (SEC)
    accounting conservatism, 44
    federal regulations, 21–22
    publicly traded companies, 74
Security market line, 176
Selling short, 184
Semi-strong market efficiency,
    128–131
Semi-variable investments, 216
Sempra, 207
Senior debt, 138
Seniority, 138
Service standards, 25–28, 314
Sewer, *see also*
    Wastewater utilities
    charging, 13
    cleanups, 12
    economic impact, 9
    mains, 12
Shareholder's equity, 47
Sharpe, William, 176
Shift in demand, 94
Shift in quantity demanded, 94
Short call, 182, 185
Short put, 184
Short-term debt, 72, 74,
    see also Debts
Short-term obligations, 66

Short-term performance, 314, *see also* Performance
Short-term planning, 256, 258
Show cause order, 237
Slack, 296
Sludge, 12
Small caps, 121, 134
Smart appliances, 233
Smart meters, 232
Smith, Adam, 117
Societal loss, 98
Software, 74–75, 150
Southern California Edison, 207
S&P, *see* Standard & Poor's 500
SPC, *see* Statistical process control (SPC)
Special purpose entries, 43
Speculative bubbles, 135
Spending
    allocation, 248
    business case development, 285–288
    justification, 248, 285–286
    need, 286–288
    risk-based, project prioritization, 275–276
Sprint, deregulation, 29
Stagflation, 105
Standard deviation, 154, 156–157
Standard & Poor's 500, 171–172
Standby rates, 228
Statement of cash flows, 55, 68–74
State regulation, 21
Statistical process control (SPC), 280
Statistics, 148–151
STDEV/STDEVP formulas, 150
Sticky prices, 117
Stocks
    market sectors, 172
    options, 61
    price movement, 158–160
    price-to-earnings, 135
    splits, 131

Storage difficulty, 4
Storm reserve asset, 76
Straight line depreciation, 50
Strategic direction, 314
Strike price, 182, 185
Strong market efficiency, 132
Study questions
    asset management, 317–318
    budgeting, 301
    economics, 107
    finance, 145–146
    fundamentals, 82–83
    ratemaking, 245–246
    risk, 194
    utilities, 31–32
Subordinate debt, 138
Subscriber loop, 9
Subsidiary ledger, 41
Sum of digits depreciation, 50
Sunk cost, 281
Supply, 86–90, 92–94
Supply curve, 88
Surplus, producer, 95–96
Sweetening, 5
Synthetic stock, 187
Systematic risk, 161

T

T account, 40
Taleb, Nassim, 164
Tariffs, 237
Tax accounting, 54, 75, *see also* Income taxes
Tax burden, 210
Tax shields, 140–141
Technical analysis, 126
Telecommunications Act, 29
Telephone utilities, 9–11, 24
Tertiary treatment, wastewater, 13
Test year, 238
The Baltimore Sun, 226
The Black Swan, 164

The Economics of Defense in the Nuclear Age, 309
Therms, 6
Tiered rates, 99–101
Timeline, 296
Time-of-use rates, 24, 233
Times interest earned ratio, 205
Time value of money
    annualized cost, 114–116
    compound interest, 110–111
    discount rates, 116–119
    fundamentals, 110
    net present value, 112–114
    weighted average cost of
        capital, 119–121
Tomlin, Lily, 25
Top down budgeting, 248–256, 312
Top line, 56
Top line growth, 59
Total capitalization, 49
Total value, 123
Traditional utility mind set,
    307–308
Transferable securities, 181
Transmission systems, 3, 5
Transportation utilities, 1
Trial balance, 41
Trunk lines, 10
Turkey example, confidence, 128
Twisted wires, 9

U

Unbundled rates, 230–232
Underwriters, 17
Unemployment rate, 105
Unit commitment, 2
Unit costs, 79
United Kingdom, 140, 310
United States, 140
Units of production depreciation,
    50

Unlevered beta, 176
Unsecured debts, 138
Usage, *see* Consumption
Utilities
    combined utilities, 14–15
    competitor example, 15–16
    deregulation, 28–30
    electric utilities, 2–4
    federal regulation, 21–23
    fundamentals, 1–2, 30–31
    gas utilities, 4–7
    government-owned utilities,
        18–19
    historical developments,
        306–309
    member-owned utilities, 19–20
    natural monopolies, 15–16
    ownership, 16–20
    privately-owned utilities, 17–18
    rates, 23–25
    regulation, 20–23
    service standards, 25–28
    state regulation, 21
    study questions, 31–32
    telephone utilities, 9–11
    types, 2–15
    wastewater utilities, 12–14
    water utilities, 7–9
Utility plants, 45, 52

V

Value at risk (VaR), 162, 164
VaR, *see* Value at risk (VaR)
Variance
    probability, 153–154
    statistics, 149–150
Vehicles, 78–79
Venture capital, 48
Vertically integrated utilities, 2
Vertical unbundling, 4, 229
Volatility, 150, 154

W

WACC, *see* Weighted average cost
   of capital (WACC)
Wal-Mart comparison, 15
Wastewater utilities, 9, 12–14
Water utilities, 7–14
Weak market efficiency, 126–128
Weighted average cost of capital
   (WACC)
   company valuation, 124
   cost considerations, 281
   defined, 120
   discount rates, 294
   fundamentals, 119–121
   reduction, 213
   residual income and EVA,
      211, 213
   revenue requirement and rate
      base, 219, 222

Weighted average method, 53
Wizards (persons), 135
WOOPS, *see* Worst-case outcome
   of probability small (WOOPS)
Working capital, 48, 72
Work scope, 289–290, 315, 316
Worst-case outcome of probability
   small (WOOPS), 275–276
Writing down asset value, 45, 59
Writing off assets, 51

Y

Yield curve, 118
Yorkshire Electricity, 310

For Product Safety Concerns and Information please contact our EU
representative GPSR@taylorandfrancis.com
Taylor & Francis Verlag GmbH, Kaufingerstraße 24, 80331 München, Germany

* 9 7 8 1 4 3 9 8 1 1 9 6 2 *